Universitext

Springer
Berlin
Heidelberg
New York
Hong Kong
London
Milan
Paris
Tokyo

James F. Blowey
Alan W. Craig
Tony Shardlow

Editors

Frontiers in
Numerical Analysis

Durham 2002

 Springer

James Blowey
Alan Craig

University of Durham
Department of Mathematical Sciences
South Road
DH1 3LE Durham, United Kingdom
e-mail: j.f.blowey@durham.ac.uk
       alan.craig@durham.ac.uk

Tony Shardlow

University of Manchester
Department of Mathematics
Oxford Road
M13 9PL Manchester, United Kingdom
e-mail: shardlow@maths.man.ac.uk

Library of Congress Cataloging-in-Publication Data

LMS-EPSRC Numerical Analysis Summer School (10th : 2002 : University of Durham)
   Frontiers in numerical analysis : Durham 2002 / James F. Blowey, Alan W. Craig, Tony
Shardlow, editors.
     p. cm. -- (Universitext)
   Includes bibliographical references and index.
   ISBN 3-540-44319-3 (pbk. : acid-free paper)
     1. Numerical analysis–Congresses. I. Blowey, James F. II. Craig, Alan W. III.
Shardlow, Tony. IV. Title.

   QA297.L59 2002
   519.4--dc21
                                                                    2003050594

ISBN 3-540-44319-3 Springer-Verlag Berlin Heidelberg New York

---

Mathematics Subject Classification (2000): 35-XX, 65-XX

Springer-Verlag Berlin Heidelberg New York
a member of BertelsmannSpringer Science+Business Media GmbH

http://www.springer.de

© Springer-Verlag Berlin Heidelberg 2003 Printed in Germany

The use of general descriptive names, registered names, trademarks, etc. in this publication
does not imply, even in the absence of a specific statement, that such names are exempt from
the relevant protective laws and regulations and therefore free for general use.

Cover design: *design & production* GmbH, Heidelberg
Typeset by the authors using a LaTeX macro package

Printed on acid-free paper   40/3142ck - 5 4 3 2 1 0

This volume is dedicated to the memory of Will Light who was a driving force in creating and running the first eight summer schools.

Preface

The Tenth LMS-EPSRC Numerical Analysis Summer School was held at the University of Durham, UK, from the 7th to the 19th of July 2002. This was the second of these schools to be held in Durham, having previously been hosted by the University of Lancaster and the University of Leicester. The purpose of the summer school was to present high quality instructional courses on topics at the forefront of numerical analysis research to postgraduate students. The speakers were Franco Brezzi, Gerd Dziuk, Nick Gould, Ernst Hairer, Tom Hou and Volker Mehrmann.

This volume presents written contributions from all six speakers which are more comprehensive versions of the high quality lecture notes which were distributed to participants during the meeting. At the time of writing it is now more than two years since we first contacted the guest speakers and during that period they have given significant portions of their time to making the summer school, and this volume, a success. We would like to thank all six of them for the care which they took in the preparation and delivery of their material.

Instrumental to the school were two groups: The five tutors who ran a very successful tutorial programme (Philip Davies, Sven Leyffer, Matthew Piggott, Giancarlo Sangalli and Vanessa Styles); the two "local experts", that is distinguished UK academics who, during the meeting, ran the academic programme on our behalf leaving us free to deal with administrative and domestic matters. These were Charlie Elliott (University of Sussex) and Sebastian Reich (Imperial College). In addition to chairing the main sessions the local experts also ran a successful programme of contributed talks from academics and students in the afternoons. The UKIE section of SIAM contributed prizes for the best talks given by graduate students. The local experts took on the bulk of the task of judging these talks. After careful and difficult consideration, and after canvassing opinion from other academics present, the prizes were awarded to Angela Mihai (Durham) and Craig Brand (Strathclyde). The general quality of the student presentations was impressively high promising a vibrant future for the subject.

The audience covered a broad spectrum, seventy-three participants ranging from research students to academics from within the UK and from abroad. A new feature of this meeting was that, thanks to the generosity of the LMS, we were able to fund a small number of students from continental Europe. As always, one of the most important aspects of the summer school was providing a forum for EU and UK numerical analysts, both young and old, to meet for an extended period and exchange ideas.

We would also like to thank the Durham postgraduates who together with those who had attended the previous Summer School ran the social

programme, Fionn Craig for dealing with registration, Rachel Duke, Tanya
Ewart, Fiona Giblin, Vicky Howard and Mary Bell for their secretarial support and our families for supporting our efforts.

We thank the LMS and the Engineering and Physical Sciences Research
Council for their financial support which covered all the costs of the main
speakers, tutors, plus the accommodation costs of the participants.

James F. Blowey, Alan W. Craig and Tony Shardlow
Durham, March 2003

Contents

Numerical Approximations to Multiscale Solutions in PDEs .. 241
Thomas Y. Hou

Numerical Methods for Eigenvalue and Control Problems 303
Volker Mehrmann

Subgrid Phenomena and Numerical Schemes

Franco Brezzi and Donatella Marini

Dipartimento di Matematica, Università di Pavia, and IMATI-C.N.R., via Ferrata 1, 27100 Pavia, Italy

Abstract. In recent times, several attempts have been made to recover some information from the subgrid scales and transfer them to the computational scales. Many stabilizing techniques can also be considered as part of this effort. We discuss here a framework in which some of these attempts can be set and analyzed.

1 Introduction

In the numerical simulation of a certain number of problems, there are physical effects that take place on a scale which is much smaller than the smallest one representable on the computational grid, but have a strong impact on the larger scales, and, therefore, cannot be neglected without jeopardizing the overall quality of the final solution.

In other cases, the discrete scheme lacks the necessary stability properties because it does not treat in a proper way the smallest scales allowed by the computational grid. As a consequence, some "smallest scale mode" appears as abnormally amplified in the final numerical results. Most types of numerical instabilities are produced in this way, such as the checkerboard pressure mode for nearly incompressible materials, or the fine-grid spurious oscillations in convection-dominated flows. See for instance [19] and the references therein for a classical overview of several types of these and other instabilities of this nature.

In the last decade it has become clear that several attempts to recover stability, in these cases, could be interpreted as a way of improving the simulation of the effects of the smallest scales on the larger ones. By doing that, the small scales can be *seen* by the numerical scheme and therefore be kept under control.

These two situations are quite different, in nature and scale. Nevertheless it is not unreasonable to hope that some techniques that have been developed for dealing with the latter class of phenomena might be adapted to deal with the former one. In this sense, one of the most promising technique seems to be the use of Residual-Free Bubbles (see e.g. [10], [18].) In the following sections, we are going to summarize the general idea behind it, trying to underline its potential and its limitations. In Section 2 we present the continuous problems in an abstract setting, and provide examples of applications, related to advection dominated flows, composite materials, and viscous incompressible flows. For application of these concepts to other problems we refer, for instance, to [13], [14], [16], [18], [24]. In Section 3 we introduce the basic features of the RFB method. Starting from a given discretization (that might possibly be unstable), we discuss the suitable *bubble space* that can be added to the original finite element space. Increasing the space with bubbles leads to the *augmented problem*, usually infinite dimensional, which, in the end, will have to be solved in some suitable approximate way. In Section 4 we give an idea of how error estimates can be deduced for the augmented problem. In Section 5 we discuss the related computational aspects, and we present several strategies that can be used to deal with the augmented problem, in order to minimize the computational cost. We shall see in particular that several other methods that are known in the literature can actually be seen as variants of the RFB procedure, in which one or another of the above strategies is employed. This includes, for advection dominated problems, the classical SUPG methods (as it was already well known, see, e.g., [4]) as well as the older Petrov-Galerkin methods based on suitable operator dependent choices of test and trial functions [25]. For composite materials, this includes both the multiscale methods of [22], [23], and the upscaling methods of [1], [2]. Finally, in Section 6 we draw some conclusions.

2   The Continuous Problem

We consider the following continuous problem

$$\begin{cases} \text{find } u \in V \text{ such that} \\ \mathcal{L}(u,v) = \langle f, v \rangle \qquad \forall\, v \in V, \end{cases} \tag{2.1}$$

where V is a Hilbert space, and V' its dual space, $\mathcal{L}(u,v)$ is a continuous bilinear form on $V \times V$, and $f \in V'$ is the forcing term. We assume that, for all $f \in V'$, problem (2.1) has a unique solution. Various problems arising from physical applications can be written in the variational form (2.1), according to different choices of the space V and the bilinear form $\mathcal{L}$. Typical choices for V, when V is a space of scalar functions, are the following: if $\mathcal{O} \subset \mathbb{R}^d$, $(d = 1, 2, 3)$ denotes a generic domain, V could be, for instance, $L^2(\mathcal{O})$, $H^1(\mathcal{O})$, $H_0^1(\mathcal{O})$, $H^2(\mathcal{O})$ or $L_0^2(\mathcal{O})$, the last one being the space of L^2−functions having zero mean value. In the case where V is a space of vector valued functions, a first choice could be to take the Cartesian product of the previous scalar spaces. Other typical choices for V can be:

$$H(\text{div}; \mathcal{O}) := \{ \tau \in (L^2(\mathcal{O}))^d \text{ such that } \nabla \cdot \tau \in L^2(\mathcal{O}) \},$$
$$H_0(\text{div}; \mathcal{O}) := \{ \tau \in H(\text{div}; \mathcal{O}) \text{ such that } \tau \cdot \mathbf{n} = 0 \text{ on } \partial\mathcal{O} \},$$

or also, for a generic domain $\mathcal{O} \subset \mathbb{R}^3$,

$$H(\mathbf{curl}; \mathcal{O}) := \{ \tau \in (L^2(\mathcal{O}))^3 \text{ such that } \nabla \wedge \tau \in (L^2(\mathcal{O}))^3 \}$$
$$H_0(\mathbf{curl}; \mathcal{O}) := \{ \tau \in H(\mathbf{curl}; \mathcal{O}) \text{ such that } \tau \wedge \mathbf{n} = 0 \text{ on } \partial\mathcal{O} \}.$$

Product spaces are also frequently used: for instance, $H(\mathrm{div}; \mathcal{O}) \times L^2(\mathcal{O})$, or $(H^1_0(\mathcal{O}))^d \times L^2_0(\mathcal{O})$, etc. Next, we provide some classical examples of problems and we indicate the corresponding space V, the bilinear form $\mathcal{L}$, and the variational formulation.

Example 2.1. Advection-dominated scalar equations:

$$-\varepsilon\Delta u + \mathbf{c} \cdot \nabla u = f \quad \text{in } \Omega; \quad u = 0 \quad \text{on } \partial\Omega$$

$$V := H^1_0(\Omega); \quad \mathcal{L}(u,v) := \int_\Omega \varepsilon\nabla u \cdot \nabla v \, dx + \int_\Omega \mathbf{c} \cdot \nabla u \, v \, dx; \quad \langle f,v \rangle := \int_\Omega fv \, dx$$

$$\mathcal{L}(u,v) = \langle f,v \rangle \quad \forall \, v \in V.$$

Example 2.2. Linear elliptic problems with composite materials:

$$-\nabla \cdot (\alpha(x)\nabla u) = f \quad \text{in } \Omega; \quad u = 0 \quad \text{on } \partial\Omega$$

$$V := H^1_0(\Omega); \quad \mathcal{L}(u,v) := \int_\Omega \alpha(x)\nabla u \cdot \nabla v \, dx; \quad \langle f,v \rangle := \int_\Omega fv \, dx$$

$$\mathcal{L}(u,v) = \langle f,v \rangle \quad \forall \, v \in V$$

(where $\alpha(x) \geq \alpha_0 > 0$ might have a very fine structure).

Example 2.3. Composite materials in mixed form, i.e., the same problem of the previous example, but now with:

$$\sigma = -\alpha\nabla\psi \quad \text{in } \Omega; \quad \nabla \cdot \sigma = f \quad \text{in } \Omega; \quad \psi = 0 \quad \text{on } \partial\Omega$$

$$V := \Sigma \times \Phi; \quad \Sigma := H(\mathrm{div}; \Omega); \quad \Phi := L^2(\Omega)$$

$$a_0(\sigma, \tau) := \int_\Omega \alpha^{-1}\sigma \cdot \tau \, dx, \quad b(\tau, \varphi) := \int_\Omega \nabla \cdot \tau \, \varphi \, dx$$

$$\mathcal{L}((\sigma,\psi),(\tau,\varphi)) := a_0(\sigma, \tau) - b(\tau, \psi) + b(\sigma, \varphi); \quad \langle f,(\tau, \varphi) \rangle := \int_\Omega f\varphi \, dx$$

$$\mathcal{L}((\sigma,\psi),(\tau,\varphi)) = \langle f,(\tau,\varphi) \rangle \quad \forall \, (\tau, \varphi) \in V.$$

Example 2.4. Stokes problem for viscous incompressible fluids:

$$-\Delta\mathbf{u} + \nabla p = \mathbf{f} \quad \text{in } \Omega; \quad \nabla \cdot \mathbf{u} = 0 \quad \text{in } \Omega; \quad \mathbf{u} = 0 \quad \text{on } \partial\Omega$$

$$V := \mathbf{U} \times Q; \quad \mathbf{U} := (H^1_0(\Omega))^d; \quad Q := L^2_0(\Omega)$$

$$a_1(\mathbf{u}, \mathbf{v}) := \int_\Omega \nabla\mathbf{u} : \nabla\mathbf{v} \, dx \quad b(\mathbf{v}, q) := \int_\Omega \nabla \cdot \mathbf{v} \, q \, dx$$

$$\mathcal{L}((\mathbf{u},p),(\mathbf{v},q)) := a_1(\mathbf{u}, \mathbf{v}) - b(\mathbf{v}, p) + b(\mathbf{u}, q); \quad \langle f,(\mathbf{v},q) \rangle := \int_\Omega \mathbf{f} \cdot \mathbf{v} \, dx$$

$$\mathcal{L}((\mathbf{u},p),(\mathbf{v},q)) = \langle f,(\mathbf{v},q) \rangle \quad \forall \, (\mathbf{v},q) \in V.$$

3 From the Discrete Problem to the Augmented Problem

Let $\mathcal{T}_h$ be a decomposition of the computational domain Ω, with the usual nondegeneracy conditions [12], and let $V_h \subset V$ be a finite element space. The original discrete problem is then:

$$\begin{cases} \text{find } u_h \in V_h \text{ such that} \\ \mathcal{L}(u_h, v_h) = \langle f, v_h \rangle \quad \forall \, v_h \in V_h. \end{cases} \tag{3.1}$$

Note that we do not assume that (3.1) has a unique solution. Indeed, the stabilization that we are going to introduce can, in some cases, take care of problems originally ill-posed. Our aim is, essentially, to solve eventually a final linear system having as many equations as the number of degrees of freedom of V_h. Apart from that, we are ready to pay some extra price, in order to have a better method. In some cases, the total amount of additional work will be small. In other cases, it can be huge. However, we want to be able to perform the extra work independently in each element so that we can do it, as a pre-processor, *in parallel*. This implies that we are ready to add as many degrees of freedom as we want at the interior of each element. For that, to V and $\mathcal{T}_h$ we associate the *maximal space of bubbles*

$$B(V; \mathcal{T}_h) = \prod_K B_V(K), \quad \text{with } B_V(K) = \{v \in V : \text{ supp}(v) \subseteq \overline{K}\}.$$

Let us give some examples of the dependence of $B_V(K)$ on V.

- if $V = H_0^1(\Omega)$ then $B_V(K) = H_0^1(K)$
- if $V = H^1(\Omega)$ then $B_V(K) = \{v \in H^1(K), v = 0 \text{ on } \partial K \cap \Omega\}$
- if $V = L^2(\Omega)$ then $B_V(K) = L^2(K)$
- if $V = L_0^2(\Omega)$ then $B_V(K) = L_0^2(K)$
- if $V = H_0^2(\Omega)$ then $B_V(K) = H_0^2(K)$
- if $V = H_0(\text{div}; \Omega)$ then $B_V(K) = H_0(\text{div}; K)$
- if $V = H(\text{div}; \Omega)$ then $B_V(K) = \{\tau \in H(\text{div}; K), \tau \cdot \mathbf{n} = 0 \text{ on } \partial K \cap \Omega\}$

Similar definitions and properties hold for the spaces $H(\mathbf{curl}; \mathcal{O})$, but we are not going to use them here.

Let us now turn to the choice of the local bubble space $B_h(K)$. If possible, we would like to augment the space V_h by adding, in each element K, the whole $B_V(K)$. This would change V_h into $V_h + B(V; \mathcal{T}_h)$. However, some conditions are needed, as we shall see below. This might forbid, in some cases, taking the whole $B_V(K)$ in the augmentation process: some components of $B_V(K)$ have to be discarded. This will become more clear in the examples below. At this very abstract and general level, we assume that, in each $K \in \mathcal{T}_h$, we choose a subspace $B_h(K) \subseteq B_V(K)$ and, for the moment, "the bigger the better". A first condition that we require is that, for every $g \in V'$, the auxiliary problem

$$\begin{cases} \text{find } w_{B,K} \in B_h(K) \text{ such that} \\ \mathcal{L}(w_{B,K}, v) = \langle g, v \rangle \quad \forall v \in B_h(K) \end{cases} \tag{3.2}$$

has a unique solution. We point out that the choice "the bigger the better" for $B_h(K)$ is made (so far) in order to understand the full potential of the method. As we shall see, in practice we will need to solve (3.2) a few times in each K. This implies that a finite dimensional choice for $B_h(K)$ will be, in the end, necessary.

Having chosen $B_h(K)$, we can now write the *augmented problem*. For that, let

$$V_A := V_h + \Pi_K B_h(K). \qquad (3.3)$$

Two requirements have to be fulfilled: first of all, in (3.3) we must have a direct sum, and, second, for every $f \in V'$, the augmented problem

$$\begin{cases} \text{find } u_A \in V_A \text{ such that} \\ \mathcal{L}(u_A, v_A) = \langle f, v_A \rangle \qquad \forall\, v_A \in V_A \end{cases} \qquad (3.4)$$

must have a unique solution. To summarize, in the augmentation process three conditions have to be fulfilled:

1) the local problems (3.2) must have a unique solution;

2) in (3.3) we must have a direct sum;

3) the augmented problem (3.4) must have a unique solution.

These are then the requirements that can guide us in choosing $B_h(K)$ in the various cases.

Examples of choices of $B_h(K)$.

Example 3.1. Referring to Examples 2.1 and 2.2 of the previous section, suppose that V_h is made of continuous piecewise linear functions. In this case it is easy to check that the choice $B_h(K) = B_V(K) \equiv H_0^1(K)$ verifies all of the three conditions.

Example 3.2. Suppose now that, always referring to Examples 2.1 and 2.2, V_h is made of continuous piecewise cubic functions. The choice $B_h(K) = B_V(K)$ is not viable anymore, as clearly condition 2) is violated: V_h contains functions of $B_V(K)$. In situations like this we should then choose a different $B_h(K)$, but we could also *reduce* the original space V_h. This is actually the simplest strategy, and we are going to follow it. Here, for instance, we can just remove the cubic bubble from $V_{h|K}$ and take a reduced space, still denoted by V_h with an abuse of notation, as a space of any serendipity cubic element (see, for instance, the element described in [12], page 50). Or we might take V_h as the space of functions v_h that are polynomials of degree ≤ 3 at the interelement boundaries and verify $Lv_h = 0$ separately in each K. Notice that these two choices produce the same augmented space V_A, and hence the same solution u_A to (3.4).

Example 3.3. Let us consider the problem of Example 2.3, and assume that $V_h = \Sigma_h \times U_h$ is made by lowest order Raviart-Thomas elements (see for instance [3]). For this problem we have

$$B_V(K) = \{\tau \in H(\text{div}; K),\ \tau \cdot \mathbf{n} = 0 \text{ on } \partial K \cap \Omega\} \times L^2(K).$$

we notice now that taking $B_h(K) = B_V(K)$ would not guarantee that problem (3.2) has a unique solution. Indeed, for internal elements K, the inf-sup condition is not satisfied, since $\int_K \operatorname{div} \tau v \, dx = 0 \; \forall \; v$ constant on K. Condition 2) would also be violated by the choice $B_h(K) = B_V(K)$: in fact, U_h being the space of piecewise constants, $U_{h|K}$ contains bubbles of $L^2(K)$. A possible remedy in this case is to take

$$B_h(K) = H_0(\operatorname{div}; K) \times L_0^2(K) \subset B_V(K).$$

With this choice V_h remains the same, and B_h is the space of all pairs $(\tau, v) \in V$ such that τ has zero normal component at the boundary of each element, and v has zero mean value in each element. The same choice for B_h would be suitable also in the case of higher order Raviart-Thomas spaces (or, say, for BDM spaces; see always [3]), but then V_h should lose all internal degrees of freedom, apart from the piecewise constant scalars.

Example 3.4. Let us now examine the Stokes problem of Example 2.4, and assume that V_h is made of piecewise quadratic velocities in $(H_0^1(\Omega))^d$, and discontinuous piecewise linear pressures in $L_0^2(\Omega)$, a choice which is known not to be stable, but can be stabilized with the present technique. Actually, in this case one can see that $B_V(K) = (H_0^1(K))^d \times L_0^2(K)$. Taking $B_h(K) = B_V(K)$ would violate condition 2), but we can reduce the space V_h, taking it to be the space of quadratic velocities and *constant* pressures. It is easy to check that with this last choice we have a direct sum in (3.3). Moreover, problem (3.4) has a unique solution, because the inf-sup condition is now verified in V_A.

Example 3.5. Let us again consider the Stokes problem of Example 2.4, but now with $V_h = U_h \times Q_h$ made of piecewise linear continuous velocities in $(H_0^1(\Omega))^d$, and piecewise constant pressures in $L_0^2(\Omega)$. It is well known that for this choice the inf-sup condition does not hold. Moreover, if we augment V_h with bubble functions, in any way, the augmented problem (3.4) will **never** verify the inf-sup condition. To see that, augment the velocity space: $U_A = U_h + \Pi_K(H_0^1(K))^d$ as much as you can, and augment the pressure space: $Q_A = Q_h + \{0\}$ as little as you can. For every $v \in (H_0^1(K))^d$ and for every constant q in K, we clearly have $(\operatorname{div} v, q) = 0$. Hence, for $q \in Q_h$:

$$\sup_{v \in V_A} \frac{(\operatorname{div} v, q)}{\|v\|_1} = \sup_{v \in U_h} \frac{(\operatorname{div} v, q)}{\|v\|_1},$$

and we know that the last quantity cannot bound $\|q\|_0$ for all $q \in Q_h$. We clearly see that, in cases like this, our strategy is totally useless, and should not be applied.

4   An Example of Error Estimates

To give an idea of how to proceed to obtain error estimates, let us consider, as an example, a general singular perturbation problem where

$$\mathcal{L}(u, v) := \varepsilon a_1(u, v) + a_0(u, v)$$

with

$$a_1(v,v) \geq \alpha\|v\|_V^2 \quad \forall\, v \in V, \qquad a_1(u,v) \leq \|u\|_V\|v\|_V \quad \forall\, u,v \in V \quad (4.1)$$

$$a_0(v,v) \geq 0 \quad \forall\, v \in V, \qquad a_0(u,v) \leq M\,\|u\|_V\,\|v\|_H \quad \forall\, u,v \in V \quad (4.2)$$

where H is a space such that $V \subset H$ with continuous embedding. We set $e := u - u_A$ and $\eta := u - u_I$, u_I being some interpolant of u in V_h. Proceeding as usual we have

$$\varepsilon\alpha\|e\|_V^2 \leq \mathcal{L}(e,e) = \mathcal{L}(e,\eta) = \varepsilon a_1(e,\eta) + a_0(e,\eta), \qquad (4.3)$$

and the term $a_0(e,\eta)$ is the source of all difficulties, since it does not contain ε as an explicit factor. In order to estimate it, let $\eta = \eta_B + \eta_H$ be any decomposition of η with $\eta_B \in B_h$ and $\eta_H \in H$. Notice that $\eta_B \in B_h \subset V_A$, so that, by Galerkin orthogonality,

$$\varepsilon a_1(e,\eta_B) = -a_0(e,\eta_B). \qquad (4.4)$$

Using this and the bounds (4.1)-(4.2) we can proceed as in [9] and deduce:

$$
\begin{aligned}
a_0(e,\eta) &= a_0(e,\eta_B) + a_0(e,\eta_H) = -\varepsilon a_1(e,\eta_B) + a_0(e,\eta_H) \\
&\leq \varepsilon\|e\|_V\|\eta_B\|_V + M\|e\|_V\|\eta_H\|_H \\
&\leq \varepsilon^{1/2}\left(\varepsilon^{1/2}\|e\|_V\|\eta_B\|_V + M\varepsilon^{-1/2}\|e\|_V\|\eta_H\|_H\right) \\
&\leq \varepsilon^{1/2}(1+M)\|e\|_V\left(\varepsilon^{1/2}\|\eta_B\|_V + \varepsilon^{-1/2}\|\eta_H\|_H\right).
\end{aligned}
\qquad (4.5)
$$

Taking now the supremum over all possible decompositions $\eta = \eta_B + \eta_H$, and then over $\varepsilon > 0$ we obtain

$$
\begin{aligned}
&a_0(e,\eta) \\
&\leq \varepsilon^{1/2}(1+M)\|e\|_V \sup_{\varepsilon>0}\left[\sup_{\eta_B+\eta_H=\eta}\left(\varepsilon^{1/2}\|\eta_B\|_V + \varepsilon^{-1/2}\|\eta_H\|_H\right)\right]. \quad (4.6)
\end{aligned}
$$

By definition (see [7]) the double supremum is the norm of η in a suitable interpolation space, usually denoted by $[B_h,H]_{\frac{1}{2},\infty}$, that for brevity we shall denote by F. Hence, (4.6) becomes

$$a_0(e,\eta) \leq \varepsilon^{1/2}(1+M)\|e\|_V\|\eta\|_F. \qquad (4.7)$$

Inserting (4.7) in (4.3) gives

$$\varepsilon\alpha\|e\|_V^2 \leq \varepsilon a_1(e,\eta) + a_0(e,\eta) \leq \varepsilon^{1/2}\|e\|_V(\varepsilon^{1/2}\|\eta\|_V + (1+M)\|\eta\|_F),$$

and finally

$$\varepsilon^{1/2}\alpha\|u - u_A\|_V \leq \varepsilon^{1/2}\|u - u_I\|_V + (1+M)\|u - u_I\|_F. \qquad (4.8)$$

Notice that an estimate for $\varepsilon^{1/2}\|u-u_A\|_V$ is not as bad as we are used to. For instance, with an argument similar to the one used before, using (4.4)-(4.5), from (4.8) we can see that

$$\|A_0(u-u_A)\|_{F'} := \sup_\varphi \frac{a_0(u-u_A,\varphi)}{\|\varphi\|_F}$$

$$= \sup_\varphi \frac{a_0(u-u_A,\varphi_B) + a_0(u-u_A,\varphi_H)}{\|\varphi\|_F}$$

$$= \sup_\varphi \frac{-\varepsilon a_1(u-u_A,\varphi_B) + a_0(u-u_A,\varphi_H)}{\|\varphi\|_F}$$

$$\leq (1+M)\varepsilon^{1/2}\|u-u_A\|_V \sup_\varphi \frac{\varepsilon^{1/2}\|\varphi_B\|_V + \varepsilon^{-1/2}\|\varphi_H\|_H}{\|\varphi\|_F}$$

$$\leq (1+M)\varepsilon^{1/2}\|u-u_A\|_V \leq C(\varepsilon^{1/2}\|u-u_I\|_V + \|u-u_I\|_F),$$

which is a typical estimate that can be obtained with stabilized methods (see, e.g., [22], [27]). We refer to [6], [9], [28] for the error analysis for residual-free bubbles methods for advection dominated problems.

5   Computational Aspects

Let us now examine the structure of the abstract augmented problem (3.4). Since we constructed the space V_A as a direct sum:

$$V_A := \Pi_K B_h(K) \oplus V_h$$

we then have the unique splittings: $u_A = u_B + u_h$, $v_A = v_B + v_h$. The augmented problem can then be written as

$$\begin{cases} \text{find } u_A = u_B + u_h \in V_A \text{ such that} \\ \mathcal{L}(u_B + u_h, v_B + v_h) = \langle f, v_B + v_h \rangle \quad \forall\, v_B \in B_h, \forall\, v_h \in V_h. \end{cases} \quad (5.1)$$

The associated system will therefore have the form:

$$\begin{pmatrix} L_{B,B} & L_{B,h} \\ L_{h,B} & L_{h,h} \end{pmatrix} \begin{pmatrix} u_B \\ u_h \end{pmatrix} = \begin{pmatrix} f_B \\ f_h \end{pmatrix} \qquad \text{with } L_{B,B} \text{ block diagonal.}$$

There are different strategies for solving the (still infinite dimensional) problem (5.1). All of them are based on the (approximate) solution of the problems

$$\begin{cases} \text{find } w_B^i \in B_h \text{ such that} \\ \mathcal{L}(w_B^i, v_B) = \mathcal{L}(v_i, v_B) \equiv \langle Lv_i, v_B \rangle \quad \forall\, v_B \in B_h, \end{cases} \quad (5.2)$$

where the $\{v_i\}$'s are a basis for V_h, plus, if necessary, the solution of the problem

$$\begin{cases} \text{find } w_B^f \in B_h \text{ such that} \\ \mathcal{L}(w_B^f, v_B) = \langle f, v_B \rangle \quad \forall\, v_B \in B_h. \end{cases} \quad (5.3)$$

As we shall see, what is actually needed, for all strategies, is the computation (for $i, j = 1, ..., \dim(V_h)$) of the quantities

$$S_{j,i} := \mathcal{L}(w_B^i, v_j) \equiv \langle w_B^i, L^* v_j \rangle, \quad \text{and} \quad T_j := \mathcal{L}(w_B^f, v_j) \equiv \langle w_B^f, L^* v_j \rangle, \quad (5.4)$$

where L^* is the adjoint operator of L. In turn, the computation of the solution of the problems (5.2) amounts to solving, in each K, the local bubble problem

$$\begin{cases} \text{find } w_{B,K}^i \in B_h(K) \text{ such that} \\ \mathcal{L}(w_{B,K}^i, b) = \langle Lv_i, b \rangle \quad \forall\, b \in B_h(K). \end{cases} \quad (5.5)$$

The same is obviously true for (5.3). Moreover, f can often be approximated, in each K, by elements of $LV_{h|K}$, so that the solution of (5.3) can be easily obtained from the solutions of the problems (5.2).

A careful inspection of the local problems (5.5) suggests several observations that are computationally relevant.

- For each v_i, the computation of w_B^i can be done in parallel.
- In each element K, the dimension of $\text{span}\{Lv_{i|K}\}$ will be small. In general, it will be less than or equal to the number of degrees of freedom of V_h in K.
- Finally, as we already pointed out, only the quantities $S_{j,i} = \langle w_B^i, L^* v_j \rangle$ are actually needed. Hence, only some averages of w_B^i will be used, and therefore a rough approximation might often be sufficient.
- The same considerations clearly hold for the contributions T_j to the right-hand side.

5.1 First Strategy

Let us see in more detail how the whole procedure can be applied in practice. For this, consider problem (5.1) and note that u_B is the solution of

$$\mathcal{L}(u_B, v_B) = -\mathcal{L}(u_h, v_B) + \langle f, v_B \rangle \quad \forall\, v_B \in B_h,$$

and can be seen as an (affine) function of u_h and f:

$$u_B = L_{B,B}^{-1}(f - Lu_h).$$

Substituting into (5.1), and taking now v_h as a test function, gives

$$\mathcal{L}(u_h, v_h) + \mathcal{L}(L_{B,B}^{-1}(f - Lu_h), v_h) = \langle f, v_h \rangle \quad \forall\, v_h \in V_h, \quad (5.6)$$

which is an equation in terms of u_h alone, where the additional term

$$\mathcal{L}(L_{B,B}^{-1}(f - Lu_h), v_h) \equiv \mathcal{L}(u_B, v_h) \quad (5.7)$$

represents the effect of the small scales onto the coarse ones. To see how to compute the additional term (5.7) let us write $u_h := \sum_i U_i v_i$ and take v_j as a test function. We have

$$\mathcal{L}(u_B, v_j) = \mathcal{L}(L_{B,B}^{-1}(f - Lu_h), v_j) = \mathcal{L}(L_{B,B}^{-1}f, v_j) - \sum_i \mathcal{L}(L_{B,B}^{-1}Lv_i, v_j)U_i$$

$$= \mathcal{L}(w_B^f, v_j) - \sum_i \mathcal{L}(w_B^i, v_j)U_i = T_j - \sum_i S_{j,i}U_i,$$

that clearly shows the use of the auxiliary terms T_j and $S_{j,i}$. Indeed, setting

$$K_{j,i} = \mathcal{L}(v_i, v_j), \quad \text{and} \quad F_j = \langle f, v_j \rangle, \tag{5.8}$$

we have from (5.6) that the U_i's can be obtained as the solution of the following linear system of equations:

$$\sum_i (K_{j,i} - S_{j,i}) U_i = F_j - T_j \qquad j = 1, ..., \dim(V_h). \tag{5.9}$$

Example 5.1. To see how this strategy can be applied, let us go back to the advection-dominated equation, that we recall here:

$$-\varepsilon \Delta u + \mathbf{c} \cdot \nabla u = f \quad \text{in } \Omega; \quad u = 0 \text{ on } \partial\Omega,$$

$$V := H_0^1(\Omega); \quad \mathcal{L}(u, v) := \int_\Omega \varepsilon \nabla u \cdot \nabla v \, dx + \int_\Omega \mathbf{c} \cdot \nabla u \, v \, dx.$$

Assume that the original finite element space V_h is made of piecewise linear continuous functions. Assume moreover that both the source term f and the convective term $\mathbf{c}$ are piecewise constant. Then, it is easy to see that for all v_i the terms Lv_i and L^*v_i are constant in each K. Consequently, all the w_B^i can be computed by solving a *single* problem in each K, that is

$$\begin{cases} \text{find } b_K \in H_0^1(K) \text{ such that} \\ \mathcal{L}(b_K, b) = \langle 1, b \rangle \qquad \forall b \in H_0^1(K). \end{cases} \tag{5.10}$$

With some computations, the problem becomes now (see, e.g., [4]):

$$\begin{cases} \text{find } u_h \in V_h \text{ such that, for all } v_h \in V_h : \\ \mathcal{L}(u_h, v_h) - \sum_K \frac{\int_K b_K \, dx}{|K|} \int_K (f - \mathbf{c} \cdot \nabla u_h) \mathbf{c} \cdot \nabla v_h \, dx = \langle f, v_h \rangle. \end{cases} \tag{5.11}$$

This coincides with the *SUPG* method with $\tau_K = \dfrac{\int_K b_K \, dx}{|K|}$ (see [11], [16]).

5.2   Alternative Computational Strategies

Another possibility is to change the space V_h: for every basis function $v_i \in V_h$, define

$$\tilde{v}_i := v_i - w_B^i, \tag{5.12}$$

and remember that w_B^i was defined by

$$\mathcal{L}(w_B^i, v_B) = \mathcal{L}(v_i, v_B) \qquad \forall \, v_B \in B_h. \tag{5.13}$$

Therefore,

$$\mathcal{L}(\tilde{v}_i, v_B) = 0 \qquad \forall \, v_B \in B_h. \tag{5.14}$$

Set now $\tilde{V}_h = \text{span}\{\tilde{v}_i\}$, and notice that, again, $V_A = \tilde{V}_h \oplus B_h$. Split u_A into $u_A = \tilde{u}_h + \tilde{u}_B$, with $\tilde{u}_h$ in $\tilde{V}_h$, and $\tilde{u}_B$ in B_h. Then, thanks to (5.14), $\tilde{u}_B$ is the solution of

$$\mathcal{L}(\tilde{u}_B, v_B) \equiv \mathcal{L}(u_A, v_B) = \langle f, v_B \rangle \quad \forall \, v_B \in B_h. \tag{5.15}$$

Hence $\tilde{u}_B$ equals w_B^f, the solution of (5.3), and can be computed *before* knowing $\tilde{u}_h$. Finally, $\tilde{u}_h$ can be computed as the solution of

$$\mathcal{L}(\tilde{u}_h, v_h) + \mathcal{L}(\tilde{u}_B, v_h) = \langle f, v_h \rangle \quad \forall \, v_h \in V_h, \tag{5.16}$$

with the same number of unknowns and equations as the dimension of V_h. It is interesting to observe that the difference between this and the first strategy is mainly psychological. Indeed, setting $\tilde{u}_h := \sum_i \tilde{U}_i \tilde{v}_i$, we have from (5.12), (5.8), and (5.4)

$$\mathcal{L}(\tilde{u}_h, v_j) = \sum_i \mathcal{L}(\tilde{v}_i, v_j)\tilde{U}_i = \sum_i \mathcal{L}(v_i - w_B^i, v_j)\tilde{U}_i = \sum_i (K_{j,i} - S_{j,i})\,\tilde{U}_i,$$

$$\mathcal{L}(\tilde{u}_B, v_j) = \mathcal{L}(w_B^f, v_j) = T_j, \tag{5.17}$$

so that, inserting (5.17) into (5.16) we obtain

$$\sum_i (K_{j,i} - S_{j,i})\,\tilde{U}_i = F_j - T_j \qquad j = 1, ..., \dim(V_h), \tag{5.18}$$

which is exactly (5.9).

A third possibility would be, assuming that the adjoint problem of (5.13) is uniquely solvable, to define $\hat{w}_B^i$ solution of

$$\mathcal{L}(v_B, \hat{w}_B^i) = \mathcal{L}(v_B, v_i) \qquad \forall \, v_B \in B_h, \tag{5.19}$$

and to associate to any v_i, basis function in V_h, the function

$$\hat{v}_i = v_i - \hat{w}_B^i. \tag{5.20}$$

Therefore, $\hat{v}_i$ is the solution of

$$\mathcal{L}(v_B, \hat{v}_i) \equiv \langle v_B, L^*\hat{v}_i \rangle = 0 \qquad \forall\, v_B \in B_h. \tag{5.21}$$

Set then $V_h^* = \text{span}\{\hat{v}_i\}$, and notice that, in general, V_h^* will be different from $\tilde{V}_h$, unless the bilinear form $\mathcal{L}$ is symmetric. We again have $V_A = V_h^* + B_h$, always with a direct sum. Take now in (5.1) for u_A the same splitting as before, that is, $u_A = \tilde{u}_h + \tilde{u}_B$, with $\tilde{u}_h \in \tilde{V}_h$, $\tilde{u}_B \in B_h$, and for v_A take instead the splitting $v_A = \hat{v}_h + v_B$, with $\hat{v}_h \in V_h^*$, $v_B \in B_h$, always without changing the final solution u_A. Substituting in (5.1) shows that $\tilde{u}_B$ is again the solution of (5.15). Hence, as before, $\tilde{u}_B$ equals w_B^f, and can be computed before knowing $\tilde{u}_h$. Finally, $\tilde{u}_h$ can be computed as the solution of

$$\mathcal{L}(\tilde{u}_h, \hat{v}_h) = \langle f, \hat{v}_h \rangle \qquad \forall\, \hat{v}_h \in V_h^*. \tag{5.22}$$

The matrix associated with (5.22) is however given by

$$\mathcal{L}(\tilde{v}_i, \hat{v}_j) = \mathcal{L}(\tilde{v}_i, v_j - \hat{w}_B^j) = \mathcal{L}(\tilde{v}_i, v_j) = K_{j,i} - S_{j,i} \tag{5.23}$$

(having used (5.20), (5.14), and (5.17)). On the other hand,

$$\langle f, \hat{v}_j \rangle = \langle f, v_j - \hat{w}_B^j \rangle = F_j - \langle f, \hat{w}_B^j \rangle, \tag{5.24}$$

and, using (5.3), (5.19), and (5.4),

$$\langle f, \hat{w}_B^j \rangle = \mathcal{L}(w_B^f, \hat{w}_B^j) = \mathcal{L}(w_B^f, v_j) = T_j. \tag{5.25}$$

We are therefore back to the system (5.18). It is somehow remarkable that the solution of (5.22) can be computed without actually computing the functions $\hat{v}_j$.

Remark 5.1. Although the above strategies, as we have seen, do coincide in practice, this is not often recognized in the literature. For instance, formulations (5.16) and (5.22), when applied to advection dominated problems coincide with the classical so-called Petrov-Galerkin methods in which suitable trial and test functions, depending on the operator, were used (see [25], and see, in Figure 5.1, the typical shape of the basis functions in $\tilde{V}_h$ and V_h^*). The above computation shows that these methods coincide with SUPG when the choice of the stabilization parameter τ_K is made as in (5.11). On the other hand, when applied to problems related to composite materials, as in Example 2.2 (respectively, Example 2.3), the formulation (5.22) reproduces the multiscale methods of [22], [23] and the upscaling method of [1], [2], respectively.

So far, we assumed that we were able to compute the solutions of the local bubble problems (5.2). As anticipated, these solutions cannot be computed exactly, but require some suitable approximation. Let us see, in the particular case of advection dominated problems, how this approximate solutions can be carried out in practice.

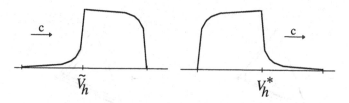

Fig. 5.1. Typical shape of the basis functions in $\tilde{V}_h$ and V_h^*.

We recall that, in this case, solving (5.15) amounts in practice to compute, in each K, the "unitary bubble" b_K, solution of

$$-\varepsilon \Delta b_K + \mathbf{c} \cdot \nabla b_K = 1 \quad \text{in each } K. \tag{5.26}$$

Actually, what we really need is its mean value in each K (see (5.11)). Several tricks can be used to compute $\int_K b_K \, dx$.

- A possibility is to solve by hand the pure convective problem, as advocated in [10]:

$$\begin{cases} \text{find } \tilde{b}_K \in H^1(K) \text{ such that} \\ \mathbf{c} \cdot \nabla \tilde{b}_K = 1 \text{ in } K, \\ \tilde{b}_K = 0 \text{ on } \partial K^- (= \text{inflow}) \end{cases}$$

Notice that the integral of $\tilde{b}_K$ on K is just the volume of a pyramid, as

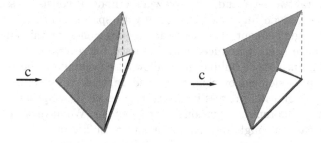

Fig. 5.2. Possible shapes of $\tilde{b}_K$; here $\mathbf{c} = (1,0)$.

shown in Figure 5.2.
- Another possibility is to solve (5.26) on a subgrid with very few degrees of freedom, but well chosen (e.g., Pseudo RFB [8], Shishkin [17], etc, see Figure 5.3. Typically few nodes in the element boundary layer are needed.
- As an alternative, one could use subgrid artificial viscosity; that means solving, instead of (5.26), the problem

$$-(\varepsilon + \varepsilon_A)\Delta b_K + \mathbf{c} \cdot \nabla b_K = 1 \quad \text{in each } K$$

on a very rough grid (typically, one node), where ε_A is a suitably chosen artificial viscosity, in general $\simeq h_K$ (see [20]). Unfortunately, the prob-

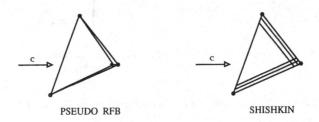

PSEUDO RFB   SHISHKIN

Fig. 5.3. Example of meshes.

lem of the optimal choice for ε_A is rather delicate. Indeed, using a one-dimensional space $B_h(K) = \text{span}\{\beta_K(x)\}$ results in an SUPG method with

$$\tau_K = \frac{(\int_K \beta \, dx)^2}{|K|(\varepsilon + \varepsilon_A) \int_K |\nabla \beta|^2 \, dx},$$

as shown in [5]. This implies that the bigger is ε_A the smaller is τ_K, that is, we add artificial viscosity for stabilizing and we decrease the stabilization parameter.

6 Conclusions

The Residual Free Bubble approach offers a unified framework for setting and analyzing several two-level and/or stabilized methods. It consists, essentially, in augmenting a given finite element space with spaces of functions having support in a single element. The necessary requirements for this augmentation process have been introduced and discussed for several examples. The disconnected nature of the bubble space allows us to eliminate the additional unknowns with an element by element procedure, that can be carried out in parallel. The elimination process involves, in general, the approximate solution of a partial differential equation in each element. We have seen however that in many cases a rough approximation can be sufficient.

The use of this type of approach for stabilizing unstable finite element formulations were already well known. Here we presented the method in a very general setting, and this allowed us to show that several other methods for stabilizing and, mostly, for dealing with subgrid phenomena, can actually be seen as a particular case of the RFB approach. This includes old methods like the Petrov Galerkin methods with special, operator dependent, trial and test functions for advection dominated problems, as well as more recent approaches like the multiscale method or the upscaling method for problems with composite materials.

Other developments and applications to different problems are surely worth further investigation, as well as some recent variants like the use of non-conforming bubbles, the possibility of adding edge-bubbles, or the connections with domain decomposition methods.

References

1. T. Arbogast, " Numerical subgrid upscaling of two-phase flow in porous media,"in "Multiphase flows and transport in porous media: State of the art", (Z. Chen, R.E. Ewing, and Z.-C. Shi eds.), Lecture Notes in Physics, Springer, Berlin, 2000.
2. T. Arbogast, S.E. Minkoff, and P.T. Keenan, "An operator-based approach to upscaling the pressure equation," in: Computational Methods in Water Resources XII, v.1, V.N. Burganos et als., eds., Computational Mechanics Publications, Southampton, U.K., 1998.
3. F. Brezzi, M. Fortin, "Mixed and Hybrid Finite Element Methods," Springer Verlag, New York, Springer Series in Computational Mathematics **15**, 1991.
4. F. Brezzi, L.P. Franca, T.J.R. Hughes, and A. Russo, " $b = \int g$," Comput. Methods Appl. Mech. Engrg. **145**, 329-339 (1997). Methods Appl. Mech. Engrg. **166**, 25-33 (1998).
5. F. Brezzi, P. Houston, L.D. Marini, and E. Süli, "Modeling subgrid viscosity for advection-diffusion problems," Comput. Methods Appl. Mech. Engrg. **190**, 1601-1610 (2000).
6. F. Brezzi, T.J.R. Hughes, L.D. Marini, A. Russo, and E. Süli, "A priori error analysis of a finite element method with residual-free bubbles for advection-dominated equations," SIAM J. Num. Anal. **36**, 1933-1948 (1999)
7. J. Bergh, J. Löfström "Interpolation Spaces" Springer Verlag, Berlin, 1976.
8. F. Brezzi, D. Marini, and A. Russo, "Applications of pseudo residual-free bubbles to the stabilization of convection-diffusion problems," Comput. Methods Appl. Mech. Engrg. **166**, 51-63 (1998).
9. F. Brezzi, D. Marini, and E. Süli, "Residual-free bubbles for advection-diffusion problems: the general error analysis," Numer. Math. **85**, 31-47 (2000).
10. F. Brezzi, A. Russo, "Choosing bubbles for advection-diffusion problems," Math. Mod. and Meth. in Appl. Sci. **4**, 571-587 (1994).
11. A.N. Brooks, T.J.R. Hughes, "Streamline Upwind/Petrov-Galerkin formulations for convection dominated flows with particular emphasis on the incompressible Navier-Stokes equations," Comput. Methods Appl. Mech. Engrg. **32**, 199-259 (1982).
12. Ph.G. Ciarlet, "The finite element method for elliptic problems," North-Holland, 1978.
13. C. Farhat, A. Macedo, and M. Lesoinne, "A two-level domain decomposition method for the iterative solution of high frequency exterior Helmholtz problems," Numer. Math. **85**, 283-308 (2000)
14. L.P. Franca, C. Farhat, A.P. Macedo and M. Lesoinne, "Residual-Free Bubbles for the Helmholtz Equation," Int. J. Num. Meth. in Eng. **40**, 4003-4009 (1997).
15. L.P. Franca, S.L. Frey and T.J.R. Hughes, "Stabilized finite element methods: I. Applications to advective-diffusive model," Comput. Methods Appl. Mech. Engrg. **95**, 253-276 (1992).
16. L.P. Franca, A.P. Macedo, "A Two-Level Finite Element Method and its Application to the Helmholtz Equation," Int. J. Num. Meth. in Eng. **43**, 23-32 (1998).
17. L.P. Franca, A. Nesliturk and M. Stynes, "On the Stability of Residual-Free Bubbles for Convection-Diffusion Problems and Their Approximation by a Two-Level Finite Element Method," Comput. Methods Appl. Mech. Engrg. **166**, 35-49 (1998).

18. L.P. Franca, A. Russo, "Deriving upwinding, mass lumping and selective reduced integration by residual-free bubbles." Appl. Math. Lett. **9**, 83-88 (1996).
19. D.F. Griffiths, A.R. Mitchell, " Spurious behavior and nonlinear instability in discretised partial differential equations," In: The dynamics of numerics and the numerics of dynamics. Inst. Math. Appl. Conf. Ser., New Ser. **34**, 215-242 (1992).
20. J.L. Guermond, "Stabilization of Galerkin approximations of transport equations by subgrid modeling," Math. Mod. Num. Anal. **33**(6), 1293-1316 (1999).
21. P. Hansbo, C. Johnson, "Streamline diffusion finite element methods for fluid flow," von Karman Institute Lectures, 1995.
22. T.Y. Hou, X.H. Wu, "A multiscale finite element method for elliptic problems in composite materials and porous media," J. Comput. Phys. **134**, 169-189 (1997).
23. T.Y. Hou, X.H. Wu, and Z. Cai, "Convergence of a multiscale finite element method for elliptic problems with rapidly oscillating coefficients," Math. of Comp. **68**, 913-943 (1999).
24. T.J.R. Hughes, "Multiscale phenomena: Green's functions, the Dirichlet to Neumann formulation, subgrid scale models, bubbles and the origins of stabilized methods," Comput. Methods Appl. Mech. Engrg. **127**, 387-401 (1995).
25. A.R. Mitchell, D.F. Griffiths, "Generalised Galerkin methods for second order equations with significant first derivative terms," In: Proc. bienn. Conf., Dundee 1977, Lect. Notes Math **630**, 90-104 (1978).
26. H.-G. Roos, M. Stynes, and L. Tobiska, "Numerical methods for singularly perturbed differential equations: convection diffusion and flow problems," Springer-Verlag, 1996.
27. A. Russo, "A posteriori error estimators via bubble functions," Math. Models Methods Appl. Sci. **6**, 353-360 (1997).
28. G. Sangalli, "Global and local error analysis for the Residual Free Bubble method applied to advection-dominated problems," Submitted to Numer. Math.

Stability of Saddle-Points in Finite Dimensions

Franco Brezzi

Dipartimento di Matematica, Università di Pavia, and IMATI-C.N.R., via Ferrata 1, 27100 Pavia, Italy

Abstract. The stability theory for mixed formulations uses, as its starting point, some basic results in functional analysis (and in particular is based on the Banach Closed Range Theorem [10]). This could represent a major obstacle for people having a limited mathematical background. Here we derive the theory in finite dimensional spaces, making as little use as possible of any type of mathematical prerequisites. The treatment is absolutely self-contained. Several variants occurring in applications are considered, and great care is devoted to the dependence of the stability constant upon the other parameters occurring in the problem. Many examples show the optimality of the results with respect to the assumptions. The poverty of instruments makes the discussion long, but all the steps are dealt with in detail. The paper is therefore recommended for readers very little mathematical background but a strong desire to learn.

1 Introduction

We present a simple introduction of the abstract theory that rules most applications of mixed and hybrid finite element methods. Such an abstract theory, in order to be developed with the sufficient amount of generality, would require a few basic instruments in functional analysis that could be unfamiliar to some readers, in particular those with a limited mathematical background. In order to help these readers, we present the theory here in a more simplified form. In particular, we shall start with simple examples of small linear systems having the block structure

$$\begin{pmatrix} A & B^T \\ B & 0 \end{pmatrix} \begin{pmatrix} \mathbf{x} \\ \mathbf{y} \end{pmatrix} = \begin{pmatrix} \mathbf{f} \\ \mathbf{g} \end{pmatrix}, \tag{1.1}$$

and discuss their *solvability* in terms of the properties of the matrices A and B. It is well known that in finite dimension, for a nonsingular linear system, we always have continuous dependence of the solution upon the data. This means that there exists a constant c such that, for every given $\mathbf{f}$ and $\mathbf{g}$

$$\|\mathbf{x}\| + \|\mathbf{y}\| \le c(\|\mathbf{f}\| + \|\mathbf{g}\|) \tag{1.2}$$

where $\mathbf{x}, \mathbf{y}$ is the solution of (1.1). Formula (1.2) deserves a very important comment. We did not specify which norms we were using for the various vectors $\mathbf{x}, \mathbf{y}, \mathbf{f}, \mathbf{g}$. We have the right to do so, since in finite dimension all norms are equivalent. Hence, the change of one norm with another would only result in a change in the numerical value of the constant c, but it would not change the basic fact that such a constant exists. In dealing with linear systems resulting from the discretization of a partial differential equation, we face however a slightly different situation. Indeed, if we want to analyze the behaviour of a given *method* when the meshsize becomes smaller and smaller, we must consider, ideally, a sequence of linear systems whose dimension increases, and tends to infinity when the meshsize tends to zero. As is well known (and can also be easily verified), the constants involved in the equivalence of different norms depend on the dimension of the space. For instance, in $\mathbb{R}^n$, the two norms

$$\|\mathbf{x}\|_1 := \sum_{i=1}^{n} |x_i| \quad \text{and} \quad \|\mathbf{x}\|_2 := (\sum_{i=1}^{n} |x_i|^2)^{1/2}, \quad (1.3)$$

are indeed equivalent, in the sense that there exist two positive constants c_1 and c_2 such that

$$c_1\|\mathbf{x}\|_1 \leq \|\mathbf{x}\|_2 \leq c_2\|\mathbf{x}\|_1, \quad (1.4)$$

for all $\mathbf{x}$ in $\mathbb{R}^n$. However, it can easily be checked that the *best* constants one can choose in (1.4) are

$$\|\mathbf{x}\|_2 \leq \|\mathbf{x}\|_1 \leq \sqrt{n}\|\mathbf{x}\|_2. \quad (1.5)$$

In particular the equality can be reached in the first inequality (for instance) when $\mathbf{x} = (1,0,\ldots,0)$ and in the second inequality (for instance) when $\mathbf{x} = (1,\ldots,1)$.

When considering a sequence of problems with increasing dimension, we have to take into account that n becomes unbounded. It is then natural to ask the following question: Is it possible, for a given choice of the norms $\|\mathbf{x}\|$, $\|\mathbf{y}\|$, $\|\mathbf{f}\|$, and $\|\mathbf{g}\|$, to find a constant c *independent of the meshsize* that makes (1.2) hold true for all meshsizes?

The above question will bring us to the study of the *stability* of a given method. However, in these notes, we shall not start from a differential problem and a differential method. We shall just consider several examples of academic sequences of linear systems, and analyze their stability with respect to different choices of the norms.

In order to read this chapter only a rudimentary background in linear algebra will be needed, but we hope that the basic ideas will still be clear enough. For a detailed treatment of some of these problems in the framework of functional spaces we refer mostly to [6], [7] and [3], while a shorter treatment can be found in [4] and [9]. Applications of these types of results to various engineering problems can be found in the above references, and mostly in [2], [8].

We summarize the outline: we first recall some elementary facts in linear algebra. The main goal for that is to fix the notation, and to refresh the memory of people with little mathematical background. Then we consider the unique solvability of problems of the type (1.1), and we describe necessary and sufficient conditions in terms of properties of matrices A and B. At this level, all norms are considered to be equivalent. In the following section we start dealing with *big matrices*, and for this we introduce different norms, together with the problem of *stability* of a sequence of problems for a given choice of the sequences of norms. In particular, different formulations are analyzed and compared for the two crucial stability conditions: the so-called *inf-sup* and *elker* conditions. In the following and final section the stability theorems are proved. In particular we shall concentrate on the *dependence* of the final stability constant upon the parameters that appear in the basic stability assumptions. As we shall see the results change according to the various assumptions that can be made on the matrices (symmetry, semidefiniteness, etc.). Several examples and counterexamples will be given, to discuss the optimality of the estimates in the various cases.

2   Notation, and Basic Results in Linear Algebra

Let r and s be integer positive numbers, and $M : \mathbb{R}^r \to \mathbb{R}^s$ an $s \times r$ real matrix. We denote by M^T the transposed matrix of M, given by

$$M^T_{i,j} = M_{j,i} \qquad i = 1,\ldots,r,\ j = 1,\ldots,s. \tag{2.1}$$

It is clear that M^T is an $r \times s$ matrix, and therefore $M^T : \mathbb{R}^s \to \mathbb{R}^r$. It is also immediate to check that

$$(M^T)^T \equiv M \tag{2.2}$$

and that

$$\mathbf{y}^T M \mathbf{x} \equiv \mathbf{x}^T M^T \mathbf{y} \qquad \forall\, \mathbf{x} \in \mathbb{R}^r,\, \forall\, \mathbf{y} \in \mathbb{R}^s. \tag{2.3}$$

Throughout this section, which is very elementary, we shall denote by 0_r and 0_s the zero vectors in $\mathbb{R}^r$ and in $\mathbb{R}^s$ respectively. This notation will be abandoned in later sections, with only a few exceptions. We define the kernel and the range of M and M^T as follows:

   i) $\ker M := \{\mathbf{x} \in \mathbb{R}^r \text{ such that } M\mathbf{x} = 0_s\}$;

   ii) $\ker M^T := \{\mathbf{y} \in \mathbb{R}^s \text{ such that } M^T\mathbf{y} = 0_r\}$;

   iii) $\operatorname{Im} M := \{\mathbf{y} \in \mathbb{R}^s \text{ such that } M\mathbf{x} = \mathbf{y} \text{ for some } \mathbf{x} \in \mathbb{R}^r\}$;

   iv) $\operatorname{Im} M^T := \{\mathbf{x} \in \mathbb{R}^r \text{ such that } M^T\mathbf{y} = \mathbf{x} \text{ for some } \mathbf{y} \in \mathbb{R}^s\}$.

$$\tag{2.4}$$

If Z is a linear subspace of $\mathbb{R}^r$, the image of the restriction of M to Z will be denoted by $M(Z)$. Hence

$$M(Z) := \{\mathbf{y} \in \mathbb{R}^s \text{ such that } M\mathbf{z} = \mathbf{y} \text{ for some } \mathbf{z} \in Z\}. \tag{2.5}$$

It is clear that $M(\mathbb{R}^r) \equiv \text{Im}(M)$. Let M be an $s \times r$ matrix. Let Z be a subspace of $\mathbb{R}^r$ and W a subspace of $\mathbb{R}^s$. We say that M, restricted to Z, is **injective** if

$$\forall \mathbf{z}^1 \in Z, \forall \mathbf{z}^2 \in Z \text{ we have: } M\mathbf{z}^1 = M\mathbf{z}^2 \Rightarrow \mathbf{z}^1 = \mathbf{z}^2. \qquad (2.6)$$

We say that M, from Z to W, is *surjective* if

$$\forall \mathbf{w} \in W, \exists \mathbf{z} \in Z \text{ such that } M\mathbf{z} = \mathbf{w}. \qquad (2.7)$$

It is easy to see that, if for instance $Z \equiv \mathbb{R}^r$, M is injective if and only if $\ker M = 0_r$. More generally, M restricted to Z is injective if and only if $\ker M \cap Z = 0_r$. On the other hand, if $W \equiv \mathbb{R}^s$, M is surjective if and only if $M(Z) = \mathbb{R}^s$. More generally, M is surjective from Z to W if and only if $W \subset M(Z)$. From now on, if we say that an $s \times r$ matrix M is injective, or surjective, without specifying the subspaces Z and W, we mean that $\ker M = 0_r$ or $\text{Im } M = \mathbb{R}^s$, respectively. In other words, by default we have that $Z = \mathbb{R}^r$ and $W = \mathbb{R}^s$.

The dimension of a linear space will be denoted by dim. Hence, for instance, $\dim(\mathbb{R}^r) = r$. The *rank* of M is defined as the dimension of its range:

$$\text{rank}(M) := \dim(\text{Im } M). \qquad (2.8)$$

Example 2.1. In order to familiarize ourselves with the notation, it will be convenient to consider an elementary example, constructed from the family of matrices

$$M_\alpha = \begin{pmatrix} 0 & 0 & 1 & 0 & 0 \\ 0 & 0 & 0 & 1 & 0 \\ 0 & 0 & 0 & 0 & \alpha \end{pmatrix}, \qquad (2.9)$$

where α is a real parameter. We have clearly $r = 5$ and $s = 3$. For our present purposes, only the cases $\alpha = 0$ and $\alpha = 1$ will be relevant. The transposed matrix will be

$$M_\alpha^T = \begin{pmatrix} 0 & 0 & 0 \\ 0 & 0 & 0 \\ 1 & 0 & 0 \\ 0 & 1 & 0 \\ 0 & 0 & \alpha \end{pmatrix}. \qquad (2.10)$$

It is immediate to check that for $\alpha = 0$ we have:

$$\begin{aligned} &\ker M_0 = \{\mathbf{x} \in \mathbb{R}^5 \text{ such that } x_3 = x_4 = 0\} \quad \dim(\ker M_0) = 3; \\ &\ker M_0^T = \{\mathbf{y} \in \mathbb{R}^3 \text{ such that } y_1 = y_2 = 0\} \quad \dim(\ker M_0^T) = 1; \\ &\text{Im } M_0 = \{\mathbf{y} \in \mathbb{R}^3 \text{ such that } y_3 = 0\} \quad \dim(\text{Im } M_0) = 2; \\ &\text{Im } M_0^T = \{\mathbf{x} \in \mathbb{R}^5 \text{ such that } x_1 = x_2 = x_5 = 0\} \quad \dim(\text{Im } M_0^T) = 2; \end{aligned} \qquad (2.11)$$

while for $\alpha = 1$, instead, we have:

$\ker M_1 = \{\mathbf{x} \in \mathbb{R}^5 \text{ such that } x_3 = x_4 = x_5 = 0\}$ $\dim(\ker M_1) = 2$;

$\ker M_1^T = \{\mathbf{y} \in \mathbb{R}^3 \text{ such that } y_1 = y_2 = y_3 = 0\} \equiv 0_3$ $\dim(\ker M_1^T) = 0$;

$\operatorname{Im} M_1 = \{ \text{ the whole } \mathbb{R}^3 \}$ $\dim(\operatorname{Im} M_1) = 3$;

$\operatorname{Im} M_1^T = \{\mathbf{x} \in \mathbb{R}^5 \text{ such that } x_1 = x_2 = 0\}$ $\dim(\operatorname{Im} M_1^T) = 3$.

$$(2.12)$$

In particular, M_1 is surjective from $\mathbb{R}^5$ to $\mathbb{R}^3$, and M_1^T is injective from $\mathbb{R}^3$ to $\mathbb{R}^5$. The same properties are not true for M_0 and M_0^T respectively.

These simple cases might also be useful to check the properties that will be discussed in the rest of the section.

For a given linear subspace Z of $\mathbb{R}^r$ we define its orthogonal subspace $Z^\perp$ as follows

$$Z^\perp := \{\mathbf{x} \in \mathbb{R}^r \text{ such that } \mathbf{x}^T\mathbf{z} = 0 \ \forall \ \mathbf{z} \in Z\}. \qquad (2.13)$$

It is clear that

$$(Z^\perp)^\perp \equiv Z. \qquad (2.14)$$

Example 2.2. For instance, with the notation of the previous example, if $Z = \ker M_\alpha$, we have, for $\alpha = 0$

$$(\ker M_0)^\perp = \{\mathbf{x} \in \mathbb{R}^5 \text{ such that } x_1 = x_2 = x_5 = 0\}$$
$$\dim((\ker M_0)^\perp) = 2, \qquad (2.15)$$

and for $\alpha = 1$

$$(\ker M_1)^\perp = \{\mathbf{x} \in \mathbb{R}^5 \text{ such that } x_1 = x_2 = 0\}$$
$$\dim((\ker M_1)^\perp) = 3. \qquad (2.16)$$

Always referring to the previous example, we have instead, in $\mathbb{R}^3$: for $\alpha = 0$

$$(\ker M_0^T)^\perp = \{\mathbf{y} \in \mathbb{R}^3 \text{ such that } y_3 = 0\} \quad \dim((\ker M_0^T)^\perp) = 2, \quad (2.17)$$

and for $\alpha = 1$

$$(\ker M_1^T)^\perp = \{ \text{ the whole } \mathbb{R}^3 \} \quad \dim((\ker M_1^T)^\perp) = 3. \qquad (2.18)$$

Notice that the definition of the orthogonal subspace relies on the choice of the whole space. For instance, it is quite common to accept that $\mathbb{R}^r \subset \mathbb{R}^{r+1}$, by identifying $(x_1, \ldots, x_r)$ with $(x_1, \ldots, x_r, 0)$. In this case we could consider Z both to be a subspace of $\mathbb{R}^r$ and a subspace of $\mathbb{R}^{r+1}$. Clearly its orthogonal subspace in $\mathbb{R}^r$ and its orthogonal subspace in $\mathbb{R}^{r+1}$ will be different. However this type of confusion will not occur here.

If, as in (2.13), Z is a subspace of $\mathbb{R}^r$, then we also have easily

$$\dim(Z^\perp) + \dim(Z) = r. \qquad (2.19)$$

We need now to prove an easy proposition.

Proposition 2.1. *The restriction of M to $(\ker M)^{\perp}$ is a one-to-one mapping between $(\ker M)^{\perp}$ and $\operatorname{Im} M$.*

Proof. The result is almost obvious, but let us detail the proof as an exercise. Let us see first that M, restricted to $(\ker M)^{\perp}$, is injective (see (2.6)). Indeed, if $\mathbf{z}^1$ and $\mathbf{z}^2$ belong to $(\ker M)^{\perp}$, and $M\mathbf{z}^1 = M\mathbf{z}^2$, then setting $\widetilde{\mathbf{x}} := \mathbf{z}^1 - \mathbf{z}^2$ we have $M\widetilde{\mathbf{x}} = 0$ and hence $\widetilde{\mathbf{x}} \in \ker M$. On the other hand the vector $\widetilde{\mathbf{x}}$, being the difference between two elements of $(\ker M)^{\perp}$, must also be in $(\ker M)^{\perp}$. Hence $\widetilde{\mathbf{x}} = 0$, which means $\mathbf{z}^1 = \mathbf{z}^2$, as we wanted. Let us now see that M, as a mapping from $(\ker M)^{\perp}$ to $\operatorname{Im} M$, is surjective (see (2.7)). For this, let $\mathbf{w}$ be an element of $\operatorname{Im} M$. By definition, there exists an $\mathbf{x} \in \mathbb{R}^r$ such that $M\mathbf{x} = \mathbf{w}$. Split this $\mathbf{x}$ into its components along $\ker M$ and $(\ker M)^{\perp}$. Let $\mathbf{x} = \mathbf{x}_K + \mathbf{z}$ be the splitting, with $\mathbf{x}_K \in \ker M$ and $\mathbf{z} \in (\ker M)^{\perp}$. By definition $M\mathbf{x}_K = 0$, so that $M\mathbf{z} = M\mathbf{x}_K + M\mathbf{z} = M\mathbf{x} = \mathbf{w}$, as we wanted. $\square$

Using Proposition 2.1 we have immediately that

$$\dim((\ker M)^{\perp}) = \dim(\operatorname{Im} M), \qquad (2.20)$$

and applying the same argument to M^T

$$\dim((\ker M^T)^{\perp}) = \dim(\operatorname{Im} M^T). \qquad (2.21)$$

Notice that (2.20) and (2.21) are in agreement with the previous examples. Check for instance (2.20), comparing (2.15) with (2.11), and comparing (2.16) with (2.12). Check similarly (2.21), comparing (2.17) with (2.11), and (2.18) with (2.12).

Using (2.20) and (2.19) we immediately have

$$\dim(\operatorname{Im} M) + \dim(\ker M) = r, \qquad (2.22)$$

which can also be easily checked on examples (2.11) and (2.12).

In the special case when $r = s$, that is, when M is a square matrix, property (2.22) corresponds to the well known property: *M is injective (that is $\ker M = 0_r$) if and only if it is surjective (that is $\operatorname{Im} M \equiv \mathbb{R}^r$)*. In this case, the square matrix M is said to be *nonsingular*. In other words,

Theorem 2.1. *Let M be a square $r \times r$ matrix. The system*

$$M\mathbf{x} = \mathbf{f} \qquad (2.23)$$

has a unique solution for every right-hand side $\mathbf{f} \in \mathbb{R}^r$ if and only if the homogeneous system $M\mathbf{x} = 0_r$ has $\mathbf{x} = 0_r$ as unique solution.

We recall now some additional elementary facts in linear algebra. We start from the following crucial result.

Theorem 2.2. *With the above notation, we have*

$$\operatorname{Im} M = (\ker M^T)^{\perp}. \qquad (2.24)$$

Proof. We start by noticing that, for every $\mathbf{y} = M\mathbf{x} \in \operatorname{Im} M$ and for every $\mathbf{z} \in \ker M^T$ we have

$$\mathbf{z}^T \mathbf{y} = \mathbf{z}^T M \mathbf{x} \equiv \mathbf{x}^T M^T \mathbf{z} = 0, \tag{2.25}$$

implying that
$$\operatorname{Im} M \subset (\ker M^T)^\perp, \tag{2.26}$$

and therefore
$$\dim(\operatorname{Im} M) \le \dim((\ker M^T)^\perp). \tag{2.27}$$

Applying the same argument to M^T we obtain instead

$$\dim(\operatorname{Im} M^T) \le \dim((\ker M)^\perp). \tag{2.28}$$

Applying (2.21), (2.28), (2.20) we have

$$\dim((\ker M^T)^\perp) = \dim(\operatorname{Im} M^T) \le \dim((\ker M)^\perp) = \dim(\operatorname{Im} M), \tag{2.29}$$

which combined with (2.26) concludes the proof. $\square$

Remark 2.1. The above theorem is the finite dimensional equivalent of the **Banach Closed Range Theorem**. We succeeded in giving an elementary proof of it only because we are dealing with finite dimensional spaces. And, indeed, the dimensional count played a crucial role in the proof. In infinite dimensional spaces the theorem requires additional assumptions, and its proof is nontrivial (see e.g. [10]).

From (2.24) we can easily deduce several useful properties. To start with, we immediately have

$$\operatorname{Im} M \equiv \mathbb{R}^s \Leftrightarrow \ker M^T = 0_s, \tag{2.30}$$

and, using (2.14),
$$\ker M^T = (\operatorname{Im} M)^\perp. \tag{2.31}$$

On the other hand, exchanging M and M^T in (2.22), (2.24), (2.30), and (2.31), and using (2.2), we immediately have

i) $\dim(\ker M^T) + \dim(\operatorname{Im} M^T) = s.$

ii) $\operatorname{Im} M^T = (\ker M)^\perp.$

iii) $\operatorname{Im} M^T \equiv \mathbb{R}^r \Leftrightarrow \ker M = 0_r.$

$$\tag{2.32}$$

iv) $\ker M = (\operatorname{Im} M^T)^\perp.$

All the above properties can also be easily checked on the example of matrices M_α in (2.9) and their transpositions.

Later it will be convenient to be able to express properties of the matrix M in terms of properties of the associated *bilinear form* (from $\mathbb{R}^r \times \mathbb{R}^s$ to $\mathbb{R}$):

$$(\mathbf{x}, \mathbf{y}) \to \mathbf{y}^T M \mathbf{x}. \tag{2.33}$$

For instance the property: *M is injective* (see (2.6)) can be expressed as

$$\forall\, \mathbf{x} \in \mathbb{R}^r \text{ with } \mathbf{x} \neq 0_r, \quad \exists\, \mathbf{y} \in \mathbb{R}^s \text{ such that } \mathbf{y}^T M \mathbf{x} \neq 0. \quad (2.34)$$

To express instead the property that *M is surjective* (that is Im $M = \mathbb{R}^s$) we can write:

$$\forall\, \mathbf{y} \in \mathbb{R}^s \text{ with } \mathbf{y} \neq 0_s, \quad \exists\, \mathbf{x} \in \mathbb{R}^r \text{ such that } \mathbf{y}^T M \mathbf{x} \neq 0. \quad (2.35)$$

Actually, (2.35) is obviously equivalent to

$$\forall\, \mathbf{y} \in \mathbb{R}^s \text{ with } \mathbf{y} \neq 0_s, \quad \exists\, \mathbf{x} \in \mathbb{R}^r \text{ such that } \mathbf{x}^T M^T \mathbf{y} \neq 0, \quad (2.36)$$

which says that M^T is injective, see (2.24). Similarly, (2.34) is equivalent to

$$\forall\, \mathbf{x} \in \mathbb{R}^r \text{ with } \mathbf{x} \neq 0_r, \quad \exists\, \mathbf{y} \in \mathbb{R}^s \text{ such that } \mathbf{x}^T M^T \mathbf{y} \neq 0, \quad (2.37)$$

which says that M^T is surjective, see (2.32ii, iv).

Note that, in the case of a square matrix ($r = s$), (2.34) and (2.35) are equivalent, as already pointed out in Theorem 2.1.

Example 2.3. Always referring to the cases of Example 2.1 we have that the bilinear form in question is now

$$(\mathbf{x}, \mathbf{y}) \to y_1 x_3 + y_2 x_4 + \alpha y_3 x_5, \quad \mathbf{x} \in \mathbb{R}^5, \mathbf{y} \in \mathbb{R}^3. \quad (2.38)$$

Remember that M_α is surjective (and M_α^T injective) for $\alpha \neq 0$. Indeed, both (2.35) and (2.36) fail for $\alpha = 0$, as can be seen by taking $\mathbf{y} = (0,0,1)^T$. On the other hand, M_α is never injective (no matter what the value of α), as can also be seen in (2.34) by taking for instance $\mathbf{x} = (1,1,0,0,0)^T$. Taking the same $\mathbf{x}$ in (2.37) we see that M_α^T is never surjective. This is clearly due to the fact that $5 > 3$.

Finally, for a given subspace Z, say, of $\mathbb{R}^s$ we introduce the *orthogonal projection* $\pi_Z \colon \mathbb{R}^s \to Z$ as follows. For a given $\mathbf{x} \in \mathbb{R}^s$, the vector $\mathbf{x}_Z := \pi_Z \mathbf{x}$ is the unique solution, in Z, of the equation

$$\mathbf{z}^T \mathbf{x}_Z \equiv \mathbf{z}^T \pi_Z \mathbf{x} = \mathbf{z}^T \mathbf{x}, \quad \forall\, \mathbf{z} \in Z. \quad (2.39)$$

As is well known, $\mathbf{x}_Z$ could also be seen as the minimizer, in Z, of the quantity $\|\mathbf{x} - \mathbf{z}\|$, that is

$$\|\mathbf{x} - \mathbf{x}_Z\| \leq \|\mathbf{x} - \mathbf{z}\|, \quad \forall\, \mathbf{z} \in Z. \quad (2.40)$$

Clearly a similar definition holds for subspaces of $\mathbb{R}^r$.

Example 2.4. Always referring to the cases of Example 2.1, if, for instance, $Z = \ker M_0$ and $\mathbf{x} = (1,2,3,4,5)^T$, then $\pi_Z \mathbf{x} = (1,2,0,0,5)^T$.

Given a subspace W of $\mathbb{R}^r$ and a subspace Z of $\mathbb{R}^s$ we might want to characterize, in terms of the bilinear form (2.33), the cases when the mapping $\mathbf{x} \to \pi_Z M\mathbf{x}$, from W into Z is injective or surjective. Using (2.34) and (2.35), and the obvious consequence of (2.39):

$$\mathbf{z}^T \pi_Z M\mathbf{x} = \mathbf{z}^T M\mathbf{x}, \quad \forall\, \mathbf{z} \in Z \;\forall\, \mathbf{x} \in \mathbb{R}^r, \tag{2.41}$$

we have that $\pi_Z M$ is injective if and only if

$$\forall\, \mathbf{w} \in W \text{ with } \mathbf{w} \neq 0_r, \quad \exists\, \mathbf{z} \in Z \text{ such that } \mathbf{z}^T M\mathbf{w} \neq 0, \tag{2.42}$$

and $\pi_Z M$ is surjective if and only if

$$\forall\, \mathbf{z} \in Z \text{ with } \mathbf{z} \neq 0_s, \quad \exists\, \mathbf{w} \in W \text{ such that } \mathbf{z}^T M\mathbf{w} \neq 0. \tag{2.43}$$

Remark 2.2. It is clear that, when $r = s$ and $W = Z$, then (2.42) and (2.43) are equivalent, that is: $\pi_Z M$, from W to itself, is injective if and only if it is surjective. Indeed, in this case $\pi_Z M\mathbf{w} = \mathbf{z}$ corresponds to a square system (see Theorem 2.1).

3 Existence and Uniqueness of Solutions: the Solvability Problem

We go back to our general form (1.1). We assume that $\mathbf{f}$ and $\mathbf{g}$ are given in $\mathbb{R}^n$ and $\mathbb{R}^m$ respectively (n and m being given integer numbers ≥ 1,) and that $\mathbf{x}$ and $\mathbf{y}$ are also sought in $\mathbb{R}^n$ and $\mathbb{R}^m$, respectively. This implies that A must be a square matrix $n \times n$ and B a rectangular matrix $m \times n$. Our present aim is to give necessary and sufficient conditions on A and B in order that (1.1) has a unique solution. We have the following basic result.

Theorem 3.1. *Let n and m be two integers ≥ 1. Let A and B be an $n \times n$ matrix and an $m \times n$ matrix, respectively. Let K be the kernel of B as in (2.4), and let π_K be the orthogonal projection operator from $\mathbb{R}^n$ onto K, as in (2.39). Then the matrix*

$$\begin{pmatrix} A & B^T \\ B & 0 \end{pmatrix} \tag{3.1}$$

is nonsingular if and only if the following two conditions are both satisfied:

$$\text{the map } \pi_K A : K \to K \text{ is surjective (or, equivalently, is injective),} \tag{3.2}$$
$$\text{the rank of } B \text{ is } m \text{ (or, equivalently, } \ker B^T = 0_m). \tag{3.3}$$

Proof. We start by noticing that the equivalence claimed in (3.2) was already made clear in Remark 2.2, while the equivalence claimed in (3.3) is an easy consequence of (2.8) and (2.30).

To prove the theorem, assume first that (3.1) is nonsingular, that is to say that the system (1.1) has a unique solution for every right-hand side $(\mathbf{f}, \mathbf{g}) \in \mathbb{R}^n \times \mathbb{R}^m$. In particular, looking at the second equation (that is, $B\mathbf{x} = \mathbf{g}$) we see that it must have a solution for every $\mathbf{g} \in \mathbb{R}^m$, and hence $\mathrm{Im}\,B \equiv \mathbb{R}^m$. This, using (2.8) gives that the rank of B is m, that is (3.3). Moreover, for every $\mathbf{f} = \mathbf{f}_K \in K$ the system

$$\begin{pmatrix} A & B^T \\ B & 0 \end{pmatrix} \begin{pmatrix} \mathbf{x} \\ \mathbf{y} \end{pmatrix} = \begin{pmatrix} \mathbf{f}_K \\ 0_m \end{pmatrix} \tag{3.4}$$

must have a solution. For every such solution we have clearly $B\mathbf{x} = 0$, that is $\mathbf{x} \in K$. We also note that for every $\mathbf{y} \in \mathbb{R}^m$, thanks to (2.32ii), $B^T\mathbf{y}$ must be orthogonal to K (kernel of B), and therefore $\pi_K B^T \mathbf{y} = 0$. Hence, taking the projection π_K of the first equation of (3.4) yields:

$$\pi_K A\mathbf{x} = \mathbf{f}_K. \tag{3.5}$$

In other words, solving (3.4) we have that: for every $\mathbf{f} = \mathbf{f}_K \in K$ there exists an $\mathbf{x} \in K$ such that (3.5) holds. Hence (3.2) holds.

Assume, conversely, that (3.2) and (3.3) hold. We want to show that the matrix (3.1) is nonsingular. This will follow if we show that the system (1.1), with $\mathbf{g} = 0$ and $\mathbf{f} = 0$ has $\mathbf{x} = 0$, $\mathbf{y} = 0$ as unique solution. Indeed, from $B\mathbf{x} = 0$ we get first that $\mathbf{x} \in K$. Taking the projection π_K of the first equation (and noting again that $\pi_K B^T \mathbf{y} = 0$) we have then $\pi_K A\mathbf{x} = 0$. This, together with $\mathbf{x} \in K$ implies $\mathbf{x} = 0$ thanks to (3.2). Finally, the first equation becomes now $B^T\mathbf{y} = 0$, and this gives $\mathbf{y} = 0$ thanks to (3.3). □

Remark 3.1. It follows easily from (3.3), using for instance (2.22), that *a necessary condition* for the solvability is $n \geq m$. This was obvious from the very beginning, but it could be a valuable first simple check for users lacking a sufficient mathematical background (see e.g. [11]).

In the remainder it will often be more convenient to express conditions (3.2) and (3.3) in terms of the *bilinear forms* $\mathbf{x}^T A\mathbf{x}$ and $\mathbf{y}^T B\mathbf{x}$ instead of the matrices A and B. For this we recall first that, thanks to Remark 2.2, condition (3.2) can be expressed equivalently by (2.42) or by (2.43). On the other hand, (3.3) can be expressed using (2.35) or (2.36). We obtain therefore the following proposition.

Proposition 3.1. *In the assumptions of Theorem 3.1, conditions (3.2) and (3.3) can be equivalently expressed as*

$$\forall\, \mathbf{x} \in K \text{ with } \mathbf{x} \neq 0 \quad \exists\, \mathbf{z} \in K \text{ such that } \mathbf{z}^T A\mathbf{x} \neq 0, \tag{3.6}$$

and

$$\forall\, \mathbf{y} \in \mathbb{R}^m \text{ with } \mathbf{y} \neq 0 \quad \exists\, \mathbf{x} \in \mathbb{R}^n \text{ such that } \mathbf{y}^T B\mathbf{x} \neq 0, \tag{3.7}$$

respectively.

The problem of checking whether (3.2) (or the equivalent (3.6)) holds or not could be avoided in some particular cases, as we shall see in the following corollary. We recall that, in general, a square $r \times r$ matrix M is said to be *positive definite* if

$$\mathbf{x}^T M \mathbf{x} > 0 \qquad \forall \, \mathbf{x} \in \mathbb{R}^r \text{ with } \mathbf{x} \neq 0. \tag{3.8}$$

More generally, if Z is a subspace of $\mathbb{R}^r$, we say that M *is positive definite on Z* if

$$\mathbf{x}^T M \mathbf{x} > 0 \qquad \forall \, \mathbf{x} \in Z \text{ with } \mathbf{x} \neq 0. \tag{3.9}$$

Then we have the following useful result:

Corollary 3.1. *Let A be an $n \times n$ matrix, and B an $m \times n$ matrix with* rank$(B) = m$. *If A is positive definite on the kernel K of B, then the matrix (3.1) is nonsingular.*

Proof. The proof follows easily from Proposition 3.1 and Theorem 3.1. Condition (3.6) follows by taking $\mathbf{z} = \mathbf{x}$ and using the fact that A is positive definite on K. Condition (3.7) follows from the fact that rank$(B) = m$. $\square$

The following corollary has more restrictive assumptions, but its use is even more simple.

Corollary 3.2. *Let A be an $n \times n$ positive definite matrix, and B an $m \times n$ matrix with* rank$(B) = m$. *Then the matrix (3.1) is nonsingular.*

The proof is immediate. The advantage of Corollary 3.2 (when we can use it!) is that there is no need to characterize the kernel K, that, in some cases, can be a nontrivial task.

We notice that the result of Theorem 3.1 could have been obtained in a different, somewhat more algebraic way [1]. As the result is particularly important, we report this alternative way as well, in the hope that two different points of view could provide a deeper understanding of the whole result.

For this, together with the kernel K of B we consider now its orthogonal complement $K^\perp$ in $\mathbb{R}^n$, that we call H. Let n_K be the dimension of K, and n_H the dimension of H. From (2.19) we have

$$n_K + n_H = n. \tag{3.10}$$

We take now a basis $\{\mathbf{x}_1^H, \ldots, \mathbf{x}_{n_H}^H\}$ in H, and a basis $\{\mathbf{x}_1^K, \ldots, \mathbf{x}_{n_K}^K\}$ in K. It is clear that

$$\{\mathbf{x}_1^H, \ldots, \mathbf{x}_{n_H}^H, \mathbf{x}_1^K, \ldots, \mathbf{x}_{n_K}^K\} \tag{3.11}$$

will be a basis for $\mathbb{R}^n$. *With respect to this basis*, we can re-write the matrices A, B, and B^T as follows:

$$A = \begin{pmatrix} A_{HH} & A_{HK} \\ A_{KH} & A_{KK} \end{pmatrix} \qquad B = \begin{pmatrix} B_H & B_K \end{pmatrix} \qquad B^T = \begin{pmatrix} B_H^T \\ B_K^T \end{pmatrix}. \tag{3.12}$$

Now, from the definition (2.4) of K we immediately have that $B_K = 0$ (that is the zero $m \times n_K$ matrix,) so that $B_K^T = 0$ as well. Splitting $\mathbf{x}$ and $\mathbf{f}$ into their orthogonal components $\mathbf{x}_H$ and $\mathbf{x}_K$, and $\mathbf{f}_H$ and $\mathbf{f}_K$, respectively, we can now write the original system (1.1) as follows

$$
\begin{pmatrix} A_{HH} & A_{HK} & B_H^T \\ A_{KH} & A_{KK} & 0 \\ B_H & 0 & 0 \end{pmatrix} \begin{pmatrix} \mathbf{x}_H \\ \mathbf{x}_K \\ \mathbf{y} \end{pmatrix} = \begin{pmatrix} \mathbf{f}_H \\ \mathbf{f}_K \\ \mathbf{g} \end{pmatrix}.
\tag{3.13}
$$

We also notice that B_H is an $m \times n_H$ matrix, where, using (2.20), $n_H = \dim(\mathrm{Im}\, B) = \mathrm{rank}(B) \leq m$. On the other hand, $\ker B_H = 0$, due to proposition 2.1, since B_H is the restriction of B to $\ker(B)^{\perp}$. We conclude that B_H is always injective, and therefore it is a nonsingular square matrix if and only if $n_H = m$.

On the other hand the system (3.13) is block-triangular, and the matrix A_{KK} is obviously square. If B_H (and hence B_H^T) is also a square matrix, then (3.13) is uniquely solvable if and only if B_H and A_{KK} are nonsingular. Let us collect the results.

- If $\mathrm{rank}(B) = m$, then B_H is square and nonsingular. Then, if we also have that A_{KK} is nonsingular, we conclude that (3.13) is uniquely solvable.
- If (3.13) is uniquely solvable then $\mathrm{rank}(B_H) = m$ (and hence $\mathrm{rank}(B) = m$). This implies that B_H is square (and hence nonsingular.) Using again the unique solvability of (3.13) we have that A_{KK} is also nonsingular.

In other words, (3.13) is nonsingular if and only if $\mathrm{rank}(B) = m$ *and* A_{KK} is nonsingular. Finally, we observe that A_{KK} is just the restriction of $\pi_K A$ to K, so that the above result coincides exactly with Theorem 3.1.

Let us now see some examples. We start by pointing out that the part of A that *must* be nonsingular is actually A_{KK}, and **not** A itself. Take for instance, for $n = 2$ and $m = 1$, the matrices

$$
A = \begin{pmatrix} 1 & 1 \\ 1 & 0 \end{pmatrix} \qquad B = \begin{pmatrix} 1 & 0 \end{pmatrix} \qquad B^T = \begin{pmatrix} 1 \\ 0 \end{pmatrix}.
\tag{3.14}
$$

Then the rank of B is 1 ($= m$), and (3.3) holds true. On the other hand we have that $K = \ker B = \{\mathbf{x} \in \mathbb{R}^2 \text{ such that } x_1 = 0\}$. Hence, in this case, the *new basis* (3.11) coincides with the original one, and the matrices are in the form (3.12) already. It is then easy to check that $A_{KK} = (0)$ and (3.2) does not hold, although A, itself, is nonsingular. Indeed the whole matrix is

$$
\begin{pmatrix} 1 & 1 & 1 \\ 1 & 0 & 0 \\ 1 & 0 & 0 \end{pmatrix}
\tag{3.15}
$$

which is clearly singular.

On the other hand, consider the choice

$$A = \begin{pmatrix} 0 & 0 \\ 0 & 1 \end{pmatrix} \qquad B = (1 \quad 0) \qquad B^T = \begin{pmatrix} 1 \\ 0 \end{pmatrix}, \qquad (3.16)$$

where A is singular. Since K is the same as before, the new coordinates (3.11) coincide again with the old ones, and we have easily that $A_{KK} = (1)$. This is clearly nonsingular, so that (3.2) is now satisfied. Indeed the whole matrix is now

$$\begin{pmatrix} 0 & 0 & 1 \\ 0 & 1 & 0 \\ 1 & 0 & 0 \end{pmatrix} \qquad (3.17)$$

which is clearly nonsingular. Along the same lines, referring to Corollary 3.2 we notice that it would not be enough to require that A is positive *semi*definite (that is $x^T A x \geq 0$ for all $x \in \mathbb{R}^n$). Indeed, for the choice

$$A = \begin{pmatrix} 1 & 0 \\ 0 & 0 \end{pmatrix} \qquad B = (1 \quad 0) \qquad B^T = \begin{pmatrix} 1 \\ 0 \end{pmatrix}, \qquad (3.18)$$

we have that A is positive semidefinite, we have that (3.3) is verified, but the whole matrix

$$\begin{pmatrix} 1 & 0 & 1 \\ 0 & 0 & 0 \\ 1 & 0 & 0 \end{pmatrix} \qquad (3.19)$$

is clearly singular.

In many cases, however, it is not immediate to see, at the first glance, what the matrix A_{KK} is. Consider for instance the case

$$A = \begin{pmatrix} a & b \\ c & d \end{pmatrix} \qquad B = (1 \quad -1) \qquad B^T = \begin{pmatrix} 1 \\ -1 \end{pmatrix}. \qquad (3.20)$$

We have in this case

$$K = \{x \in \mathbb{R}^2 \text{ such that } x_1 - x_2 = 0\}. \qquad (3.21)$$

Hence, K can be presented as the one-dimensional subset of $\mathbb{R}^2$ made of vectors of the type $(\alpha, \alpha)^T$ with $\alpha \in \mathbb{R}$. In its turn, H can now be presented as the one-dimensional subset of $\mathbb{R}^2$ made of vectors of the type $(\beta, -\beta)^T$ with $\beta \in \mathbb{R}$. In order to reach the form (3.12) we have now to express the matrix A in the new basis $\{x_1^H, \ldots, x_{n_H}^H, x_1^K, \ldots, x_{n_K}^K\}$ that is now simply $\{x_1^H, x_1^K\}$ where, if we want an orthonormal basis, we can take $x_1^H = (1/\sqrt{2}, -1/\sqrt{2})^T$ and $x_1^K = (1/\sqrt{2}, 1/\sqrt{2})^T$. After some classical computations, we can see that, *in this new basis*, the matrix A takes the form

$$\tilde{A} = \frac{1}{2} \begin{pmatrix} a-b-c+d & a+b-c-d \\ a-b+c-d & a+b+c+d \end{pmatrix}. \qquad (3.22)$$

From (3.22) we have that A_{KK} is the 1×1 matrix $(\frac{1}{2}(a+b+c+d))$, which is nonsingular if and only if $a+b+c+d \neq 0$.

However, the fact that the condition $a+b+c+d \neq 0$ is necessary and sufficient for the matrix

$$
\begin{pmatrix}
a & b & 1 \\
c & d & -1 \\
1 & -1 & 0
\end{pmatrix}
\tag{3.23}
$$

to be nonsingular can be checked in a shorter time directly, without using the above theory. In cases like that (which are the majority) it would be simpler to deal directly with the restriction of $\pi_K A$ to K, which is A_{KK} in the original variables. This would require us to apply the (original) matrix

$$
\begin{pmatrix}
a & b \\
c & d
\end{pmatrix}
\tag{3.24}
$$

to the general vector (in the original coordinates) $\mathbf{x}_K = (\alpha, \alpha)^T$ in K, obtaining the vector

$$
A\mathbf{x}_K = \begin{pmatrix}
\alpha(a+b) \\
\alpha(c+d)
\end{pmatrix}.
\tag{3.25}
$$

Then we have to check whether the component of $A\mathbf{x}_K$ in K (that is $\pi_K Ax$) is different from zero. As K is one-dimensional, this amounts to taking the scalar product

$$
(\mathbf{x}_1^K)^T A\mathbf{x}_K = (1/\sqrt{2} \quad 1/\sqrt{2}) A\mathbf{x}_K = \frac{\alpha}{\sqrt{2}}(a+b+c+d),
\tag{3.26}
$$

and seeing if it is different from zero when α is different from zero. We clearly obtain the same condition, but, this time, in a quicker way.

Sometimes, the matrix A itself has a block structure of the type

$$
\begin{pmatrix}
C & D^T \\
D & 0
\end{pmatrix}.
\tag{3.27}
$$

Then again one has to be careful and require the nonsingularity of A just on the kernel of B. In some cases, together with an A with the structure (3.27) we have a B with the structure $B = (E \quad 0)$ or $B = (0 \quad E)$, so that the whole matrix has the block structure

$$
\begin{pmatrix}
C & D^T & E^T \\
D & 0 & 0 \\
E & 0 & 0
\end{pmatrix}
\quad \text{or} \quad
\begin{pmatrix}
C & D^T & 0 \\
D & 0 & E^T \\
0 & E & 0
\end{pmatrix},
\tag{3.28}
$$

respectively. In these cases, it can be a useful exercise to rewrite conditions (3.2) and (3.3) in terms of properties of the matrices C, D, and E.

To fix ideas, let us assume that, in the first case of (3.28), C is an $r \times r$ matrix, D is an $s \times r$ matrix, and E an $k \times r$ matrix. It is clear that we can directly use Theorem 3.1, with $n = r$ and $m = s + k$. It is not difficult to see that condition (3.3) can now be expressed as: $\mathrm{rank}(D) = s$ and $\mathrm{rank}(E) = k$ (or, equivalently, $\ker D^T = 0_s$ and $\ker E^T = 0_k$). On the other hand, condition (3.2) requires now that *the restriction C_{KK} of C to K is nonsingular*, where $K = \ker D \cap \ker E$.

To deal with the second case of (3.28) we assume instead that C is an $r \times r$ matrix, D is an $s \times r$ matrix, and E a $k \times s$ matrix. Possibly the easiest way to apply Theorem 3.1 consists in performing first an exchange of rows and columns, to reach the form

$$\begin{pmatrix} C & 0 & D^T \\ 0 & 0 & E \\ D & E^T & 0 \end{pmatrix}. \tag{3.29}$$

Then we can take $n = r + k$ and $m = s$ with

$$A = \begin{pmatrix} C & 0 \\ 0 & 0 \end{pmatrix} \qquad B = (D \;\; E^T) \qquad B^T = \begin{pmatrix} D^T \\ E \end{pmatrix}. \tag{3.30}$$

If we apply Theorem 3.1 to this form, condition (3.3) (that is: $\ker B^T = 0_m = 0_s$) requires now $\ker D^T \cap \ker E = 0_s$. Then we have to look at the kernel of B, and require the nonsingularity of A. It is clear that the kernel of B, in this case, is given by

$$K = \{(\mathbf{x}, \mathbf{z}) \in \mathbb{R}^r \times \mathbb{R}^k \text{ such that } D\mathbf{x} + E^T\mathbf{z} = 0_s\}. \tag{3.31}$$

This includes all pairs of the form $(0_r, \tilde{\mathbf{z}})$, with $\tilde{\mathbf{z}} \in \ker E^T$. When we apply the matrix A to one of these vectors, we obviously obtain the zero vector. Hence, if we want that the restriction of A to K is nonsingular, we must first require that these pairs are reduced to $(0_r, 0_k)$, that is we must require that $\ker E^T = 0_k$. But K might also contain pairs $(\mathbf{x}, \mathbf{z})$ with $\mathbf{x} \neq 0_r$, provided $D\mathbf{x} \in \mathrm{Im}\, E^T$. This subset of $\mathbb{R}^r$ can be characterized, also using (2.32ii), as

$$\begin{aligned} \tilde{K} &= \{\mathbf{x} \in \mathbb{R}^r \text{ such that } D\mathbf{x} = E^T\mathbf{z} \text{ for some } \mathbf{z} \in \mathbb{R}^k\} \\ &\equiv \{\mathbf{x} \in \mathbb{R}^r \text{ such that } \tilde{\mathbf{z}}^T D\mathbf{x} = 0 \quad \forall \tilde{\mathbf{z}} \in \ker E\}. \end{aligned} \tag{3.32}$$

On the space $\tilde{K}$ we have to require the nonsingularity of C, that is, using for instance (3.6),

$$\forall \tilde{\mathbf{x}} \in \tilde{K} \quad \exists \mathbf{x} \in \tilde{K} \text{ such that } \tilde{\mathbf{x}}^T C\mathbf{x} \neq 0. \tag{3.33}$$

Hence the conditions can be summarized as:

$$\begin{aligned} &\ker D^T \cap \ker E = 0_s, \\ &\ker E^T = 0_k, \\ &\pi_{\tilde{K}} C \text{ is nonsingular } \tilde{K} \to \tilde{K}. \end{aligned} \tag{3.34}$$

It is not difficult to verify that conditions (3.34) are necessary and sufficient for the nonsingularity of the whole matrix.

Remark 3.2. There is another equivalent way to apply Theorem 3.1, which amounts to considering directly $n = r + s$, $m = k$ and

$$A = \begin{pmatrix} C & D^T \\ D & 0 \end{pmatrix} \qquad B = E \quad \text{directly.} \qquad (3.35)$$

As we are dealing with necessary and sufficient conditions, we would find exactly the same conditions as before, possibly with a longer argument.

A different, more interesting problem arises when we consider the case of systems of the type

$$\begin{pmatrix} A & B^T \\ B & C \end{pmatrix} \begin{pmatrix} \mathbf{x} \\ \mathbf{y} \end{pmatrix} = \begin{pmatrix} \mathbf{f} \\ \mathbf{g} \end{pmatrix}, \qquad (3.36)$$

where again A and B are $n \times n$ and $m \times n$ matrices, respectively, and C is an $m \times m$ matrix. The name of the game here is to see C as a perturbation of the original problem (1.1). We therefore assume that matrices A and B satisfy (3.2) and (3.3), plus, possibly, some minor additional requirement, and we look for conditions on C in order to have the unique solvability of (3.36). A first result is more or less obvious.

Proposition 3.2. *Assume that (3.2) and (3.3) are satisfied. Then there exists an $\varepsilon > 0$ such that, for every $m \times m$ matrix C satisfying*

$$\|C\mathbf{y}\| \le \varepsilon \|\mathbf{y}\|, \quad \forall \, \mathbf{y} \in \mathbb{R}^m, \qquad (3.37)$$

problem (3.36) is uniquely solvable.

The proof is based on the following obvious fact: if the determinant of a matrix is different from zero, and if we perturb the matrix by a small enough quantity, the determinant will still be different from zero. We omit the mathematical details.

The following theorem is more interesting, and more relevant for applications. In order to prove it, however, we are going to need the following lemma, that will also be useful in other occasions.

Lemma 3.1. *Assume that A is a symmetric $n \times n$ matrix satisfying*

$$\mathbf{x}^T A \mathbf{x} \ge 0, \quad \forall \, \mathbf{x} \in \mathbb{R}^n. \qquad (3.38)$$

Then for every $\mathbf{x} \in \mathbb{R}^n$ we have

$$\mathbf{x}^T A \mathbf{x} = 0 \quad \Rightarrow \quad A\mathbf{x} = 0. \qquad (3.39)$$

We recall the proof of Lemma 3.1 for the convenience of the reader. Using (3.38) we easily have that, for any $\mathbf{z} \in \mathbb{R}^n$ and for any real number s,

$$(\mathbf{x} + s\mathbf{z})^T A(\mathbf{x} + s\mathbf{z}) \geq 0. \tag{3.40}$$

Expanding (3.40) in powers of s and using the symmetry of A we have

$$\mathbf{x}^T A\mathbf{x} + 2s\mathbf{z}^T A\mathbf{x} + s^2 \mathbf{z}^T A\mathbf{z} \geq 0, \tag{3.41}$$

implying that the equation (in s) $\mathbf{x}^T A\mathbf{x} + 2s\mathbf{z}^T A\mathbf{x} + s^2 \mathbf{z}^T A\mathbf{z} = 0$ cannot have distinct real roots, and therefore

$$\Delta \equiv (2\mathbf{z}^T A\mathbf{x})^2 - 4(\mathbf{x}^T A\mathbf{x})(\mathbf{z}^T A\mathbf{z}) \leq 0, \tag{3.42}$$

and, dividing by 4,

$$(\mathbf{z}^T A\mathbf{x})^2 \leq (\mathbf{x}^T A\mathbf{x})(\mathbf{z}^T A\mathbf{z}). \tag{3.43}$$

Hence $\mathbf{x}^T A\mathbf{x} = 0$ implies that $\mathbf{z}^T A\mathbf{x} = 0$ *for all* $\mathbf{z} \in \mathbb{R}^n$, and therefore $A\mathbf{x} = 0$.

Remark 3.3. The proof of Lemma 3.1 could have been shortened by using the fact that a symmetric matrix can be diagonalized. Indeed, if A is diagonal, then $\mathbf{x}^T A\mathbf{x} = \sum_{i=1}^n \alpha_i(x_i)^2$. If this is zero, and A is positive semidefinite (implying $\alpha_i \geq 0$ for all i's), we immediately obtain that $\alpha_i(x_i)^2 = 0$ for all i's, and then $\alpha_i x_i = 0$ for all i's (that is $A\mathbf{x} = 0_n$). However, the proof reported before generalizes more easily to infinite dimensional spaces, where the passage to the diagonal form is far less trivial.

We are now ready to present the theorem.

Theorem 3.2. *Assume that B is an $m \times n$ matrix verifying (3.3), and A is an $n \times n$ symmetric matrix, verifying (3.38) and (3.2) with $K = \ker B$. Then for every $m \times m$ matrix C satisfying*

$$\mathbf{y}^T C\mathbf{y} \leq 0, \quad \forall\, \mathbf{y} \in \mathbb{R}^m \tag{3.44}$$

problem (3.36) is uniquely solvable.

Proof. The proof can be done easily by showing that the homogeneous version of (3.36) (that is when $\mathbf{f}$ and $\mathbf{g}$ are both equal to zero) has $\mathbf{x} = 0$, $\mathbf{y} = 0$ as the unique solution. For this, let $(\mathbf{x}, \mathbf{y})$ be the solution of the homogeneous system. Taking the scalar product of $\mathbf{x}^T$ times the first equation of (3.36) we get

$$\mathbf{x}^T A\mathbf{x} + \mathbf{x}^T B^T \mathbf{y} = 0, \tag{3.45}$$

while taking the scalar product of $\mathbf{y}^T$ times the second equation of (3.36) we obtain

$$\mathbf{y}^T B\mathbf{x} + \mathbf{y}^T C\mathbf{y} = 0. \qquad (3.46)$$

Subtracting (3.46) from (3.45), and using (2.3) we have therefore

$$\mathbf{x}^T A\mathbf{x} - \mathbf{y}^T C\mathbf{y} = 0. \qquad (3.47)$$

Using (3.44) and (3.38) in (3.47) we have then

$$\mathbf{x}^T A\mathbf{x} = \mathbf{y}^T C\mathbf{y} = 0. \qquad (3.48)$$

We can now use the symmetry of A and Lemma 3.1 and deduce that $A\mathbf{x} = 0$. Using this in the first equation we obtain now $B^T\mathbf{y} = 0$ which, thanks to (3.3), implies $\mathbf{y} = 0$. This, in turn, gives $C\mathbf{y} = 0$, so that, from the second equation, $B\mathbf{x} = 0$. Hence $\mathbf{x}$ belongs to ker B, and having already $A\mathbf{x} = 0$ we can use (3.2) to conclude that $\mathbf{x}$ is also equal to zero.   □

Remark 3.4. Looking at the proof of Theorem 3.2 we also see that we can trade the symmetry assumption on A with the condition that A is positive definite on the whole $\mathbb{R}^n$. Indeed, the symmetry was only used in Lemma 3.1, to show that $\mathbf{x}^T A\mathbf{x} = 0$ implies $A\mathbf{x} = 0$. If A is supposed to be positive definite, from $\mathbf{x}^T A\mathbf{x} = 0$ we have immediately $\mathbf{x} = 0$ and then $\mathbf{y} = 0$ as before.

Theorem 3.2 has a counterpart, in which the symmetry assumption is shifted from A to C.

Theorem 3.3. *Assume that B is an $m \times n$ matrix verifying (3.3), and A is an $n \times n$ matrix, verifying (3.2) with $K = \ker B$, and (3.38). Then for every symmetric $m \times m$ matrix C satisfying (3.44) problem (3.36) is uniquely solvable.*

Proof. We proceed exactly as in the proof of Theorem 3.2. Let $(\mathbf{x}, \mathbf{y})$ be a solution of the homogeneous system: taking the scalar products of $\mathbf{x}^T$ times the first equation, then the scalar problem of $\mathbf{y}^T$ times the second equation and finally taking the difference we reach again (3.47) and (3.48). This time we apply Lemma 3.1 to the matrix C, obtaining $C\mathbf{y} = 0$. Then we can go back to Theorem 3.1 and using (3.2) and (3.3) we obtain $\mathbf{x} = 0_n$ and $\mathbf{y} = 0_m$.   □

Remark 3.5. The results of Theorems 3.2 and 3.3 could be summarized as follows. Assume that A and B verify the assumptions (3.2) and (3.3) (that would give the result for $C = 0$.) Assume further that A is positive semidefinite (3.38), and C is negative semidefinite (3.44). Then, if *at least one* of the two matrices A and C is symmetric, then problem (3.36) is uniquely solvable.

We shall now see that the symmetry assumptions in Theorem 3.2 or in Theorem 3.3 cannot be easily reduced. Indeed, if we consider the case

$$A = \begin{pmatrix} 1 & 0 & 0 \\ 0 & 1 & 1 \\ 0 & -1 & 0 \end{pmatrix} \quad B = \begin{pmatrix} 0 & 1 & 0 \\ 0 & 0 & 1 \end{pmatrix} \quad C = \begin{pmatrix} 0 & 1 \\ -1 & -1 \end{pmatrix}, \quad (3.49)$$

we see that A is nonnegative, C is nonpositive, $\ker B^T = 0_2$ and A is nonsingular when restricted to the $\ker B \equiv \{\mathbf{x} \in \mathbb{R}^3 \text{ such that } x_2 = x_3 = 0\}$. Hence all the assumptions of Theorem 3.2 except the symmetry assumption are satisfied. It is easy to see that the whole matrix

$$\begin{pmatrix} 1 & 0 & 0 & 0 & 0 \\ 0 & 1 & -1 & 1 & 0 \\ 0 & 1 & 0 & 0 & 1 \\ 0 & 1 & 0 & 0 & 1 \\ 0 & 0 & 1 & -1 & -1 \end{pmatrix} \quad (3.50)$$

is singular, since the third and fourth rows are equal. Indeed, we cannot apply Theorem 3.2 nor Theorem 3.3 since neither A nor C is symmetric.

On the other hand, it is obvious that we cannot give up the assumption that A and C have, in some weak sense, opposite signs, because the elementary choice

$$A = (1) \quad B = (1) \quad C = (1) \quad (3.51)$$

gives rise to the singular matrix

$$\begin{pmatrix} 1 & 1 \\ 1 & 1 \end{pmatrix}. \quad (3.52)$$

Similarly, we cannot even give up one of the two signs (of A or of C), since, for instance, the choice

$$A = \begin{pmatrix} 0 & 1 \\ 1 & 1 \end{pmatrix} \quad B = \begin{pmatrix} 1 & 0 \end{pmatrix} \quad C = (-1) \quad (3.53)$$

with C symmetric and negative definite and A symmetric but indefinite, produces the singular matrix

$$\begin{pmatrix} 0 & 1 & 1 \\ 1 & 1 & 0 \\ 1 & 0 & -1 \end{pmatrix}. \quad (3.54)$$

4   The Case of Big Matrices. The Inf-Sup Condition

We now place ourselves again in the context of problems of the type (1.1). However, this time, we suppose that we are actually given a *sequence* of problems, with increasing dimensions. It is clear that this will be the case when we are going to consider discretizations of a given, say, partial differential equation, with a sequence of finer and finer meshes. Consider therefore for $k = 1, 2, \ldots$ the problems

$$\begin{pmatrix} A_k & B_k^T \\ B_k & 0 \end{pmatrix} \begin{pmatrix} \mathbf{x}_k \\ \mathbf{y}_k \end{pmatrix} = \begin{pmatrix} \mathbf{f}_k \\ \mathbf{g}_k \end{pmatrix}, \tag{4.1}$$

where A_k is an $n_k \times n_k$ matrix, B_k an $m_k \times n_k$ matrix, and the dimensions n_k and m_k tend to infinity when k goes to infinity. We are interested in conditions that ensure not only the unique solvability of each problem (4.1), but also a stability estimate of the type (1.2):

$$\|\mathbf{x}_k\| + \|\mathbf{y}_k\| \leq c(\|\mathbf{f}_k\| + \|\mathbf{g}_k\|) \tag{4.2}$$

where the constant c *does not depend on* k. This requirement is obviously meaningless unless we specify the norms that we intend to use. As anticipated in the Introduction, the choice of the norms, in this case, is not irrelevant: although they are all equivalent, the *constants* involved in the equivalence may (and, in general, do) depend on the dimensions, that we are assuming to be going to infinity.

On the other hand, if we want to use these abstract results in order to provide *a priori* error bounds for some realistic discretization of a differential problem, we are not totally free in the choice of the norms.

In general, in the finite element context, the norms to be used will be the norms in some functional space where the differential problem itself is set. Hence, in practice, we are going to have little choice. At the present level, however, we have no functional spaces yet (nor, for what matters, a differential problem). Hence, we are going to consider norms, or, rather, families of norms, that are defined independently from functional spaces and discretization schemes. In the present section, we shall reconsider several aspects that were discussed in Section 2, but, this time, introducing norms, and analyzing the behaviour of the various constants in dependence of the chosen norms.

In particular, we go back to the notation of Section 2, and we consider, for every $k = 1, 2, \ldots$ an $r(k) \times s(k)$ matrix M_k. For the sake of simplicity, from now on we shall drop the index k unless it will really be necessary, and we shall just *remember* that r, s, and M depend on k.

We now recall that, in the previous sections, several nonsingularity (or injectivity, or surjectivity) conditions were expressed by formulae of the type

$$\forall \, \mathbf{y} \in \mathbb{R}^s \text{ with } \mathbf{y} \neq 0, \quad \exists \, \mathbf{x} \in \mathbb{R}^r \text{ such that } \mathbf{y}^T M \mathbf{x} \neq 0. \tag{4.3}$$

Taking this as an example (and assuming that $r \geq s$, otherwise (4.3) can never hold,) we shall now write (4.3) in a different way.

We assume that, for each k, we are given a norm $\|\cdot\|_X$ on $\mathbb{R}^r$ and a norm $\|\cdot\|_Y$ on $\mathbb{R}^s$. As stated, we shall just *remember* that these norms actually depend on k, although this will not appear in our notation.

To start with, we observe that (4.3) can also be written as

$$\forall \, \mathbf{y} \in \mathbb{R}^s \text{ with } \mathbf{y} \neq 0, \quad \exists \, \mathbf{x} \in \mathbb{R}^r \text{ such that } \mathbf{y}^T M \mathbf{x} > 0. \tag{4.4}$$

Indeed, if a certain $\mathbf{x}$ makes $\mathbf{y}^T M \mathbf{x}$ negative, taking $\widetilde{\mathbf{x}} = -\mathbf{x}$ we have that $\mathbf{y}^T M \widetilde{\mathbf{x}}$ is positive. We also remark that, for fixed $\mathbf{y} \neq 0_s$ and $\mathbf{x} \neq 0_r$, condition $\mathbf{y}^T M \mathbf{x} > 0$ is equivalent to

$$\frac{\mathbf{y}^T M \mathbf{x}}{\|\mathbf{x}\|_X \, \|\mathbf{y}\|_Y} > 0. \tag{4.5}$$

Hence we can set, for $\mathbf{x}$ and $\mathbf{y}$ different from zero,

$$q(\mathbf{x}, \mathbf{y}) := \frac{\mathbf{y}^T M \mathbf{x}}{\|\mathbf{x}\|_X \, \|\mathbf{y}\|_Y}, \tag{4.6}$$

and write (4.3) once more in the equivalent form

$$\forall \, \mathbf{y} \in \mathbb{R}^s \text{ with } \mathbf{y} \neq 0, \quad \exists \, \mathbf{x} \in \mathbb{R}^r \text{ with } \mathbf{x} \neq 0, \text{ such that } q(\mathbf{x}, \mathbf{y}) > 0. \tag{4.7}$$

It is not difficult to see that (4.7) is equivalent to

$$\forall \, \mathbf{y} \in \mathbb{R}^s \text{ with } \mathbf{y} \neq 0, \quad \sup_{\mathbf{x} \in \mathbb{R}^r \setminus \{0\}} q(\mathbf{x}, \mathbf{y}) > 0. \tag{4.8}$$

Indeed, for every given and fixed $\mathbf{y}$ we have: if $q(\mathbf{x}, \mathbf{y})$ is positive for at least one $\mathbf{x}$, then its supremum (in $\mathbf{x}$) is positive, and, conversely, if the supremum in $\mathbf{x}$ of $q(\mathbf{x}, \mathbf{y})$ is positive, then $q(\mathbf{x}, \mathbf{y})$ itself will be positive for at least one $\mathbf{x}$. Let us look at (4.8): it says that a certain quantity

$$\sigma(\mathbf{y}) := \sup_{\mathbf{x} \in \mathbb{R}^r \setminus \{0\}} q(\mathbf{x}, \mathbf{y}) \equiv \sup_{\mathbf{x} \in \mathbb{R}^r \setminus \{0\}} \frac{\mathbf{y}^T M \mathbf{x}}{\|\mathbf{x}\|_X \, \|\mathbf{y}\|_Y}, \tag{4.9}$$

is positive for every $\mathbf{y} \in \mathbb{R}^s \setminus \{0\}$. We would like to show that this is equivalent to saying that:

$$\inf_{\mathbf{y} \in \mathbb{R}^s \setminus \{0\}} \sigma(\mathbf{y}) > 0. \tag{4.10}$$

As the function σ from $\mathbb{R}^s \setminus \{0\}$ into $\mathbb{R}$ is clearly continuous, we would like to apply a basic theorem in elementary calculus that says that every continuous function on a closed bounded subset of $\mathbb{R}^s$ has a minimum. Unfortunately, $\mathbb{R}^s \setminus \{0\}$ is not closed, nor bounded. Hence, we will have first to restrict ourselves to the unit sphere $S_1 \equiv \{\mathbf{y} \in \mathbb{R}^s, \quad \|\mathbf{y}\|_Y = 1\}$, apply the theorem there, and then go back to $\mathbb{R}^s \setminus \{0\}$. For doing this, we notice that σ is homogeneous, in the sense that $\sigma(\lambda \mathbf{y}) = \sigma(\mathbf{y})$ for all $\lambda \in \mathbb{R}$ and for all $\mathbf{y} \in \mathbb{R}^s$. Hence, $\sigma(\mathbf{y})$ will be positive for every $\mathbf{y} \in \mathbb{R} \setminus \{0\}$ if and only if it is positive for every $\mathbf{y} \in S_1$. On S_1 the function σ must have a minimum (due to the above mentioned theorem in basic calculus,) and hence the condition

$$\sigma(\mathbf{y}) > 0 \quad \forall \, \mathbf{y} \in S_1 \tag{4.11}$$

is equivalent to

$$\inf_{\mathbf{y} \in S_1} \sigma(\mathbf{y}) > 0. \tag{4.12}$$

Using again the fact that σ is homogeneous, we now see that, in its turn, condition (4.12) is equivalent to (4.10), since:

$$\inf_{\mathbf{y} \in \mathbb{R}^s \setminus \{0\}} \sigma(\mathbf{y}) = \inf_{\mathbf{y} \in \mathbb{R}^s \setminus \{0\}} \sigma(\frac{\mathbf{y}}{\|\mathbf{y}\|_Y}) = \inf_{\mathbf{w} \in S_1} \sigma(\mathbf{w}). \tag{4.13}$$

With some pain (if you are not too strong in basic calculus, and some boredom if you are) we saw that (4.8) is equivalent to (4.10). Collecting (4.10) and (4.9) (and all our chain of equivalences) we conclude that our original condition (4.3) is actually equivalent to

$$\inf_{\mathbf{y} \in \mathbb{R}^s \setminus \{0\}} \sup_{\mathbf{x} \in \mathbb{R}^r \setminus \{0\}} \frac{\mathbf{y}^T M \mathbf{x}}{\|\mathbf{x}\|_X \|\mathbf{y}\|_Y} > 0. \tag{4.14}$$

We notice now that the quantity which appears at the left-hand side of (4.14) does not depend on $\mathbf{x}$ or on $\mathbf{y}$ anymore, but depends only on the matrix M. In order to simplify the notation, we assume *from now on* that when taking infimums and supremums we implicitly discard the value 0. Hence we can set

$$\mu(M) := \inf_{\mathbf{y} \in \mathbb{R}^s} \sup_{\mathbf{x} \in \mathbb{R}^r} \frac{\mathbf{y}^T M \mathbf{x}}{\|\mathbf{x}\|_X \|\mathbf{y}\|_Y}, \tag{4.15}$$

and we remember that, if we have a sequence of matrices M_k, then $\mu(M_k)$ clearly depends on k. If we want a condition that is actually independent of k we must then require that there exists a $\mu > 0$, independent of k, such that

$$\inf_{\mathbf{y} \in \mathbb{R}^s} \sup_{\mathbf{x} \in \mathbb{R}^r} \frac{\mathbf{y}^T M \mathbf{x}}{\|\mathbf{x}\|_X \|\mathbf{y}\|_Y} \geq \mu > 0 \tag{4.16}$$

for all matrices M in the sequence M_k. Condition (4.16) is usually called *the inf-sup condition*.

From the above discussion, going somewhat backward, it is not difficult to see that, for a given $\mu > 0$, it is equivalent to requiring that

$$\inf_{\mathbf{y} \in \mathbb{R}^s} \sup_{\mathbf{x} \in \mathbb{R}^r} \frac{\mathbf{y}^T M \mathbf{x}}{\|\mathbf{x}\|_X \|\mathbf{y}\|_Y} \geq \mu, \tag{4.17}$$

or that

$$\forall \, \mathbf{y} \in \mathbb{R}^s \text{ with } \mathbf{y} \neq 0, \quad \sup_{\mathbf{x} \in \mathbb{R}^r} \frac{\mathbf{y}^T M \mathbf{x}}{\|\mathbf{x}\|_X} \geq \mu \|\mathbf{y}\|_Y, \tag{4.18}$$

or finally that

$$\forall \, \mathbf{y} \in \mathbb{R}^s \setminus \{0\}, \quad \exists \, \mathbf{x} \in \mathbb{R}^r \setminus \{0\} \text{ such that } \mathbf{y}^T M \mathbf{x} \geq \mu \|\mathbf{x}\|_X \|\mathbf{y}\|_Y. \tag{4.19}$$

In order to analyze in a deeper way the dependence of the constant μ on M and on the norms $\|\cdot\|_X$ and $\|\cdot\|_Y$, it will be convenient to limit the possible choice of the norms. Hence, we assume that for each k we are given an $r(k) \times r(k)$ symmetric positive definite matrix Ξ_k with entries ξ_{ij}, and an $s(k) \times s(k)$ symmetric positive definite matrix H_k with entries η_{ij}. Again, for the sake of simplicity, we shall drop the index k and just *remember* that Ξ and H depend on k. To the matrices Ξ and H we associate the norms $\|\cdot\|_X$ and $\|\cdot\|_Y$ defined through

$$\|\mathbf{x}\|_X^2 := \mathbf{x}^T \Xi \mathbf{x}, \tag{4.20}$$

and

$$\|\mathbf{y}\|_Y^2 := \mathbf{y}^T \Xi \mathbf{y}. \tag{4.21}$$

We shall also need the *dual norms* of $\|\cdot\|_X$ and $\|\cdot\|_Y$. At a general level, the dual norm $\|\cdot\|_F$ of a given norm $\|\cdot\|_X$ is defined as the smallest possible norm that makes the Cauchy-Schwarz type inequality

$$\mathbf{x}^T \mathbf{f} \leq \|\mathbf{x}\|_X \|\mathbf{f}\|_F, \quad \forall \, \mathbf{x} \in \mathbb{R}^r \; \forall \, \mathbf{f} \in \mathbb{R}^r \tag{4.22}$$

hold true. This is achieved by taking, for all $\mathbf{f} \in \mathbb{R}^r$,

$$\|\mathbf{f}\|_F := \sup_{\mathbf{x} \in \mathbb{R}^r \setminus \{0\}} \frac{\mathbf{x}^T \mathbf{f}}{\|\mathbf{x}\|_X}. \tag{4.23}$$

The following lemma helps us in finding the choice of $\mathbf{x}$ that realizes the supremum in (4.23).

Lemma 4.1. *Let Ξ be an $r \times r$ symmetric and positive definite matrix, and let the norm $\|\cdot\|_X$ be defined as in (4.20). Then, for every given $\mathbf{f} \in \mathbb{R}^r$ the supremum in (4.23) is achieved when*

$$\mathbf{x} = \Xi^{-1} \mathbf{f}. \tag{4.24}$$

Proof. With the change of variable

$$\mathbf{z} := \Xi^{1/2} \mathbf{x} \tag{4.25}$$

(see the Remark 4.1 below, if you are not familiar with the notation) the supremum in (4.23) becomes

$$\sup_{\mathbf{z} \in \mathbb{R}^r \setminus \{0\}} \frac{\mathbf{z}^T \Xi^{-1/2} \mathbf{f}}{(\mathbf{z}^T \mathbf{z})^{1/2}} \tag{4.26}$$

and the supremum is clearly reached when $\mathbf{z} = \Xi^{-1/2} \mathbf{f}$. Using (4.25) we obtain (4.24), that is the desired result. □

Remark 4.1. In (4.25) we used the square root of the symmetric and positive definite matrix Ξ. This can be introduced in several ways. Possibly the simplest one is to diagonalize the matrix Ξ, by writing it as $\Xi = T^{-1}\Lambda T$ with T an orthogonal matrix (that means $T^T \equiv T^{-1}$), and Λ a diagonal matrix with positive diagonal entries. Now $\Lambda^{1/2}$ can easily be defined by taking the square root of every diagonal entry of Λ. Finally we can define $\Xi^{1/2} := T^{-1}\Lambda^{1/2}T$.

Using Lemma 4.1, and using the $\mathbf{x}$ given in (4.24) it is easy to see that (4.23) implies

$$\|\mathbf{f}\|_F = \frac{\mathbf{x}^T\mathbf{f}}{\mathbf{x}^T\Xi\mathbf{x}} = \frac{(\Xi^{-1}\mathbf{f})^T\mathbf{f}}{((\Xi^{-1}\mathbf{f})^T\Xi(\Xi^{-1}\mathbf{f}))^{1/2}} = \frac{\mathbf{f}^T\Xi^{-1}\mathbf{f}}{(\mathbf{f}^T\Xi^{-1}\mathbf{f})^{1/2}}, \qquad (4.27)$$

which simplifies to

$$\|\mathbf{f}\|_F^2 = \mathbf{f}^T\Xi^{-1}\mathbf{f}, \quad \forall\, \mathbf{f} \in \mathbb{R}^r. \qquad (4.28)$$

In particular using (4.28) we have, for every $\mathbf{z} \in \mathbb{R}^r$,

$$\|\Xi\mathbf{z}\|_F^2 = (\Xi\mathbf{z})^T\Xi^{-1}\Xi\mathbf{z} = \mathbf{z}^T\Xi\mathbf{z} = \|\mathbf{z}\|_X^2 \qquad (4.29)$$

so that

$$\|\Xi\mathbf{z}\|_F = \|\mathbf{z}\|_X. \qquad (4.30)$$

In a similar way, we can define the dual norm of $\|\cdot\|_Y$ by

$$\|\mathbf{g}\|_G^2 := \mathbf{g}^T H^{-1}\mathbf{g}, \quad \forall\, \mathbf{g} \in \mathbb{R}^s, \qquad (4.31)$$

as the smallest norm that gives

$$\mathbf{y}^T\mathbf{g} \le \|\mathbf{y}\|_Y \|\mathbf{g}\|_G \qquad \forall\, \mathbf{y} \in \mathbb{R}^s \qquad \forall\, \mathbf{g} \in \mathbb{R}^s. \qquad (4.32)$$

Note that, as in (4.30), we have, for all $\mathbf{z} \in \mathbb{R}^s$,

$$\|H\mathbf{z}\|_G = \|\mathbf{z}\|_Y. \qquad (4.33)$$

We are now ready to see how the value of the *inf-sup* appearing in (4.17) can actually be related to properties of matrix M. To see this, we assume, for a given matrix M and for a given constant $\mu > 0$, that we have

$$\mu \le \inf_{\mathbf{y}\in\mathbb{R}^s} \sup_{\mathbf{x}\in\mathbb{R}^r} \frac{\mathbf{y}^T M\mathbf{x}}{\|\mathbf{y}\|_Y \|\mathbf{x}\|_X}, \qquad (4.34)$$

and look for other interesting properties that can be related to M and μ. We do that in the next two propositions.

Proposition 4.1. *Assume that μ is a positive constant, and M is an $s \times r$ matrix satisfying (4.34). Let moreover $\|\cdot\|_F$ be the dual norm of $\|\cdot\|_X$ as defined in (4.23) and (4.31). Then we have*

$$\mu\|\mathbf{y}\|_Y \le \|M^T\mathbf{y}\|_F \quad \forall\, \mathbf{y} \in \mathbb{R}^s. \qquad (4.35)$$

Proof. For every $\mathbf{y} \in \mathbb{R}^s$ we have from (4.34) that

$$\sup_{\mathbf{x} \in \mathbb{R}^r} \frac{\mathbf{y}^T M \mathbf{x}}{\|\mathbf{x}\|} \geq \mu \|\mathbf{y}\|_Y, \tag{4.36}$$

since μ is the *infimum* over $\mathbf{y}$ (as we actually already saw in (4.18).) Using the elementary relation (2.3) we have immediately that

$$\sup_{\mathbf{x} \in \mathbb{R}^r} \frac{\mathbf{x}^T M^T \mathbf{y}}{\|\mathbf{x}\|_X} \geq \mu \|\mathbf{y}\|_Y. \tag{4.37}$$

Considering the $\mathbf{x}$ that realizes the supremum in (4.37) we have then

$$\frac{\mathbf{x}^T M^T \mathbf{y}}{\|\mathbf{x}\|_X} \geq \mu \|\mathbf{y}\|_Y. \tag{4.38}$$

Finally from (4.37) and (4.22)

$$\mu \|\mathbf{y}\|_Y \leq \|M^T \mathbf{y}\|_F. \quad \square \tag{4.39}$$

Proposition 4.2. *Assume that μ is a positive constant, and M is an $s \times r$ matrix satisfying (4.34). Let moreover $\| \cdot \|_G$ be the dual norm of $\| \cdot \|_Y$ as defined in and (4.31). Then there exists a linear operator R_M from $\mathbb{R}^s$ to $\mathbb{R}^r$ such that*

$$M(R_M \mathbf{g}) = \mathbf{g} \qquad \forall \, \mathbf{g} \in \mathbb{R}^s, \tag{4.40}$$

and

$$\mu \|R_M \mathbf{g}\|_X \leq \|\mathbf{g}\|_G \qquad \forall \, \mathbf{g}_H \in \mathbb{R}^s. \tag{4.41}$$

Proof. For every $\mathbf{g} \in \mathbb{R}^s$ we consider $(\mathbf{x}_g, \mathbf{y}_g)$ as the solution of

$$\begin{aligned} \Xi \mathbf{x}_g + M^T \mathbf{y}_g &= 0, \\ M \mathbf{x}_g &= \mathbf{g}, \end{aligned} \tag{4.42}$$

where Ξ is the symmetric and positive definite $r \times r$ matrix used to define $\| \cdot \|_X$ as in (4.20). Using (4.34) and the properties of Ξ we can apply Theorem 3.1 and obtain that (4.42) has a unique solution. We now multiply $\mathbf{x}_g^T$ times the first equation of (4.42) and $\mathbf{y}_g^T$ times the second equation. Taking the difference we obtain

$$\mathbf{x}_g^T \Xi \mathbf{x}_g = -\mathbf{y}_g^T \mathbf{g}. \tag{4.43}$$

Using the definition of $\| \cdot \|_X$ (see (4.20)) and (4.32) we have then from (4.43)

$$\|\mathbf{x}_g\|_X^2 \leq \|\mathbf{y}_g\|_Y \|\mathbf{g}\|_G. \tag{4.44}$$

As we assumed (4.34) we can apply Proposition 4.1: taking (4.35) with $\mathbf{y} = \mathbf{y}_g$, and then using the first equation of (4.42), we obtain

$$\mu \|\mathbf{y}_g\|_Y \leq \|M^T \mathbf{y}_g\|_F \equiv \|\Xi \mathbf{x}_g\|_F. \tag{4.45}$$

42   Franco Brezzi

Using (4.30) with $\mathbf{z} = \mathbf{x}_g$ we have now

$$\|\Xi\mathbf{x}_g\|_F^2 = \|\mathbf{x}_g\|_X^2, \tag{4.46}$$

so that combining (4.44), (4.45), and (4.46) we obtain

$$\|\mathbf{x}_g\|_X^2 \leq \|\mathbf{y}_g\|_Y \|\mathbf{g}\|_G \leq \mu^{-1}\|\Xi\mathbf{x}_g\|_F \|\mathbf{g}\|_G \leq \mu^{-1}\|\mathbf{x}_g\|_X \|\mathbf{g}\|_G, \tag{4.47}$$

and finally

$$\mu\|\mathbf{x}_g\|_X \leq \|\mathbf{g}\|_G. \tag{4.48}$$

Setting now

$$R_M\mathbf{g} := \mathbf{x}_g \quad \forall\, \mathbf{g} \in \mathbb{R}^s, \tag{4.49}$$

we have easily that R_M is a linear operator that satisfies (4.40) and (4.41). □

We conclude our analysis of the *inf-sup* condition (4.34) with the following proposition, which collects several equivalent formulations.

Proposition 4.3. *Let M be an $s \times r$ matrix, and let $\mu > 0$ be a real number. Let moreover the norms $\|\cdot\|_X$, $\|\cdot\|_Y$, $\|\cdot\|_F$, and $\|\cdot\|_G$ be defined as in (4.20), (4.21), (4.28), and (4.31), respectively, with Ξ and H symmetric and positive definite. Then the following conditions are equivalent.*

IS1 $\forall\, \mathbf{y} \in \mathbb{R}^s \setminus \{0\}\ \exists\, \mathbf{x} \in \mathbb{R}^r \setminus \{0\}$ such that $\mathbf{y}^T M\mathbf{x} \geq \mu\|\mathbf{x}\|_X \|\mathbf{y}\|_Y$, (4.50)

IS2 $\forall\, \mathbf{y} \in \mathbb{R}^s \setminus \{0\}\quad \sup\limits_{\mathbf{x}\in\mathbb{R}^r} \dfrac{\mathbf{y}^T M\mathbf{x}}{\|\mathbf{x}\|_X} \geq \mu\|\mathbf{y}\|_Y,$ (4.51)

IS3 $\inf\limits_{\mathbf{y}\in\mathbb{R}^s} \sup\limits_{\mathbf{x}\in\mathbb{R}^r} \dfrac{\mathbf{y}^T M\mathbf{x}}{\|\mathbf{x}\|_X \|\mathbf{y}\|_Y} \geq \mu,$ (4.52)

IS4 $\forall\, \mathbf{y} \in \mathbb{R}^s\quad \mu\|\mathbf{y}\|_Y \leq \|M^T\mathbf{y}\|_F,$ (4.53)

IS5 $\forall\, \mathbf{g} \in \mathbb{R}^s\ \exists\, \mathbf{x}_g \in \mathbb{R}^r$ such that $M\mathbf{x}_g = \mathbf{g}$ and $\mu\|\mathbf{x}_g\|_X \leq \|\mathbf{g}\|_G.$ (4.54)

Proof. The equivalence of (4.50), (4.51), and (4.52) has been seen in (4.18), (4.19) and the discussion before them. We already saw, in Propositions 4.1 and 4.2, that (4.52) implies (4.53) and (4.54). Hence it remains to prove that (4.53) implies (4.50), and that also (4.54) implies (4.50). For this, assume that (4.53) holds. For every $\mathbf{y} \in \mathbb{R}^s$ we take

$$\mathbf{x} := \Xi^{-1}M^T\mathbf{y}. \tag{4.55}$$

Then, using (4.55) and (4.28) we have

$$\mathbf{y}^T M\mathbf{x} = \mathbf{y}^T M\Xi^{-1}M^T\mathbf{y} = \|M^T\mathbf{y}\|_F^2 \tag{4.56}$$

and using (4.30) and (4.55)

$$\|\mathbf{x}\|_X = \|\Xi\mathbf{x}\|_F = \|M^T\mathbf{y}\|_F, \tag{4.57}$$

so that from (4.56) and (4.57)

$$\mathbf{y}^T M \mathbf{x} = \|\mathbf{x}\|_X \, \|M^T \mathbf{y}\|_F \tag{4.58}$$

and (4.50) follows easily from (4.53) and (4.58).

Assume now that (4.54) holds. For every $\mathbf{y} \in \mathbb{R}^s$ set

$$\mathbf{g} := H\mathbf{y}. \tag{4.59}$$

Notice that, using (4.33), we have

$$\mathbf{y}^T \mathbf{g} = \mathbf{y}^T H \mathbf{y} = \|\mathbf{y}\|_Y^2 = \|\mathbf{g}\|_G^2. \tag{4.60}$$

If (4.54) holds, we can find an $\mathbf{x}_g \in \mathbb{R}^r$ such that $M\mathbf{x}_g = \mathbf{g}$. Then from (4.60) we have

$$\mathbf{y}^T M \mathbf{x}_g = \mathbf{y}^T \mathbf{g} = \|\mathbf{y}\|_Y^2, \tag{4.61}$$

while, using (4.54) and again (4.60)

$$\mu \|\mathbf{x}_g\|_X \leq \|\mathbf{g}\|_G = \|\mathbf{y}\|_Y. \tag{4.62}$$

Combining (4.61) and (4.62) we have

$$\mathbf{y}^T M \mathbf{x}_g = \|\mathbf{y}\|_Y^2 \geq \mu \|\mathbf{y}\|_Y \|\mathbf{x}_g\|_X \tag{4.63}$$

and (4.50) follows. $\square$

Example 4.1. Assume for instance that both Ξ and H coincide with the identity matrix. Assume that $r > s$ and that M has the simple structure

$$M = \begin{pmatrix} m_1 & 0 & \cdot & \cdot & 0 & 0 & 0 \cdot 0 \\ 0 & m_2 & \cdot & \cdot & 0 & 0 & 0 \cdot 0 \\ \cdot & \cdot & \cdot & & & \cdot & \cdot \cdot \cdot \\ \cdot & \cdot & & \cdot & & \cdot & \cdot \cdot \cdot \\ 0 & 0 & \cdot & & \cdot m_{s-1} & 0 & 0 \cdot 0 \\ 0 & 0 & \cdot & & \cdot & 0 & m_s \, 0 \cdot 0 \end{pmatrix}. \tag{4.64}$$

Then clearly one has

$$\mu = \min\{m_1, \ldots, m_s\}. \tag{4.65}$$

On the other hand, assume also that Ξ and H are diagonal matrices:

$$\Xi = \begin{pmatrix} \xi_1 & 0 & \cdot & \cdot & \cdot & 0 & 0 \\ 0 & \xi_2 & \cdot & \cdot & \cdot & 0 & 0 \\ \cdot & \cdot & \cdot & \cdot & \cdot & \cdot & \cdot \\ \cdot & \cdot & \cdot & \cdot & \cdot & \cdot & \cdot \\ 0 & 0 & \cdot & \cdot & \cdot \xi_{r-1} & 0 \\ 0 & 0 & \cdot & \cdot & \cdot & 0 & \xi_r \end{pmatrix} \qquad H = \begin{pmatrix} \eta_1 & 0 & \cdot & \cdot & 0 & 0 \\ 0 & \eta_2 & \cdot & \cdot & 0 & 0 \\ \cdot & \cdot & \cdot & \cdot & \cdot & \cdot \\ \cdot & \cdot & \cdot & \cdot & \cdot & \cdot \\ 0 & 0 & \cdot & \cdot \eta_{s-1} & 0 \\ 0 & 0 & \cdot & \cdot & 0 & \eta_s \end{pmatrix}.$$

Now, for a given $\mathbf{y} \in \mathbb{R}^s$ we can take

$$x_i := y_i \eta_i^{1/2} \xi_i^{-1/2} \text{ for } 1 \leq i \leq s \quad \text{and } x_i = 0 \text{ for } s < i \leq r, \qquad (4.66)$$

that gives easily, using (4.20), (4.66), and (4.21)

$$\|\mathbf{x}\|_X^2 = \sum_{i=1}^s \xi_i x_i^2 = \sum_{i=1}^s \xi_i y_i^2 \eta_i \xi_i^{-1} = \sum_{i=1}^s \eta_i y_i^2 = \|\mathbf{y}\|_Y^2. \qquad (4.67)$$

Setting now

$$\mu := \min\{m_1 \xi_1^{-1/2} \eta_1^{-1/2}, m_2 \xi_2^{-1/2} \eta_2^{-1/2}, \dots, m_s \xi_s^{-1/2} \eta_s^{-1/2}\}, \qquad (4.68)$$

we easily have, using (4.66), (4.68), and then (4.21) and (4.67)

$$\mathbf{y}^T M \mathbf{x} = \sum_{i-1}^s y_i m_i y_i \eta_i^{1/2} \xi_i^{-1/2} = \sum_{i=1}^s \eta_i y_i^2 m_i \eta_i^{-1/2} \xi_i^{-1/2}$$

$$\geq \sum_{i=1}^s \eta_i y_i^2 \mu = \mu \|\mathbf{y}\|_Y^2 = \mu \|\mathbf{x}\|_X \|\mathbf{y}\|_Y. \qquad (4.69)$$

This proves that the choice (4.68) for μ satisfies (4.50). It is also easy to check that (4.68) gives the best (that, is, the biggest) possible choice for μ (if j is the index realizing the minimum in (4.68), take $y_i = \delta ij$).

This shows that the actual value of the inf-sup constant μ depends not only on the matrix M but also on the matrices Ξ and H used to define the norms $\|\cdot\|_X$ and $\|\cdot\|_Y$.

5   The Case of Big Matrices. The Problem of Stability

We now go back to the framework of Section 3, and try to use the results of the previous section in order to achieve uniform stability estimates for *sequences* of problems of the type (4.1), that we repeat for convenience of the reader;

$$\begin{pmatrix} A_k & B_k^T \\ B_k & 0 \end{pmatrix} \begin{pmatrix} \mathbf{x}_k \\ \mathbf{y}_k \end{pmatrix} = \begin{pmatrix} \mathbf{f}_k \\ \mathbf{g}_k \end{pmatrix}. \qquad (5.1)$$

Clearly we have to make suitable assumptions on the matrices A_k and B_k, and on the norms that we are going to use. We start with the precise assumptions on the spaces and norms.

Assumption A1 *We assume that we are given, for each $k \in \mathbb{N}$:*

- *two positive integers $n(k)$, and $m(k)$;*
- *a symmetric and positive definite $n(k) \times n(k)$ matrix Ξ, that we use to define the norms $\| \cdot \|_X$ and $\| \cdot \|_F$ in $\mathbb{R}^{n(k)}$ as in (4.20) and (4.28);*
- *a symmetric and positive definite $m(k) \times m(k)$ matrix H that we use to define the norms $\| \cdot \|_Y$ and $\| \cdot \|_G$ in $\mathbb{R}^{m(k)}$ as in (4.21) and (4.31).*

We consider now the matrices A_k and B_k.

Assumption A2 *We assume that we are given, for each $k \in \mathbb{N}$:*

- *an $n(k) \times n(k)$ matrix A_k;*
- *an $n(k) \times m(k)$ matrix B_k;*

and we define, for each $k \in \mathbb{N}$, the kernel K_k of B_k:

$$K_k := \{\mathbf{x} \in \mathbb{R}^n \text{ such that } B_k \mathbf{x} = 0_m\}. \tag{5.2}$$

We then require the *uniform continuity* of matrices A_k and B_k with respect to the chosen norms.

Assumption A3 *There exists an $M_a \in \mathbb{R}$, independent of k, such that*

$$\forall\, \mathbf{z} \in \mathbb{R}^n, \quad \forall\, \mathbf{x} \in \mathbb{R}^n \quad \mathbf{z}^T A_k \mathbf{x} \le M_a \|\mathbf{z}\|_X \|\mathbf{x}\|_X \tag{5.3}$$

and there exists an $M_b \in \mathbb{R}$, independent of k, such that

$$\forall\, \mathbf{y} \in \mathbb{R}^m, \quad \forall\, \mathbf{x} \in \mathbb{R}^n \quad \mathbf{y}^T B_k \mathbf{x} \le M_b \|\mathbf{y}\|_Y \|\mathbf{x}\|_X. \tag{5.4}$$

Remark 5.1. The uniform continuity assumptions made in (5.3) and (5.4) are quite natural in a finite element context. However, if you are not familiar with it, or with its mathematical aspects, just assume that this is the game that we are going to play. The rest of the chapter will show more than clearly the motivations for these choices. For the moment, we just notice that (5.3) and (5.4), together with the previous Assumption 1 on the norms, imply, for all $\mathbf{x} \in \mathbb{R}^n$ and for all $\mathbf{y} \in \mathbb{R}^m$,

$$\|A_k \mathbf{x}\|_F \equiv \sup_{\mathbf{z} \in \mathbb{R}^n} \frac{\mathbf{z}^T A_k \mathbf{x}}{\|\mathbf{z}\|_X} \le M_a \|\mathbf{x}\|_X, \tag{5.5}$$

$$\|B_k \mathbf{x}\|_G \equiv \sup_{\mathbf{z} \in \mathbb{R}^m} \frac{\mathbf{z}^T B_k \mathbf{x}}{\|\mathbf{z}\|_Y} \le M_b \|\mathbf{x}\|_X, \tag{5.6}$$

and

$$\|B_k^T \mathbf{y}\|_F \equiv \sup_{\mathbf{z} \in \mathbb{R}^n} \frac{\mathbf{z}^T B_k^T \mathbf{y}}{\|\mathbf{z}\|_X} \le M_b \|\mathbf{y}\|_Y. \tag{5.7}$$

We are now ready to present our first stability result (see [5]).

Theorem 5.1. *In the Assumptions A1, A2, and A3 listed above, assume that the sequences of matrices A_k and B_k verify*

$$\exists\, \alpha > 0 \text{ such that } \forall\, k \in \mathbb{N}, \quad \inf_{x \in K_k} \sup_{z \in K_k} \frac{z^T A_k x}{\|x\|_X \|z\|_X} \geq \alpha, \qquad (5.8)$$

(where K_k is defined in (5.2)), and

$$\exists\, \beta > 0 \text{ such that } \forall\, k \in \mathbb{N}, \quad \inf_{y \in \mathbb{R}^m} \sup_{x \in \mathbb{R}^n} \frac{y^T B_k x}{\|x\|_X \|y\|_Y} \geq \beta. \qquad (5.9)$$

Then for every $k \in \mathbb{N}$, and for every right-hand side $\mathbf{f}_k \in \mathbb{R}^n$ and $\mathbf{g}_k \in \mathbb{R}^m$, the problem

$$\begin{pmatrix} A_k & B_k^T \\ B_k & 0 \end{pmatrix} \begin{pmatrix} \mathbf{x}_k \\ \mathbf{y}_k \end{pmatrix} = \begin{pmatrix} \mathbf{f}_k \\ \mathbf{g}_k \end{pmatrix} \qquad (5.10)$$

has a unique solution $(\mathbf{x}_k, \mathbf{y}_k) \in \mathbb{R}^n \times \mathbb{R}^m$. Moreover if M_* is defined as

$$M_* := \max\{M_a, M_b, 1\} \qquad (5.11)$$

then we have

$$\|\mathbf{x}_k\|_X \leq \frac{M_*}{\alpha}\|\mathbf{f}_k\|_F + \frac{2M_*}{\alpha\beta}\|\mathbf{g}_k\|_G \qquad (5.12)$$

and

$$\|\mathbf{y}_k\|_Y \leq \frac{2M_*^2}{\alpha\beta}\|\mathbf{f}_k\|_F + \frac{2M_*^2}{\alpha\beta^2}\|\mathbf{g}_k\|_G. \qquad (5.13)$$

Proof. Existence and uniqueness of the solution of (5.10) follow easily from Theorem 3.1 (or, more immediately, from Proposition 3.1), having already established that (5.8) implies (3.6) and (5.9) implies (3.7).

Therefore we only have to prove (5.12) and (5.13). In order to have a simpler notation, we drop most of the indices k: hence the matrices will be A and B, the kernel of B will be K, and the right-hand side $(\mathbf{f}, \mathbf{g})$. We continue to denote the solution of (5.10) by $(\mathbf{x}_k, \mathbf{y}_k)$. The proof will be given in four steps.

First step. We consider the lifting operator defined in Proposition 4.2, that we now call R_B, and we apply it to the right-hand side $\mathbf{g}$. We set $\mathbf{x}_g := R_B \mathbf{g}$. According to Proposition 4.2 we have

$$B\mathbf{x}_g = \mathbf{g} \qquad (5.14)$$

and

$$\beta\|\mathbf{x}_g\|_X \leq \|\mathbf{g}\|_G \qquad (5.15)$$

Second step. We now set

$$\mathbf{x}_K := \mathbf{x}_k - \mathbf{x}_g \qquad (5.16)$$

and we would like to show that

$$\|\mathbf{x}_K\|_X \le \frac{1}{\alpha}(\|\mathbf{f}\|_F + M_a\|\mathbf{x}_g\|_X). \qquad (5.17)$$

For this, we start by using Proposition 4.3 (on the matrix A restricted to K), and in particular the fact that (4.52) implies (4.50). We have that, using (5.8), for every $\mathbf{x} \in K$ (and in particular for $\mathbf{x} = \mathbf{x}_K$), there exists a $\widetilde{\mathbf{x}} \in K$ such that

$$\widetilde{\mathbf{x}}^T A \mathbf{x}_K \ge \alpha \|\mathbf{x}_K\|_X \|\widetilde{\mathbf{x}}\|_X, \qquad (5.18)$$

that we write now as

$$\frac{\widetilde{\mathbf{x}}^T A \mathbf{x}_K}{\|\widetilde{\mathbf{x}}\|_X} \ge \alpha \|\mathbf{x}_K\|_X. \qquad (5.19)$$

Multiplying $\widetilde{\mathbf{z}}^T$ times the first equation of (5.10) and using (5.16) we obtain

$$\widetilde{\mathbf{z}}^T \mathbf{f} = \widetilde{\mathbf{z}}^T A \mathbf{x}_k + +\widetilde{\mathbf{z}}^T B^T \mathbf{y}_k = \widetilde{\mathbf{z}}^T A \mathbf{x}_g + \widetilde{\mathbf{z}}^T A \mathbf{x}_K + \widetilde{\mathbf{z}}^T B^T \mathbf{y}_k. \qquad (5.20)$$

Using the fact that $\widetilde{\mathbf{z}}^T \in K = \ker B$, so that $\widetilde{\mathbf{z}}^T B^T \mathbf{y}_k \equiv \mathbf{y}_k^T B \widetilde{\mathbf{z}} = 0$, and rearranging terms we have

$$\widetilde{\mathbf{z}}^T A \mathbf{x}_K = \widetilde{\mathbf{z}}^T \mathbf{f} - \widetilde{\mathbf{z}}^T A \mathbf{x}_g. \qquad (5.21)$$

Using now (5.19), then (5.21), and finally (4.22) and (5.3), we have

$$\alpha \|\mathbf{x}_K\|_X \le \frac{\widetilde{\mathbf{z}}^T A \mathbf{x}_K}{\|\widetilde{\mathbf{z}}\|_X} = \frac{\widetilde{\mathbf{z}}^T \mathbf{f} - \widetilde{\mathbf{z}}^T A \mathbf{x}_g}{\|\widetilde{\mathbf{z}}\|_X} \le \|\mathbf{f}\|_F + M_a\|\mathbf{x}_g\|_X, \qquad (5.22)$$

and (5.17) follows.

Third Step. Next, we want to show that

$$\|\mathbf{y}_k\|_Y \le \frac{1}{\beta}(\|\mathbf{f}\|_F + M_a\|\mathbf{x}_k\|_X). \qquad (5.23)$$

We start by using Proposition (4.1) and then the first equation of (5.10) to obtain

$$\beta \|\mathbf{y}_k\|_Y \le \|B^T \mathbf{y}_k\|_F \le \|\mathbf{f}\|_F + \|A \mathbf{x}_k\|_F. \qquad (5.24)$$

We can now use (5.5) in (5.24) and obtain

$$\beta \|\mathbf{y}_k\|_Y \le \|\mathbf{f}\|_F + M_a\|\mathbf{x}_k\|_X, \qquad (5.25)$$

and (5.23) follows immediately.

Fourth Step. The fourth step consists only in collecting, in a telescopic way, the previous estimates (5.15), (5.17), and (5.23). Notice that we followed, in a certain sense, the order suggested by (3.13), solving the system from the last to the first equation.

We first combine (5.17) and (5.15) to obtain an estimate for $\mathbf{x}_K$:

$$\|\mathbf{x}_K\|_X \le \frac{1}{\alpha}(\|\mathbf{f}\|_F + \frac{M_a}{\beta}\|\mathbf{g}\|_G). \tag{5.26}$$

Then we combine (5.16),(5.15), and (5.26) to get the estimate on $\mathbf{x}_k$:

$$\|\mathbf{x}_k\|_X \le \|\mathbf{x}_g\|_X + \|\mathbf{x}_K\|_X \le \frac{1}{\alpha}\|\mathbf{f}\|_F + \frac{1}{\beta}(\frac{M_a}{\alpha} + 1)\|\mathbf{g}\|_G. \tag{5.27}$$

Finally, we combine (5.27) and (5.23) to obtain the estimate on $\mathbf{y}_k$:

$$\|\mathbf{y}_k\|_Y \le (\frac{1}{\beta} + \frac{M_a}{\alpha})\|\mathbf{f}\|_F + \frac{M_a}{\beta^2}(\frac{M_a}{\alpha} + 1)\|\mathbf{g}\|_G. \tag{5.28}$$

Hence the theorem is proved. □

Remark 5.2. It is clear that formulae (5.12) and (5.13) give a much better insight on the dependence of the estimate (1.2) upon the constants α and β which appear in (5.8) and (5.9). Still (5.27) and (5.28) allow a better control on M_a as well.

Remark 5.3. Actually, in Theorem 5.1 we could allow the constants α and β to depend on k, by changing (5.27) and (5.28) into

$$\|\mathbf{x}_k\|_X \le \|\mathbf{x}_H\|_X + \|\mathbf{x}_K\|_X \le \frac{1}{\alpha_k}\|\mathbf{f}\|_F + \frac{1}{\beta_k}(\frac{M_a}{\alpha_k} + 1)\|\mathbf{g}\|_G, \tag{5.29}$$

and

$$\|\mathbf{y}_k\|_Y \le (\frac{1}{\beta_k} + \frac{M_a}{\alpha_k})\|\mathbf{f}\|_F + \frac{M_a}{\beta_k^2}(\frac{M_a}{\alpha_k} + 1)\|\mathbf{g}\|_G. \tag{5.30}$$

We see from (5.29) and (5.30) that if, for instance, α_k remains bounded away from zero but β_k tends to zero, then: for $\mathbf{g} \ne 0$, $\mathbf{x}_k$ will explode as $1/\beta_k$ and $\mathbf{y}_k$ will explode like $1/\beta_k^2$, while for $\mathbf{g} = 0$, $\mathbf{x}_k$ will remain bounded, and $\mathbf{y}_k$ will explode like $1/\beta_k$. On the other hand, if instead α_k tends to zero and β_k stays bounded away from zero, then both $\mathbf{x}_k$ and $\mathbf{y}_k$ will explode like $1/\alpha_k$, unless both $\mathbf{f}$ and $\mathbf{g}$ are zero (a case with little interest indeed.)

Remark 5.4. The estimates (5.29) and (5.30) cannot be improved as far as the dependence from α and β is concerned. Indeed, if one considers the system

$$\begin{pmatrix} 1 & 1 & \beta \\ 1 & \alpha & 0 \\ \beta & 0 & 0 \end{pmatrix} \cdot \begin{pmatrix} x_1 \\ x_2 \\ y \end{pmatrix} = \begin{pmatrix} f_1 \\ f_2 \\ g \end{pmatrix} \tag{5.31}$$

one easily obtains

$$x_1 = \frac{1}{\beta}g, \tag{5.32}$$

$$x_2 = \frac{1}{\alpha}f_2 - \frac{1}{\alpha\beta}g, \tag{5.33}$$

and

$$y = \frac{1}{\beta} f_1 - \frac{1}{\alpha\beta} f_2 + \frac{1-\alpha}{\alpha\beta^2} g, \tag{5.34}$$

which shows the optimality of (5.15), (5.17), and (5.23), and hence of the final estimates (5.29) and (5.30).

In many applications, the following corollary, similar to Corollary 3.1, will prove to be very useful

Corollary 5.1. *Assume, as in Theorem 5.1, that Assumptions A1, A2, and A3 are satisfied. Assume moreover that A is positive definite on the kernel K, that is*

$$\exists\, \alpha > 0 \ \text{such that}\ \mathbf{x}^T A \mathbf{x} \geq \alpha \|\mathbf{x}\|_X^2 \quad \forall\, \mathbf{x} \in K \tag{5.35}$$

and that (5.9) holds. Then the same conclusion of Theorem 5.1 holds.

The proof is trivial since (5.35) implies (5.8).

It is clear that if we know that (5.35) holds for every $\mathbf{x} \in \mathbb{R}^n$ then it will also hold for all $\mathbf{x} \in K$. In these cases there will be no need to characterize the kernel K of B. The situation clearly extends Corollary 3.2.

We notice that in the example (5.31) the matrix A is symmetric, and (5.35) holds. Therefore the additional assumptions of symmetry for A and the ellipticity in the kernel (5.35) cannot improve the quality of the bounds in their dependence on α and β. Similarly, assuming A to satisfy (5.35) for every $\mathbf{x} \in \mathbb{R}^n$ will not improve the bound either, as shown by the system

$$\begin{pmatrix} 1 & -1 & \beta \\ 1 & \alpha & 0 \\ \beta & 0 & 0 \end{pmatrix} \cdot \begin{pmatrix} x_1 \\ x_2 \\ y \end{pmatrix} = \begin{pmatrix} f \\ f \\ g \end{pmatrix}. \tag{5.36}$$

However, one might object, the matrix A in (5.33) is positive definite, but not symmetric. Is it possible to have better estimates assuming A to be symmetric, positive semidefinite in $\mathbb{R}^n$, and positive definite on K? Yes, it is. But in order to see that, we need first to extend Lemma 3.1.

Lemma 5.1. *Assume that A is an $n \times n$ symmetric positive semidefinite matrix, in the sense of (3.38). Assume that (5.3) holds, and assume that the dual norm $\|\cdot\|_F$ of $\|\cdot\|_X$ has been defined as in (4.23). Then we have*

$$\forall\, \mathbf{x} \in \mathbb{R}^n \ \forall\, \mathbf{z} \in \mathbb{R}^n \quad (\mathbf{z}^T A \mathbf{x})^2 \leq (\mathbf{x}^T A \mathbf{x})(\mathbf{z}^T A \mathbf{z}), \tag{5.37}$$

and

$$\forall\, \mathbf{x} \in \mathbb{R}^n \quad \|A\mathbf{x}\|_F^2 \leq M_a\, \mathbf{x}^T A \mathbf{x} \quad \forall\, \mathbf{x} \in \mathbb{R}^n. \tag{5.38}$$

Proof. We notice immediately that (5.37) has already been shown during the proof of Lemma 3.1 (see inequality (3.43)), so that we only need to prove (5.38). Dividing both sides of (5.37) by $\|\mathbf{z}\|_X^2$ and taking the supremum with respect to $\mathbf{z}$ on both sides we obtain

$$\left(\sup_{\mathbf{z}\in\mathbb{R}^n}\frac{(\mathbf{z}^T A\mathbf{x})}{\|\mathbf{z}\|_X}\right)^2 \leq (\mathbf{x}^T A\mathbf{x})\sup_{\mathbf{z}\in\mathbb{R}^n}\frac{(\mathbf{z}^T A\mathbf{z})}{\|\mathbf{z}\|_X^2}, \tag{5.39}$$

where we also used the fact that the supremum of the squares is the square of the supremum. Using (4.23) in the left-hand side, and (5.3) in the right-hand side of (5.39) we obtain (5.38). $\square$

We are now able to prove the following variant of Theorem 5.1.

Theorem 5.2. *In the Assumptions A1, A2, and A3 listed above, assume that the matrices A_k are symmetric and positive semidefinite (see (3.38)), and satisfy*

$$\exists\,\alpha > 0 \text{ such that } \forall\,k \in \mathbb{N} \quad \mathbf{x}^T A_k\mathbf{x} \geq \alpha\|\mathbf{x}\|_X^2 \quad \forall\,\mathbf{x} \in K_k, \tag{5.40}$$

where K_k is defined in (5.2). Assume also that B_k verifies (5.9). Then for every $k \in \mathbb{N}$, and for every right-hand side $\mathbf{f}_k \in \mathbb{R}^n$ and $\mathbf{g}_k \in \mathbb{R}^m$, the problem (5.10) has a unique solution $(\mathbf{x}_k, \mathbf{y}_k) \in \mathbb{R}^n \times \mathbb{R}^m$. Moreover if M_* is defined as in (5.11) then we have

$$\|\mathbf{x}_k\|_X \leq \frac{1}{\alpha}\|\mathbf{f}_k\|_F + \frac{2M_*^{1/2}}{\alpha^{1/2}\beta}\|\mathbf{g}_k\|_G \tag{5.41}$$

and

$$\|\mathbf{y}_k\|_Y \leq \frac{2M_*^{1/2}}{\alpha^{1/2}\beta}\|\mathbf{f}_k\|_F + \frac{2M_*}{\beta^2}\|\mathbf{g}_k\|_G. \tag{5.42}$$

Proof. We have to go back to the proof of Theorem 5.1, and try to improve it here and there. To start with, we keep the first step unchanged. Hence we have a $\mathbf{x}_g$ that satisfies (5.14) and (5.15). The major difference will be in second step. We again set $\mathbf{x}_K := \mathbf{x}_k - \mathbf{x}_g$ as in (5.16) (and we still have $\mathbf{x}_K \in K$). We want to improve (5.17) to

$$\left(\mathbf{x}_K^T A\mathbf{x}_K\right)^{1/2} \leq \frac{1}{\alpha^{1/2}}\|\mathbf{f}\|_F + \frac{M_a^{1/2}}{\beta}\|\mathbf{g}\|_G, \tag{5.43}$$

and

$$\|\mathbf{x}_K\|_X \leq \frac{1}{\alpha}\|\mathbf{f}\|_F + \frac{M_a^{1/2}}{\beta\alpha^{1/2}}\|\mathbf{g}\|_G. \tag{5.44}$$

We start by noticing that (5.40) gives immediately

$$\|\mathbf{x}_K\|_X \leq \frac{1}{\alpha^{1/2}}\left(\mathbf{x}_K^T A\mathbf{x}_K\right)^{1/2}. \tag{5.45}$$

Then we take the scalar product of $\mathbf{x}_K^T$ times the first equation of (5.10), and we notice that $\mathbf{x}_K^T B^T \mathbf{y}_k = \mathbf{y}_k B \mathbf{x}_K = 0$ since $\mathbf{x}_K \in K$, so that, using (4.22)

$$\mathbf{x}_K^T A \mathbf{x}_k = \mathbf{x}_K^T \mathbf{f} \le \|\mathbf{x}_K\|_X \|\mathbf{f}\|_F. \tag{5.46}$$

Using (5.16), (5.46) and (5.37), and finally (5.45), we obtain

$$\mathbf{x}_K^T A \mathbf{x}_K = \mathbf{x}_K^T A \mathbf{x}_k - \mathbf{x}_K^T A \mathbf{x}_g \le \|\mathbf{x}_K\|_X \|\mathbf{f}\|_F + \left(\mathbf{x}_K^T A \mathbf{x}_K\right)^{1/2} \left(\mathbf{x}_g^T A \mathbf{x}_g\right)^{1/2}$$

$$\le \left(\mathbf{x}_K^T A \mathbf{x}_K\right)^{1/2} \left(\frac{1}{\alpha^{1/2}} \|\mathbf{f}\|_F + \left(\mathbf{x}_g^T A \mathbf{x}_g\right)^{1/2}\right) \tag{5.47}$$

and (5.43) follows from (5.47), (5.3), and (5.15). Finally, (5.44) follows from (5.43) and (5.45). We anticipate the fourth step, and start collecting our estimate on $\mathbf{x}_k$. From (5.44) and (5.15) we have

$$\|\mathbf{x}_k\|_X \le \|\mathbf{x}_K\|_X + \|\mathbf{x}_g\|_X \le \frac{1}{\alpha} \|\mathbf{f}\|_F + \frac{\alpha^{1/2} + M_a^{1/2}}{\beta \alpha^{1/2}} \|\mathbf{g}\|_G. \tag{5.48}$$

We can now use our improved estimate in order to change the third step in the proof of Theorem 5.1. Indeed we can now write, instead of (5.24):

$$\beta \|\mathbf{y}_k\|_Y \le \|B^T \mathbf{y}_k\|_F \le \|\mathbf{f}\|_F + \|A \mathbf{x}_k\|_F \le \|\mathbf{f}\|_F + \|A \mathbf{x}_K\|_F + \|A \mathbf{x}_g\|_F. \tag{5.49}$$

We now use (5.38) (for $\mathbf{x} = \mathbf{x}_K$) and (5.43) to bound the second term in the right-hand side of (5.49), and (5.5) with (5.15) to bound the third term:

$$\beta \|\mathbf{y}_k\|_Y \le \|\mathbf{f}\|_F + (M_a)^{1/2} \left(\mathbf{x}_K^T A \mathbf{x}_K\right)^{1/2} + M_a \|\mathbf{x}_g\|_X$$

$$\le \|\mathbf{f}\|_F + \frac{M_a^{1/2}}{\alpha^{1/2}} \|\mathbf{f}\|_F + \frac{M_a}{\beta} \|\mathbf{g}\|_G + \frac{M_a}{\beta} \|\mathbf{g}\|_G \tag{5.50}$$

that reduces to the final estimate on $\mathbf{y}_k$:

$$\|\mathbf{y}_k\|_Y \le \frac{\alpha^{1/2} + M_a^{1/2}}{\alpha^{1/2} \beta} \|\mathbf{f}\|_F + + \frac{2 M_a}{\beta^2} \|\mathbf{g}\|_G. \tag{5.51}$$

Finally using (6.5) in (5.48) and (5.51) we obtain (5.41) and (5.42). □

Remark 5.5. We notice that the result of Theorem 5.2 is optimal, as can be seen considering the system

$$\begin{pmatrix} 2 & \sqrt{\alpha} & \beta \\ \sqrt{\alpha} & \alpha & 0 \\ \beta & 0 & 0 \end{pmatrix} \cdot \begin{pmatrix} x_1 \\ x_2 \\ y \end{pmatrix} = \begin{pmatrix} f_1 \\ f_2 \\ g \end{pmatrix}. \tag{5.52}$$

whose solution is given by

$$x_1 = \frac{g}{\beta}, \quad x_2 = \frac{f_2}{\alpha} - \frac{g}{\alpha^{1/2} \beta}, \quad y = \frac{f_1}{\beta} - \frac{f_2}{\alpha^{1/2} \beta} - \frac{1}{\beta^2}. \tag{5.53}$$

Notice as well that in this example condition (5.40) holds for every $\mathbf{x} \in \mathbb{R}^n$ (here $\mathbb{R}^2$), so that there is no hope of improving the bounds by requiring A to be symmetric and positive definite in the whole $\mathbb{R}^n$.

6 Additional Considerations

We shall now discuss briefly the case of problems of the type (3.36) where an additional matrix C is present. As a first step, we have to extend our assumptions.

Assumption A4 *We assume that we are given, for each $k \in \mathbb{N}$, an $m(k) \times m(k)$ matrix C_k, and we assume moreover that there exists a constant M_c, independent of k, such that*

$$\forall \, \mathbf{z} \in \mathbb{R}^m \, \forall \, \mathbf{y} \in \mathbb{R}^m \quad \mathbf{z}^T C_k \mathbf{y} \leq M_c \|\mathbf{z}\|_Y \|\mathbf{y}\|_Y. \qquad (6.1)$$

We notice that, as in (5.5), (5.6), and (5.7), we have now

$$\|C_k \mathbf{y}\|_G \equiv \sup_{\mathbf{z} \in \mathbb{R}^m} \frac{\mathbf{z}^T C_k \mathbf{y}}{\|\mathbf{z}\|_Y} \leq M_c \|\mathbf{y}\|_Y. \qquad (6.2)$$

We would like to extend the results of Theorem 5.1 to the system

$$\begin{pmatrix} A_k & B_k^T \\ B_k & C_k \end{pmatrix} \begin{pmatrix} \mathbf{x}_k \\ \mathbf{y}_k \end{pmatrix} = \begin{pmatrix} \mathbf{f}_k \\ \mathbf{g}_k \end{pmatrix}. \qquad (6.3)$$

Following Theorem 3.2 we shall assume that A is symmetric and nonsingular on K. It will therefore be more convenient, in order to reach optimal estimates in an easier way, to use (5.40) directly instead of (5.8). We repeat (5.40) for convenience of the reader,

$$\exists \, \alpha > 0 \text{ such that } \forall \, k \in \mathbb{N} \, \forall \, \mathbf{z} \in K_k \quad \mathbf{z}^T A_k \mathbf{z} \geq \alpha \|\mathbf{z}\|_X^2. \qquad (6.4)$$

For technical reasons, it will also be easier to deal separately with the case in which $\mathbf{f}_k = 0$ for all k and the case in which $\mathbf{g}_k = 0$ for all k. We start therefore with the following lemma.

Lemma 6.1. *Assume that the matrices A_k, B_k, and C_k, and the norms that we are going to use, verify Assumptions A1-A4. Assume that (6.4) and (5.9) are satisfied. Assume moreover that, for each k, A_k is symmetric and positive semidefinite, and C_k negative semidefinite. Assume finally that $\mathbf{f}_k = 0$ for every k. Then, for every $\mathbf{g}_k$ problem (4.1) has a unique solution. Moreover, setting*

$$M_* := \max\{M_a, M_b, M_c, 1\} \qquad (6.5)$$

we have

$$\|\mathbf{x}_k\|_X \leq \frac{4 M_*^{5/2}}{\alpha^{1/2} \beta^3} \|\mathbf{g}_k\|_G, \qquad (6.6)$$

$$\|\mathbf{y}_k\|_Y \leq \frac{M_*}{\beta^2} \|\mathbf{g}_k\|_G. \qquad (6.7)$$

Proof. As we did before, we drop all indices k with the exception of the solution that will still be $(\mathbf{x}_k, \mathbf{y}_k)$. Then, using Proposition 4.1 (and in particular (4.35)) together with the first equation of (6.3) we obtain

$$\beta\|\mathbf{y}_k\|_Y \le \|B^T\mathbf{y}_k\|_F = \|A\mathbf{x}_k\|_F. \tag{6.8}$$

Now, following the beginning of the proof of Theorem 3.2 we take the scalar product of $\mathbf{x}_k^T$ times the first equation of (6.3), then we take the scalar product of $\mathbf{y}_k^T$ times the second equation of (6.3), and we take the difference, obtaining

$$\mathbf{x}_k^T A\mathbf{x}_k - \mathbf{y}_k^T C\mathbf{y}_k = -\mathbf{y}_k^T\mathbf{g}. \tag{6.9}$$

Using then Lemma 5.1, the fact that C is negative semidefinite, and finally (4.32) we have

$$\|A\mathbf{x}_k\|_F^2 \le -M_a\mathbf{x}_k^T A\mathbf{x}_k \le M_a\mathbf{y}_k^T\mathbf{g} \le M_a\|\mathbf{y}_k\|_Y\|\mathbf{g}\|_G, \tag{6.10}$$

that, combined with (6.8) yields

$$\|A\mathbf{x}_k\|_F \le \frac{M_a}{\beta}\|\mathbf{g}\|_G \tag{6.11}$$

and using again (6.8)

$$\|\mathbf{y}_k\|_Y \le \frac{M_a}{\beta^2}\|\mathbf{g}\|_G, \tag{6.12}$$

that proves (6.7). The proof now becomes similar to the proof of Theorem 5.1. Using Proposition 4.2 we have the existence of an operator (that we call R_B) such that for every $\mathbf{w} \in \mathbb{R}^m$

$$B(R_B\mathbf{w}) = \mathbf{w}, \tag{6.13}$$

and

$$\beta\|R_B\mathbf{w}\|_X \le \|\mathbf{w}\|_G. \tag{6.14}$$

We set therefore

$$\tilde{\mathbf{x}} := R_B(\mathbf{g} - C\mathbf{y}_k) \tag{6.15}$$

and we have

$$B\tilde{\mathbf{x}} = \mathbf{g} - C\mathbf{y}_k, \tag{6.16}$$

together with

$$\beta\|\tilde{\mathbf{x}}\|_X \le \|\mathbf{g} - C\mathbf{y}_k\|_G \le (1 + \frac{M_cM_a}{\beta^2})\|\mathbf{g}\|_G, \tag{6.17}$$

where, in the last step, we used (6.2) and (6.12). From (6.17) we have then immediately

$$\|\tilde{\mathbf{x}}\|_X \le \frac{\beta^2 + M_*^2}{\beta^3}\|\mathbf{g}\|_G. \tag{6.18}$$

Setting now

$$\mathbf{x}_K := \mathbf{x}_k - \tilde{\mathbf{x}} \tag{6.19}$$

we have from (6.16) and the second equation of (6.3) that $\mathbf{x}_K \in K$ (the kernel of B). We notice then that, from the first equation of (6.3) and remembering that $\mathbf{f} = 0$:

$$\mathbf{x}_K^T A \mathbf{x}_k = -\mathbf{x}_K^T B^T \mathbf{y}_k = \mathbf{y}^T B \mathbf{x}_K = 0. \tag{6.20}$$

Moreover, using (6.19), (6.20), and (3.43) we have

$$\mathbf{x}_K^T A \mathbf{x}_K = -\mathbf{x}_K^T A \tilde{\mathbf{x}} \le (\mathbf{x}_K^T A \mathbf{x}_K)^{1/2} (\tilde{\mathbf{x}}^T A \tilde{\mathbf{x}})^{1/2}, \tag{6.21}$$

that gives easily

$$\mathbf{x}_K^T A \mathbf{x}_K \le \tilde{\mathbf{x}}^T A \tilde{\mathbf{x}}. \tag{6.22}$$

Hence we can use (6.4) and (6.22) to obtain

$$\alpha \|\mathbf{x}_K\|_X^2 \le \mathbf{x}_K^T A \mathbf{x}_K \le \tilde{\mathbf{x}}^T A \tilde{\mathbf{x}}, \tag{6.23}$$

and finally

$$\|\mathbf{x}_K\|_X \le (\frac{M_a}{\alpha})^{1/2} \|\tilde{\mathbf{x}}\|_X. \tag{6.24}$$

Notice that this improves the estimate in (5.17) with $\mathbf{f} = 0$ and $\tilde{\mathbf{x}}$ in place of $\mathbf{x}_g$. This is due to the fact that we are now using the symmetry of A. Finally we can collect (6.19), (6.18) and (6.23) and have an estimate for $\mathbf{x}$:

$$\|\mathbf{x}_k\|_X \le \|\mathbf{x}_K\|_X + \|\tilde{\mathbf{x}}\|_X \le (1 + (\frac{M_a}{\alpha})^{1/2}) \|\tilde{\mathbf{x}}\|_X$$
$$\le (1 + (\frac{M_a}{\alpha})^{1/2}) \frac{\beta^2 + M_*^2}{\beta^3} \|\mathbf{g}\|_G. \tag{6.25}$$

Using the definition (6.5) and the fact that $\beta \le M_b \le M_*$ in (6.25) we obtain (6.6) and the proof is completed. $\square$

Remark 6.1. The dependence of the constants in (6.6) and (6.7) on α and β cannot be improved. Indeed, taking for instance the problem

$$\begin{pmatrix} \alpha & \sqrt{\alpha} & -\sqrt{\alpha} & 0 & 0 \\ \sqrt{\alpha} & 2 & 1 & \beta & 0 \\ -\sqrt{\alpha} & 1 & 2 & 0 & \beta \\ 0 & \beta & 0 & 0 & 1 \\ 0 & 0 & \beta & -1 & 0 \end{pmatrix} \cdot \begin{pmatrix} x_1 \\ x_2 \\ x_3 \\ y_1 \\ y_2 \end{pmatrix} = \begin{pmatrix} 0 \\ 0 \\ 0 \\ -1 \\ -1 \end{pmatrix}, \tag{6.26}$$

we have easily, as unique solution,

$$x_1 = \frac{3}{\beta^3 \alpha^{1/2}}, \quad x_2 = -\frac{3 + \beta^2}{\beta^3}, \quad x_3 = \frac{3 - \beta^2}{\beta^3}, \tag{6.27}$$

$$y_1 = \frac{3}{\beta^2}, \quad y_2 = \frac{3}{\beta^2}, \tag{6.28}$$

that shows the (essential) optimality of (6.12) and (6.25).

We consider now the case when $\mathbf{g}_k$ is equal to zero and $\mathbf{f}_k$ is not.

Lemma 6.2. *Assume that the matrices A_k, B_k, and C_k, and the norms that we are going to use, verify Assumptions A1-A4. Assume that (6.4) and (5.9) are satisfied. Assume moreover that, for each k, A_k is symmetric and positive semidefinite, and C_k negative semidefinite. Assume finally that $\mathbf{g}_k = 0$ for every k. Then, for every $\mathbf{f}_k$ problem (4.1) has a unique solution. Moreover if M_* is defined as in (6.5) then we have*

$$\|\mathbf{x}_k\|_X \leq \frac{5M_*^4}{\alpha\beta^4}\|\mathbf{f}_k\|_F. \tag{6.29}$$

$$\|\mathbf{y}_k\|_Y \leq \frac{4M_*^{5/2}}{\alpha^{1/2}\beta^3}\|\mathbf{f}_k\|_F. \tag{6.30}$$

Proof. As we did before, we drop all indices k with the exception of the solution that will still be $(\mathbf{x}_k, \mathbf{y}_k)$. As in the previous lemma, we take the scalar product of $\mathbf{x}_k^T$ times the first equation of (6.3), then we take the scalar product of $\mathbf{y}_k^T$ times the second equation of (6.3), and we take the difference, obtaining

$$\mathbf{x}_k^T A \mathbf{x}_k - \mathbf{y}_k^T C \mathbf{y}_k = \mathbf{x}_k^T \mathbf{f}. \tag{6.31}$$

Using then Lemma 5.1, the fact that C is negative semidefinite, and finally (4.32) we have

$$\|A\mathbf{x}_k\|_F^2 \leq M_a \mathbf{x}_k^T A \mathbf{x}_k \leq M_a \mathbf{x}_k^T \mathbf{f} \leq M_a \|\mathbf{x}_k\|_X \|\mathbf{f}\|_F. \tag{6.32}$$

Next, we use Proposition 4.1 and in particular (4.35) to obtain, from the first equation of (6.3),

$$\beta\|\mathbf{y}_k\|_Y \leq \|B^T\mathbf{y}_k\|_F \equiv \|\mathbf{f}_k - A\mathbf{x}_k\|_F$$
$$\leq \|A\mathbf{x}_k\|_F + \|\mathbf{f}\|_F. \tag{6.33}$$

We consider now, as we did before, the operator R_B as defined in Proposition 4.2 and we set

$$\widetilde{\mathbf{x}} := R_B(-C\mathbf{y}_k) \tag{6.34}$$

so that

$$B\widetilde{\mathbf{x}} + C\mathbf{y}_k = 0 \tag{6.35}$$

and

$$\beta\|\widetilde{\mathbf{x}}\|_X \leq \|C\mathbf{y}_k\|_G \leq M_c\|\mathbf{y}_k\|_Y, \tag{6.36}$$

where we also used (6.2). We set now

$$\mathbf{x}_K := \mathbf{x}_k - \widetilde{\mathbf{x}}, \tag{6.37}$$

and notice that clearly $B\mathbf{x}_K = 0$, so that $\mathbf{x}_K \in K = \ker B$. Our next (and most delicate) step will be to estimate $\mathbf{x}_K$ in terms of $\widetilde{\mathbf{x}}$. We notice first that, using (6.4)

$$\alpha\|\mathbf{x}_K\|_X^2 \leq \mathbf{x}_K^T A \mathbf{x}_K, \tag{6.38}$$

which implies that

$$\|\mathbf{x}_K\|_X \leq \Big(\frac{\mathbf{x}_K^T A \mathbf{x}_K}{\alpha}\Big)^{1/2}. \tag{6.39}$$

Then we estimate $\mathbf{x}_K^T A \mathbf{x}_K$. We remember again that $\mathbf{x}_K^T B^T \mathbf{y}_k = 0$ (since $\mathbf{x}_K \in \ker B$), so that, using (6.37) and the first equation of (6.3)

$$\mathbf{x}_K^T A \mathbf{x}_K = \mathbf{x}_K^T A \mathbf{x}_k - \mathbf{x}_K^T A \tilde{\mathbf{x}} = \mathbf{x}_K^T \mathbf{f} - \mathbf{x}_K^T A \tilde{\mathbf{x}}. \tag{6.40}$$

We now use (4.22), (3.43), and then (6.39) in (6.40) obtaining

$$\begin{aligned}
\mathbf{x}_K^T A \mathbf{x}_K &\leq \|\mathbf{f}\|_F \|\mathbf{x}_K\|_X + (\mathbf{x}_K^T A \mathbf{x}_K)^{1/2}(\tilde{\mathbf{x}}^T A \tilde{\mathbf{x}})^{1/2} \\
&\leq \|\mathbf{f}\|_F \Big(\frac{\mathbf{x}_K^T A \mathbf{x}_K}{\alpha}\Big)^{1/2} + (\mathbf{x}_K A \mathbf{x}_K)^{1/2}(\tilde{\mathbf{x}}^T A \tilde{\mathbf{x}})^{1/2} \\
&\leq (\mathbf{x}_K^T A \mathbf{x}_K)^{1/2} \Big(\frac{1}{\sqrt{\alpha}}\|\mathbf{f}\|_F + (\tilde{\mathbf{x}}^T A \tilde{\mathbf{x}})^{1/2}\Big),
\end{aligned} \tag{6.41}$$

implying

$$(\mathbf{x}_K^T A \mathbf{x}_K)^{1/2} \leq \frac{1}{\sqrt{\alpha}}\|\mathbf{f}\|_F + (\tilde{\mathbf{x}}^T A \tilde{\mathbf{x}})^{1/2}. \tag{6.42}$$

Inserting (6.42) into (6.39), and then using (5.3) we have now

$$\|\mathbf{x}_K\|_X \leq \frac{1}{\alpha}\|\mathbf{f}\|_F + \Big(\frac{\tilde{\mathbf{x}}^T A \tilde{\mathbf{x}}}{\alpha}\Big)^{1/2} \leq \frac{1}{\alpha}\|\mathbf{f}\|_F + \frac{M_a^{1/2}}{\alpha^{1/2}}\|\tilde{\mathbf{x}}\|_X. \tag{6.43}$$

We can now collect (6.37), (6.36), and (6.43) to obtain an estimate for $\mathbf{x}_k$

$$\|\mathbf{x}_k\|_X \leq \|\mathbf{x}_K\|_X + \|\tilde{\mathbf{x}}\|_X \leq \frac{M_c M_a^{1/2}}{\alpha^{1/2}\beta}\|\mathbf{y}_k\|_Y + \frac{1}{\alpha}\|\mathbf{f}\|_F. \tag{6.44}$$

Now we take the square of both sides of (6.33), we use $(a+b)^2 \leq 2(a^2+b^2)$, we insert (6.32), and finally (6.44):

$$\begin{aligned}
\beta^2 \|\mathbf{y}_k\|_Y^2 &\leq 2\|A\mathbf{x}_k\|_F^2 + 2\|\mathbf{f}\|_F^2 \leq 2M_a\|\mathbf{x}_k\|_X \|\mathbf{f}\|_F + 2\|\mathbf{f}\|_F^2 \\
&\leq 2\|\mathbf{f}\|_F \Big(\frac{M_c M_a^{3/2}}{\alpha^{1/2}\beta}\|\mathbf{y}_k\|_Y + \frac{M_a}{\alpha}\|\mathbf{f}\|_F\Big) + 2\|\mathbf{f}\|_F^2. \tag{6.45}
\end{aligned}$$

We use now the fact that, for positive real numbers t, a, and b, if $t^2 \leq at + b$ then $t \leq a + \sqrt{b}$. Applied to (6.45) this gives

$$\|\mathbf{y}_k\|_Y \leq \frac{2M_c M_a^{3/2}}{\alpha^{1/2}\beta^3}\|\mathbf{f}\|_F + \frac{(2M_a + 2\alpha)^{1/2}}{\alpha^{1/2}\beta}\|\mathbf{f}\|_F. \tag{6.46}$$

Using the fact that $\alpha \leq M_a \leq M_*$ and $\beta \leq M_b \leq M_*$ we can rewrite (6.46) as

$$\|\mathbf{y}_k\|_Y \leq \frac{4M_*^{5/2}}{\alpha^{1/2}\beta^3}\|\mathbf{f}\|_F. \tag{6.47}$$

Inserting (6.47) into (6.44) we obtain the corresponding estimate for $\mathbf{x}_k$:

$$\|\mathbf{x}_k\|_X \leq \left(\frac{4M_*^4}{\alpha\beta^4} + \frac{1}{\alpha}\right)\|\mathbf{f}\|_F \leq \left(\frac{5M_*^4}{\alpha\beta^4}\right)\|\mathbf{f}\|_F, \qquad (6.48)$$

which concludes the proof. $\square$

Remark 6.2. The result (6.29)-(6.30) cannot be improved in its dependence from the constants α and β. Indeed, if we consider the system

$$\begin{pmatrix} 2\alpha & \sqrt{\alpha} & -\sqrt{\alpha} & 0 & 0 \\ \sqrt{\alpha} & 2 & 1 & \beta & 0 \\ -\sqrt{\alpha} & 1 & 2 & 0 & \beta \\ 0 & \beta & 0 & 0 & 1 \\ 0 & 0 & \beta & -1 & 0 \end{pmatrix} \cdot \begin{pmatrix} x_1 \\ x_2 \\ x_3 \\ y_1 \\ y_2 \end{pmatrix} = \begin{pmatrix} 2 \\ 0 \\ 0 \\ 0 \\ 0 \end{pmatrix}, \qquad (6.49)$$

we have easily, as unique solution,

$$x_1 = \frac{3+\beta^4}{\alpha\beta^4}, \qquad x_2 = \frac{-3-\beta^2}{\alpha^{1/2}\beta^4}, \qquad x_3 = \frac{3-\beta^2}{\alpha^{1/2}\beta^4}, \qquad (6.50)$$

$$y_1 = \frac{3-\beta^2}{\alpha^{1/2}\beta^3}, \qquad y_2 = \frac{3+\beta^2}{\alpha^{1/2}\beta^3}, \qquad (6.51)$$

that shows the (essential) optimality of (6.29) and (6.30).

We can now collect the results of the previous two lemmas.

Theorem 6.1. *Assume that the matrices A_k, B_k, and C_k, and the norms that we are going to use, verify Assumptions A1-A4. Assume that (6.4) and (5.9) are satisfied. Assume moreover that, for each k, A_k is symmetric and positive semidefinite, and C_k negative semidefinite. Then, for every $\mathbf{f}_k$ and $\mathbf{g}_k$ problem (4.1) has a unique solution. Moreover if M_* is defined as in (6.5) then we have*

$$\|\mathbf{x}_k\|_X \leq \frac{5M_*^4}{\alpha\beta^4}\|\mathbf{f}_k\|_F + \frac{4M_*^{5/2}}{\alpha^{1/2}\beta^3}\|\mathbf{g}_k\|_G \qquad (6.52)$$

and

$$\|\mathbf{y}_k\|_Y \leq \frac{4M_*^{5/2}}{\alpha^{1/2}\beta^3}\|\mathbf{f}_k\|_F + \frac{M_*}{\beta^2}\|\mathbf{g}_k\|_G. \qquad (6.53)$$

The proof follows easily by linearity.

The dependence of the constants in (6.52) and (6.53) on α and β improves noticeably if we assume that C_k is symmetric as well. As an example we can consider the particular case (relevant in applications) of systems still having the structure (6.3), where $-C$ is a symmetric and positive definite matrix verifying

$$\gamma\|\mathbf{y}\|_Y^2 \leq -\mathbf{y}^T C\mathbf{y} \leq M_c\|\mathbf{y}\|_Y^2 \quad \forall \mathbf{y} \in \mathbb{R}^m. \qquad (6.54)$$

We notice that our assumption (6.54) easily implies that

$$\frac{1}{M_c}\|\mathbf{z}\|_G^2 \le -\mathbf{z}^T C^{-1}\mathbf{z} \le \frac{1}{\gamma}\|\mathbf{z}\|_G^2 \quad \forall \mathbf{z} \in \mathbb{R}^m \tag{6.55}$$

as can easily be seen, for instance, by diagonalizing the matrix $H^{-1/2}CH^{-1/2}$. From (6.54) and (6.55) we easily obtain as well that

$$\|\mathbf{y}\|_Y \le \frac{1}{\gamma}\|C\mathbf{y}\|_G \quad \forall \mathbf{y} \in \mathbb{R}^m, \tag{6.56}$$

and

$$\|\mathbf{z}\|_G \le M_c\|C^{-1}\mathbf{z}\|_Y \quad \forall \mathbf{z} \in \mathbb{R}^m. \tag{6.57}$$

Let us see first how to make the estimate for $\mathbf{f} = 0$. Following the notation of Lemma 6.1 we still have (6.11), (6.12), and (6.24). Our target is to improve (6.18), which is suboptimal in our (stronger) assumptions. For this we restart by taking once more the the scalar product of the first equation times $\mathbf{x}^T$, getting

$$\mathbf{x}_k^T A\mathbf{x}_k + \mathbf{x}_k^T B^T \mathbf{y}_k = 0 \tag{6.58}$$

and we substitute $\mathbf{y}_k = C^{-1}(\mathbf{g} - B\mathbf{x}_k)$. Recalling that A is positive semidefinite we obtain

$$-\mathbf{x}_k^T B^T C^{-1} B\mathbf{x}_k \le -\mathbf{x}_k^T B^T C^{-1}\mathbf{g} = -\mathbf{g}^T C^{-1} B\mathbf{x}_k. \tag{6.59}$$

Using (6.55) with $\mathbf{z} = B\mathbf{x}_k$ and (6.56) with $\mathbf{y} = C^{-1}B\mathbf{x}_k$ we have from (6.59)

$$\|B\mathbf{x}_k\|_G^2 \le M_c(-\mathbf{x}_k^T B^T C^{-1} B\mathbf{x}_k) \le M_c(-\mathbf{g}^T C^{-1} B\mathbf{x}_k) \le \frac{M_c}{\gamma}\|\mathbf{g}\|_G\|B\mathbf{x}_k\|_G \tag{6.60}$$

that easily gives

$$\|B\mathbf{x}_k\|_G \le \frac{M_c}{\gamma}\|\mathbf{g}\|_G. \tag{6.61}$$

As $\beta\|\tilde{\mathbf{x}}\|_X \le \|B\tilde{\mathbf{x}}\|_G = \|B\mathbf{x}_k\|_G$ we have then

$$\|\tilde{\mathbf{x}}\|_X \le \frac{M_c}{\gamma\beta}\|\mathbf{g}\|_G. \tag{6.62}$$

We can now use this improved estimate in (6.24), and we obtain

$$\|\mathbf{x}_K\|_X \le \left(\frac{M_a}{\alpha}\right)^{1/2}\|\tilde{\mathbf{x}}\|_X \le \frac{M_c M_a^{1/2}}{\gamma\beta\alpha^{1/2}}\|\mathbf{g}\|_G. \tag{6.63}$$

We notice at this point that we have another way to obtain an estimate for $\mathbf{y}_k$, apart from (6.12) that we kept from the previous analysis; actually from (6.56), the second equation of (6.3) and (6.61):

$$\|\mathbf{y}_k\|_Y \le \frac{1}{\gamma}\|\mathbf{g} - B\mathbf{x}_k\|_G \le \left(\frac{1}{\gamma} + \frac{M_c}{\gamma^2}\right)\|\mathbf{g}\|_G = \frac{\gamma + M_c}{\gamma^2}\|\mathbf{g}\|_G. \tag{6.64}$$

With some manipulations we see that (6.12) and (6.64) can be combined into

$$\|\mathbf{y}_k\|_Y \le \frac{M_a(M_c + \gamma)}{M_a\gamma^2 + (M_c + \gamma)\beta^2}\|\mathbf{g}\|_G. \tag{6.65}$$

We collect the results for $\mathbf{f} = 0$

$$\|\mathbf{x}_k\|_X \le \left(\left(\frac{M_a}{\alpha}\right)^{1/2} + 1\right)\frac{M_c}{\gamma\beta}\|\mathbf{g}\|_G, \qquad \|\mathbf{y}_k\|_Y \le \frac{M_a(M_c + \gamma)}{M_a\gamma^2 + (M_c + \gamma)\beta^2}\|\mathbf{g}\|_G. \tag{6.66}$$

We consider now the case in which $\mathbf{g} = 0$. As before we can keep part of the previous analysis, but we can improve on it in several places. From the proof of Lemma 6.2 we keep the definition of $\widetilde{\mathbf{x}}$ and $\mathbf{x}_K$, and the estimates (6.42) and (6.43). We take now the scalar product of $\widetilde{\mathbf{x}}^T$ times the first equation of (6.3), and substitute $\mathbf{y}_k = -C^{-1}B\mathbf{x}_k$:

$$\widetilde{\mathbf{x}}^T A\mathbf{x}_k - \widetilde{\mathbf{x}}^T B^T C^{-1}B\mathbf{x}_k = \widetilde{\mathbf{x}}^T \mathbf{f}. \tag{6.67}$$

We now recall that $\mathbf{x}_k = \widetilde{\mathbf{x}} + \mathbf{x}_K$ and that $B\widetilde{\mathbf{x}} = B\mathbf{x}_k$, and rewrite (6.67) as follows

$$\widetilde{\mathbf{x}}^T A\mathbf{x}_K + \widetilde{\mathbf{x}}^T A\widetilde{\mathbf{x}} - \mathbf{x}_k^T B^T C^{-1}B\mathbf{x}_k = \widetilde{\mathbf{x}}^T \mathbf{f}. \tag{6.68}$$

We apply now (6.55) with $\mathbf{z} = B\mathbf{x}_k$ and we recall that $\beta\|\widetilde{\mathbf{x}}\|_G \le \|B\widetilde{\mathbf{x}}\|_G = \|B\mathbf{x}_k\|_G$. We obtain

$$\begin{aligned}
&\widetilde{\mathbf{x}}^T A\widetilde{\mathbf{x}} + \frac{1}{M_c}\|B\mathbf{x}_k\|_G^2 \\
&\le \widetilde{\mathbf{x}}^T A\widetilde{\mathbf{x}} - \mathbf{x}_k^T B^T C^{-1}B\mathbf{x}_k \\
&\le -\widetilde{\mathbf{x}}^T A\mathbf{x}_K + \|\mathbf{f}\|_F\|\widetilde{\mathbf{x}}\|_X \le -\widetilde{\mathbf{x}}^T A\mathbf{x}_K + \frac{1}{\beta}\|\mathbf{f}\|_F\|B\mathbf{x}_k\|_G.
\end{aligned} \tag{6.69}$$

We apply then (3.43) and (6.42):

$$\begin{aligned}
&\widetilde{\mathbf{x}}^T A\widetilde{\mathbf{x}} + \frac{1}{M_c}\|B\mathbf{x}_k\|_G^2 \\
&\le (\widetilde{\mathbf{x}}^T A\widetilde{\mathbf{x}})^{1/2}(\mathbf{x}_K^T A\mathbf{x}_K)^{1/2} + \frac{1}{\beta}\|\mathbf{f}\|_F\|B\mathbf{x}_k\|_G \\
&\le \frac{1}{\alpha^{1/2}}\|\mathbf{f}\|_F(\widetilde{\mathbf{x}}^T A\widetilde{\mathbf{x}})^{1/2} + \widetilde{\mathbf{x}}^T A\widetilde{\mathbf{x}} + \frac{1}{\beta}\|\mathbf{f}\|_F\|B\mathbf{x}_k\|_G.
\end{aligned} \tag{6.70}$$

We simplify $\widetilde{\mathbf{x}}^T A\widetilde{\mathbf{x}}$ on both sides, and we use again $\beta\|\widetilde{\mathbf{x}}\|_X \le \|B\mathbf{x}_k\|_G$:

$$\frac{1}{M_c}\|B\mathbf{x}_k\|_G^2 \le \frac{M_a^{1/2}}{\alpha^{1/2}\beta}\|\mathbf{f}\|_F\|B\mathbf{x}_k\|_G + \frac{1}{\beta}\|\mathbf{f}\|_F\|B\mathbf{x}_k\|_G \tag{6.71}$$

and we divide both sides by $\|B\mathbf{x}_k\|_G$ obtaining

$$\frac{1}{M_c}\|B\mathbf{x}_k\|_G \le \frac{M_a^{1/2}}{\alpha^{1/2}\beta}\|\mathbf{f}\|_F + \frac{1}{\beta}\|\mathbf{f}\|_F \le \frac{M_a^{1/2} + \alpha^{1/2}}{\alpha^{1/2}\beta}\|\mathbf{f}\|_F, \tag{6.72}$$

which is the basis of our improved estimates. From (6.72) we first derive

$$\|\widetilde{\mathbf{x}}\|_X \leq \frac{1}{\beta}\|B\widetilde{\mathbf{x}}\|_G \leq M_c \frac{M_a^{1/2} + \alpha^{1/2}}{\alpha^{1/2}\beta^2}\|\mathbf{f}\|_F, \tag{6.73}$$

and then we use it in (6.43)

$$\|\mathbf{x}_K\|_X \leq \frac{1}{\alpha}\|\mathbf{f}\|_F + \frac{M_a^{1/2}}{\alpha^{1/2}}\|\widetilde{\mathbf{x}}\|_X \leq \left(\frac{1}{\alpha} + \frac{M_a^{1/2}}{\alpha^{1/2}} M_c \frac{M_a^{1/2} + \alpha^{1/2}}{\alpha^{1/2}\beta^2}\right)\|\mathbf{f}\|_F$$

$$\leq \left(\frac{1}{\alpha} + \frac{M_c M_a + M_c(M_a\alpha)^{1/2}}{\alpha\beta^2}\right)\|\mathbf{f}\|_F. \tag{6.74}$$

From the second equation of (6.3), (6.56), and (6.73) we also derive our improved estimate for $\mathbf{y}_k$

$$\|\mathbf{y}_k\|_Y = \|C^{-1}B\mathbf{x}_k\|_Y \leq \frac{1}{\gamma}\|B\mathbf{x}_k\|_G \leq \frac{M_c}{\gamma}\frac{M_a^{1/2} + \alpha^{1/2}}{\alpha^{1/2}\beta}\|\mathbf{f}\|_F. \tag{6.75}$$

We collect the results for $\mathbf{g} = 0$, using the fact that $\alpha \leq M_a$,

$$\|\mathbf{x}_k\|_X \leq \frac{\beta^2 + 4M_c M_a}{\alpha\beta^2}\|\mathbf{f}\|_F, \qquad \|\mathbf{y}_k\|_Y \leq \frac{2M_c M_a^{1/2}}{\gamma\alpha^{1/2}\beta}\|\mathbf{f}\|_F. \tag{6.76}$$

We finally present (6.66) and (6.76) in one theorem.

Theorem 6.2. *Assume that the matrices A_k, B_k, and C_k, and the norms that we are going to use, verify Assumptions A1-A4. Assume that (6.4) and (5.9) are satisfied. Assume moreover that, for each k, A_k is symmetric and positive semidefinite, and C_k is symmetric and negative definite, satisfying (6.54). Then, for every $\mathbf{f}_k$ and $\mathbf{g}_k$ problem (4.1) has a unique solution. Moreover we have*

$$\|\mathbf{x}_k\|_X \leq \frac{\beta^2 + 4M_c M_a}{\alpha\beta^2}\|\mathbf{f}\|_F + \frac{2M_a^{1/2}M_c}{\alpha^{1/2}\gamma\beta}\|\mathbf{g}\|_G \tag{6.77}$$

and

$$\|\mathbf{y}_k\|_Y \leq \frac{2M_c M_a^{1/2}}{\gamma\alpha^{1/2}\beta}\|\mathbf{f}\|_F + \frac{M_a(M_c + \gamma)}{M_a\gamma^2 + (M_c + \gamma)\beta^2}\|\mathbf{g}\|_G. \tag{6.78}$$

We remark that in several applications we have $C = \varepsilon I$, so that $M_c = \gamma = \varepsilon$. In this case the estimates (6.77) and (6.78) become

$$\|\mathbf{x}_k\|_X \leq \frac{\beta^2 + 4\varepsilon M_a}{\alpha\beta^2}\|\mathbf{f}\|_F + \frac{2M_a^{1/2}}{\alpha^{1/2}\beta}\|\mathbf{g}\|_G \tag{6.79}$$

and

$$\|\mathbf{y}_k\|_Y \leq \frac{2M_a^{1/2}}{\alpha^{1/2}\beta}\|\mathbf{f}\|_F + \frac{2M_a}{M_a\varepsilon + 2\beta^2}\|\mathbf{g}\|_G. \tag{6.80}$$

We also point out that (6.79) and (6.80) are optimal, with respect to the dependency of the stability constants the parameters α, β and ε. To see this, consider the problem

$$
\begin{pmatrix}
2\alpha & \sqrt{\alpha} & -\sqrt{\alpha} & 0 & 0 \\
\sqrt{\alpha} & 2 & 1 & \beta & 0 \\
-\sqrt{\alpha} & 1 & 2 & 0 & \beta \\
0 & \beta & 0 & -\varepsilon & 0 \\
0 & 0 & \beta & 0 & -\varepsilon
\end{pmatrix}
\cdot
\begin{pmatrix}
x_1 \\ x_2 \\ x_3 \\ y_1 \\ y_2
\end{pmatrix}
=
\begin{pmatrix}
2f \\ 0 \\ 0 \\ 0 \\ 2g
\end{pmatrix},
\tag{6.81}
$$

whose solution is given by

$$
x_1 = \frac{f(\beta^2 + \varepsilon)}{\alpha\beta^2} + \frac{g}{\beta\alpha^{1/2}}, \quad x_2 = -\frac{f\varepsilon}{\alpha^{1/2}\beta^2} - \frac{3g\varepsilon}{\beta(3\varepsilon + \beta^2)}, \tag{6.82}
$$

$$
x_3 = \frac{f\varepsilon}{\alpha^{1/2}\beta^2} + \frac{g(3\varepsilon + 2\beta^2)}{\beta(3\varepsilon + \beta^2)}, \tag{6.83}
$$

$$
y_1 = -\frac{f}{\alpha^{1/2}\beta} - \frac{3g}{3\varepsilon + \beta^2}, \quad y_2 = \frac{f}{\alpha^{1/2}\beta} - \frac{3g}{3\varepsilon + \beta^2}. \tag{6.84}
$$

References

1. Arnold, D. N. Discretization by finite elements of a model parameter dependent problem. Numer. Math. 37 (1981), no. 3,405–421.
2. Bathe, K. J. Finite Element Procedures, Prentice Hall, 1996.
3. Braess, D. Finite elements. Theory, fast solvers, and applications in solid mechanics. Cambridge University Press, Cambridge, 1997.
4. Brenner, S. C. and Scott, L. R. The mathematical theory of finite element methods. Texts in Applied Mathematics, 15. Springer-Verlag, New York, 1994.
5. Brezzi, F. On the existence, uniqueness and approximation of saddle-point problems arising from Lagrangian multipliers. Rev. Française Automat. Informat. Recherche Opérationnelle Sér. Rouge 8 (1974), no. R-2, 129–151.
6. Brezzi, F. and Fortin, M. Mixed and Hybrid Finite Element Methods. Springer Verlag, New York, Springer Series in Computational Mathematics 15, 1991.
7. Girault, V. and Raviart, P.-A. Finite element methods for Navier-Stokes equations. Theory and algorithms. Springer Series in Computational Mathematics, 5. Springer-Verlag, Berlin, 1986.
8. Hughes, T. J. R. The Finite Element Method: Linear Static and Dynamic Finite Element Analysis, Prentice-Hall, Dover, Mineola, New York. 2000.
9. Quarteroni, A. and Valli, A. Numerical approximation of partial differential equations. Springer Series in Computational Mathematics, 23. Springer-Verlag, Berlin, 1994.
10. Yosida, K. Functional analysis. Die Grundlehren der Mathematischen Wissenschaften, Band 123 Academic Press, Inc., New York; Springer-Verlag, Berlin
11. Zienkiewicz, O. C. and Taylor, R. L. The finite element method. Vol. 1,2,3. Fifth edition. Butterworth-Heinemann, Oxford, 2000.

Mean Curvature Flow and Related Topics

Klaus Deckelnick[1] and Gerhard Dziuk[2]

[1] Institut für Analysis und Numerik, Universität Magdeburg, Universitätsplatz 2,
D–39106 Magdeburg, Germany
[2] Institut für Angewandte Mathematik, Universität Freiburg,
Hermann-Herder-Straße 10, D-79104 Freiburg, Germany

Abstract. The subject of these notes is the numerical approximation of hypersurfaces $\Gamma(t)$ which evolve according to the following anisotropic mean curvature flow

$$\beta(\nu)V = -\operatorname{div}\left(\nabla_\nu \gamma(x,\nu)\right) + c(x,t,\nu) \qquad \text{on } \Gamma(t).$$

Here, ν is a unit normal to $\Gamma(t)$, V is the normal velocity, γ is a given positive function, which is positively homogeneous of degree one in the second variable, and β and c are given functions. We consider three approaches in order to track the evolution of $\Gamma(t)$: the parametric approach, the description as a graph and the level set method. In each case we derive a PDE which corresponds to the above equation and recall the necessary analytical background. These PDEs are subsequently discretized in space with the help of Finite Elements and in time by the backward Euler method. We discuss the resulting schemes and describe the corresponding stability and error analysis. For each of the aforementioned approaches we also present figures of computed curves and surfaces. A detailed error analysis for the anisotropic curvature flow of graphs is included.

1 Introduction

These notes concern the numerical approximation of hypersurfaces $\Gamma(t) \subset \mathbb{R}^{n+1}$, which evolve according to a geometric law of the form

$$\beta(\nu)V = -\operatorname{div}\left(\nabla_\nu \gamma(x,\nu)\right) + c(x,t,\nu) \qquad \text{on } \Gamma(t). \qquad (1.1)$$

Here, ν is a unit normal to $\Gamma(t)$ and V is the normal velocity. The function γ is positive and positively homogeneous of degree one in the second variable and is often called the weight or anisotropy function. The expression ∇_ν refers to differentiation with respect to the second variable. Finally, β and c are given functions where β, usually referred to as the mobility, is assumed to be positive. Let us list some problems in which a law of the form (1.1) arise.

1) Geometry

We begin by describing classical evolution by mean curvature. Consider a family $(\Gamma(t))_{t\in[0,T]}$ of closed compact hypersurfaces such that $\Gamma(t)$ forms the boundary of an open bounded set $\Omega(t)$. Let us denote by ν the unit normal to $\Gamma(t)$ which points out of $\Omega(t)$, and by H the mean curvature of

$\Gamma(t)$. Our sign convention will be that H is positive for spheres. We say that $\Gamma(t)$ moves by mean curvature if

$$V = -H \quad \text{on } \Gamma(t). \tag{1.2}$$

Note that this corresponds to the choices $\beta \equiv 1$, $c \equiv 0$ and $\gamma(x,p) = |p|$, since $H = \operatorname{div} \nu$. It is well-known that this flow can be interpreted as the L^2-gradient flow for the area functional. In general, given an initial hypersurface Γ_0, smooth solutions of (1.2) will only exist locally in time. Global existence of solutions has been obtained in the case $n = 1$ (the so-called curve shortening flow) for embedded Γ_0 ([36], [38]) and in the case $n > 1$ for hypersurfaces Γ_0 which enclose a convex set ([39]). Otherwise, the flow may develop singularities like the well-known pinching-off singularity for certain dumbbell type initial surfaces. Tracking the evolution beyond such a singularity requires a lot of care in the description of $\Gamma(t)$: level set and phase field approaches have been shown to be successful.

2) Image Processing

One of the most important problems in image processing is to automatically detect contours of objects. We essentially follow the exposition in [3]. Suppose that $M \subset \mathbb{R}^{n+1}$ ($n = 1$ or 2) is a given object and let $I(x) = \chi_{\Omega \setminus M}(x)$ be the characteristic function of $\Omega \setminus M$. The function

$$g(x) = \frac{1}{1 + |\nabla I_\sigma(x)|^2},$$

where I_σ is a mollification of I, will be small near the contour of M. It is therefore natural to look for minimizers of the functional

$$J(\Gamma) = \int_\Gamma g$$

where Γ is respectively a curve in $\mathbb{R}^2$ and a surface in $\mathbb{R}^3$. The corresponding L^2-gradient flow leads to the following evolution law: find curves/surfaces (moving "snakes") $\Gamma(t)$ (t is an artificial time) such that

$$V = -\operatorname{div}(g(x)\nu) = -g H - \nabla g \cdot \nu \quad \text{on } \Gamma(t),$$

where H is again the mean curvature and ν the unit normal to $\Gamma(t)$. Also, $\Gamma(0) = \Gamma_0$, where Γ_0 contains M. We see, that this law fits into the framework (1.1) if we choose $\beta \equiv 1$, $c \equiv 0$ and $\gamma(x,p) = g(x)|p|$.

3) Stefan Problem with Kinetic Undercooling

Consider a container $\Omega \subset \mathbb{R}^{n+1}$ ($n = 1$ or 2) filled with an undercooled liquid. An initial solid seed grows into the liquid towards the undercooled walls of Ω. A mathematical model for this situation is the Stefan problem with kinetic undercooling, in which the solid–liquid phase boundary is described by a curve/surface $\Gamma(t)$ and has to be determined together with the temperature distribution. The problem then reads: for a given initial phase boundary Γ_0 and initial temperature distribution $T_0 = T_0(x)$ $(x \in \overline{\Omega})$, find the temperature $T = T(x,t)$ and the phase boundary $\Gamma(t)$ $(t > 0)$, such that the heat equation is satisfied in the bulk, i.e.

$$T_t - \Delta T = 0 \quad \text{in } \Omega \setminus \Gamma(t),$$

together with the initial value $T(\cdot, 0) = T_0$ in $\overline{\Omega}$ and boundary value $T = T_0$ on $\partial\Omega$. On the moving boundary the following two conditions are satisfied:

$$V = -\frac{1}{\varepsilon_L} \left[\frac{\partial T}{\partial \nu} \right] \quad \text{on } \Gamma(t),$$
$$T + \varepsilon_V V + \varepsilon_H H = 0 \quad \text{on } \Gamma(t). \tag{1.3}$$

Here, $\varepsilon_V, \varepsilon_H$ are given functions depending on the normal of Γ and ε_L is the constant latent heat of solidification. By $[\partial T / \partial \nu]$ we denote the jump in the normal derivative of the temperature field across the interface. Note that (1.3) can be rewritten as

$$\frac{\varepsilon_V}{\varepsilon_H} V = -H - \frac{1}{\varepsilon_H} T \quad \text{on } \Gamma(t).$$

If we consider T as being given, this equation again fits into the framework of (1.1).

4) Surface Diffusion

The following law does not fit into (1.1), but we list it as an example of a geometric evolution where the normal velocity depends on higher derivatives of mean curvature. We call the equation

$$V = \Delta_\Gamma H \quad \text{on } \Gamma \tag{1.4}$$

the surface diffusion equation. Here, Δ_Γ denotes the Laplace–Beltrami operator on Γ. The law arises in the modelling of mass diffusion within the bounding surface of a solid body, see [8].

In these notes we shall restrict ourselves to evolution laws (1.1), in which $\beta \equiv 1$, $c \equiv 0$ and $\gamma(x, p) = \gamma(p)$, i.e.

$$V = -\operatorname{div}(\nabla\gamma(\nu)) \quad \text{on } \Gamma(t). \tag{1.5}$$

We remark, that it is possible to extend our results to the more general setting (1.1) under suitable assumptions on the data (cf. [20] for the case $\gamma = \gamma(x, p)$). Equation (1.5) is complemented by an initial condition and a suitable boundary condition if necessary. In order to solve this problem analytically or numerically we need a description of $\Gamma(t)$. Here we shall focus on three possible approaches.

a) Parametric Approach

The hypersurfaces $\Gamma(t)$ are given as $\Gamma(t) = u(\cdot,t)(M)$, where M is a suitable reference manifold (fixing the topological type of $\Gamma(t)$) and $u : M \times [0,T) \to \mathbb{R}^{n+1}$ has to be determined. We shall see later, how (1.5) translates into a nonlinear parabolic system of PDEs for the vector u.

b) Graphs

We assume that $\Gamma(t)$ can be written in the form

$$\Gamma(t) = \{(x, u(x,t)) \,|\, x \in \Omega\},$$

where $\Omega \subset \mathbb{R}^n$ and the height function $u : \Omega \times [0,T) \to \mathbb{R}$ has to be found. The law (1.5) now leads to a nonlinear parabolic equation for u. Clearly, the assumption that $\Gamma(t)$ is a graph is rather restrictive; however, techniques which were developed for this case have turned out to be very helpful in understanding more general situations. Furthermore, the corresponding PDE has a formal similarity to a regularized version of the level set PDE, which proved to be pivotal in analyzing certain numerical schemes for the level set approach.

c) Level Set Method

One looks for $\Gamma(t)$ as the zero level set of an auxiliary function $u : \mathbb{R}^{n+1} \times [0,\infty) \to \mathbb{R}$, i.e.

$$\Gamma(t) = \{x \in \mathbb{R}^{n+1} \,|\, u(x,t) = 0\}.$$

The law (1.5) now translates into a nonlinear, degenerate and singular PDE for u. Contrary to parametric and graph approach, the level set method is capable of tracking topological changes of $\Gamma(t)$. This advantage however needs to be offset against the fact that the problem now becomes $(n+1)$–dimensional in space. In general, the decision for one or the other approach will depend on whether one expects topological changes (like pinching–off or merging) in the flow.

These notes are organized as follows. In Chapter 2 we present some useful notation from geometric analysis and introduce the concept of the anisotropy γ together with its relevant properties. Chapter 3 deals with the parametric approach. We start with the classical curve shortening flow and present a semidiscrete numerical scheme as well as error estimates. This approach is subsequently generalized to the anisotropic case. Next, we show how to apply the above ideas to the approximation of higher dimensional surfaces. A crucial point is to construct numerical schemes which reflect the intrinsic nature of the flow. Chapter 4 introduces the level set equation as a way of handling topological changes. We briefly discuss the framework of viscosity solutions which allows a satisfactory existence and uniqueness theory. For numerical purposes it is convenient to regularize the level set equation. We collect some properties of the regularized problem and clarify its formal similarity to the graph setting. This motivates a detailed study of the graph case in Chapters 5 and 6. In Chapter 5 we study the isotropic case and recall some analytical results as well as a convergence analysis for a discretization in space by linear finite elements. We finish this chapter by discussing the discretization in time together with the question of stability. In Chapter 6 we generalize these ideas to the anisotropic case. We derive the corresponding PDE and use a variational form in order to discretize in space by linear finite elements. A detailed error analysis is presented for the resulting scheme and we finish by addressing time discretization issues. Finally, Chapter 7 shows how to exploit results proved in the graph setting to obtain convergence for numerical approximations of the level set equation. For the convenience of the reader we have included a long list of references, which are related to the subject of these notes, but not all of which are cited in the text.

2 Some Geometric Analysis

2.1 Tangential Gradients and Curvature

Let Γ be a C^2 hypersurface in $\mathbb{R}^{n+1}$ with unit normal ν. This means that Γ can locally be represented as the graph of a C^2 function over an open subset of $\mathbb{R}^n$. For a function $u : U \to \mathbb{R}$ which is defined and smooth in a neighbourhood U of Γ, we define its tangential gradient on Γ by

$$\nabla_\Gamma u = \nabla u - \nabla u \cdot \nu\, \nu.$$

It is not difficult to show that $\nabla_\Gamma u$ only depends on the values of u on Γ. We use the notation

$$\nabla_\Gamma u = (\underline{D}_1 u, \ldots, \underline{D}_{n+1} u)$$

for the $n+1$ components of the tangential gradient. Obviously

$$\nabla_\Gamma u \cdot \nu = 0.$$

For a function u which has second derivatives we define the Laplace-Beltrami operator of u on Γ as the tangential divergence of the tangential gradient, i.e. by

$$\Delta_\Gamma u = \nabla_\Gamma \cdot \nabla_\Gamma u$$

or

$$\Delta_\Gamma u = \sum_{j=1}^{n+1} \underline{D}_j \underline{D}_j u.$$

The matrix (H_{jk}) where

$$H_{jk} = \underline{D}_j \nu_k \qquad j, k = 1, \ldots, n+1$$

has one eigenvalue which is equal to 0 with corresponding eigenvector ν. The remaining n eigenvalues are the principal curvatures of Γ. We call the trace of (H_{jk}) the mean curvature of Γ, i.e.

$$H = \sum_{j=1}^{n+1} H_{jj}.$$

Note that this coincides with the usual definition up to a factor n.

The formula for integration by parts on Γ is

$$\int_\Gamma \underline{D}_j u = \int_{\partial\Gamma} u\mu_j - \int_\Gamma uH\nu_j \qquad j = 1, \ldots, n+1.$$

Here μ is the outward pointing unit normal vector on the boundary of Γ - tangential to Γ. Obviously this implies the Green formula

$$\int_\Gamma \nabla_\Gamma u \cdot \nabla_\Gamma v = \int_{\partial\Gamma} u\nabla_\Gamma v \cdot \mu - \int_\Gamma u\Delta_\Gamma v.$$

The previous definitions and results allow us to formulate a Lemma which will be very important for the numerical treatment of mean curvature flow.

Lemma 2.1. *For smooth enough Γ we have the following relations for the coordinate function $u(x) = x_j$, $j = 1, \ldots, n+1$:*

$$-\Delta_\Gamma x_j = H\nu_j,$$
$$\nabla_\Gamma \cdot \nu = H,$$

and

$$\int_\Gamma H(x)\nu(x) \cdot \phi(x)\,dA(x) = \int_\Gamma \nabla_\Gamma x \cdot \nabla_\Gamma \phi(x)\,dA(x)$$

for every $\mathbb{R}^{n+1}$ –valued test function ϕ which vanishes on the boundary of Γ.

2.2   Moving Surfaces

We will be concerned with the curvature driven motion of interfaces. Therefore we shall need some information concerning the time derivative of volume or area integrals. We formulate this for moving graphs.

Proposition 2.1. *Assume that all quantities make sense. Then for the surface*

$$\Gamma(t) = \{(x, u(x,t))|\, x \in \underline{\Omega}\}$$

and the domain

$$\Omega(t) = \{(x, x_{n+1})|\, x \in \underline{\Omega},\, 0 < x_{n+1} < u(x,t)\}$$

we have

$$\frac{d}{dt}\int_{\Omega(t)} g = \int_{\Omega(t)} \frac{\partial g}{\partial t} - \int_{\Gamma(t)} gV,$$

where $V = -u_t/\sqrt{1+|\nabla u|^2}$ is the normal velocity of $\Gamma(t)$. Assume that $V = 0$ on $\partial\Omega$. Then

$$\frac{d}{dt}\int_{\Gamma(t)} g = \int_{\Gamma(t)} \frac{\partial g}{\partial t} + \int_{\Gamma(t)} VHg + \int_{\Gamma(t)} V\frac{\partial g}{\partial \nu}.$$

2.3   The Concept of Anisotropy

In phase transition problems it is often necessary to treat motion of interfaces which are driven by anisotropic curvature. This is induced by anisotropic surface energy, which generalizes area in the isotropic case to weighted area in the anisotropic case. Anisotropic surface energy has the form

$$E_\gamma = \int_\Gamma \gamma(\nu), \qquad (2.1)$$

where Γ is a surface with normal ν and γ is a given anisotropy function. We shall call an anisotropy function γ admissible if it satisfies the following requirements:

(i) $\gamma \in C^3(\mathbb{R}^{n+1} \setminus \{0\})$, $\gamma(p) > 0$ for $p \in \mathbb{R}^{n+1} \setminus \{0\}$.
(ii) γ is positively homogeneous of degree one, i.e.

$$\gamma(\lambda p) = |\lambda|\gamma(p) \qquad \text{for all } \lambda \neq 0, p \neq 0. \qquad (2.2)$$

(iii) there exists $\gamma_0 > 0$ such that

$$\gamma_{pp}(p)q \cdot q \geq \gamma_0|q|^2 \quad \text{for all } p, q \in \mathbb{R}^{n+1}, \ |p| = 1, \ p \cdot q = 0. \qquad (2.3)$$

Here, and in what follows, we shall denote by γ_p the gradient of γ and by γ_{pp} the matrix of second derivatives. It is not difficult to verify that (2.2) implies

$$\gamma_p(p) \cdot p = \gamma(p), \ \gamma_{pp}(p)p \cdot q = 0, \qquad (2.4)$$

$$\gamma_p(\lambda p) = \frac{\lambda}{|\lambda|}\gamma_p(p), \ \gamma_{pp}(\lambda p) = \frac{1}{|\lambda|}\gamma_{pp}(p) \qquad (2.5)$$

for all $p \in \mathbb{R}^{n+1} \setminus \{0\}$, $q \in \mathbb{R}^{n+1}$ and $\lambda \neq 0$. The convexity assumption (2.3) will be crucial for analysis and numerical methods.

Anisotropy can be visualized by using the Frank diagram $\mathcal{F}$ and the Wulff shape $\mathcal{W}$

$$\mathcal{F} = \{p \in \mathbb{R}^{n+1} \mid \gamma(p) \leq 1\},$$
$$\mathcal{W} = \{q \in \mathbb{R}^{n+1} \mid \gamma^*(q) \leq 1\}.$$

Here γ^* is the dual of γ, which is given by

$$\gamma^*(q) = \sup_{p \in \mathbb{R}^{n+1} \setminus \{0\}} \frac{p \cdot q}{\gamma(p)}.$$

Let us consider some examples:

(i) The choice $\gamma(p) = |p|$ is called the isotropic case; in particular we have that $\mathcal{F} = \mathcal{W} = \{p \in \mathbb{R}^{n+1} \mid |p| = 1\}$ is the unit sphere.

(ii) A typical choice for anisotropy is the discrete l^r-norm for $1 \le r \le \infty$,

$$\gamma(p) = \|p\|_{l^r} = \left(\sum_{k=1}^{n+1} |p_k|^r\right)^{\frac{1}{r}}, \quad 1 \le r < \infty \qquad (2.6)$$

with the obvious modification for $r = \infty$.

(iii) For a given positive definite $(n+1) \times (n+1)$ matrix G, the anisotropy function

$$\gamma(p) = \sqrt{Gp \cdot p} \qquad (2.7)$$

models an anisotropy which is defined by a (constant) Riemannian metric.

(iv) An anisotropy function, which is used in a physical context, is

$$\gamma(p) = \left(1 - A\left(1 - \frac{\|p\|_{l^4}^4}{\|p\|_{l^2}^4}\right)\right) \|p\|_{l^2} \qquad (2.8)$$

where A is a parameter. For $A < 0.25$ the Frank diagram is convex. In Figure 2.1 we show Frank diagram and Wulff shape for $A = 0.24$.

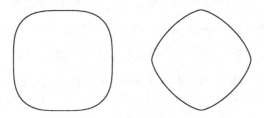

Fig. 2.1. Frank diagram (left) and Wulff shape (right) for the anisotropy function (2.8) with $A = 0.24$ and $n = 1$.

The following Lemma will be used in our error estimates for anisotropic mean curvature flow.

Lemma 2.2. *Let γ be an admissible weight function. Then there exists a constant $c_1 > 0$ such that for $p, q \in S^n := \{x \in \mathbb{R}^{n+1} \mid |x| = 1\}$ with $|p+q| > 0$*

$$\gamma(q) - \gamma_p(p) \cdot q \ge c_1 |q - p|^2. \qquad (2.9)$$

3   Parametric Mean Curvature Flow

We are interested in evolving a compact n–dimensional surface Γ_0 without a boundary in $\mathbb{R}^{n+1}$ according to mean curvature flow $V = -H$. As already mentioned above, in the parametric approach one chooses a suitable reference manifold M and then looks for maps $u(\cdot,t) : M \to \mathbb{R}^{n+1}$ $(0 \le t < T)$ such that $\Gamma(t) = u(\cdot,t)(M)$. If we require that u satisfies

$$\frac{\partial u}{\partial t} = -H(u)\nu(u) \quad \text{on } M \times (0,T),$$

then $V = -H$ on $\Gamma(t)$ follows by taking the dot product with the normal $\nu(u)$. Observing that $u(\cdot,t)$ is the identity on $\Gamma(t)$ and recalling Lemma 2.1 we derive

$$H(u)\nu(u) = -\Delta_{\Gamma(t)}u$$

with the Laplace-Beltrami operator $\Delta_{\Gamma(t)}$ on $\Gamma(t)$. Thus we are led to the following initial value problem

$$\frac{\partial u}{\partial t} - \Delta_{\Gamma(t)}u = 0 \quad \text{on } M \times (0,T)$$
$$u(\cdot,0) = u_0 \text{ on } M, \tag{3.1}$$

where $u_0 : M \to \Gamma_0$ is a parameterization of Γ_0.

The standard example for a solution of this problem is the shrinking sphere. Assume that $M = S^n$ and write

$$u(x,t) = R(t)x, \quad x \in S^n.$$

Then (3.1) is equivalent to the ODE

$$\frac{dR}{dt} + \frac{n}{R} = 0, \ R(0) = R_0$$

where $R_0 > 0$ is the radius of the initial sphere. The solution is given by

$$R(t) = \sqrt{R_0^2 - 2nt}$$

and thus the sphere shrinks down to a point at final time $T = R_0^2/(2n)$.

3.1   Curve Shortening Flow

Mean curvature evolution in the one–dimensional case is usually referred to as curve shortening flow. In the case of closed curves, a convenient choice of a reference manifold is S^1, so that the curve $\Gamma(t)$ is parametrized by $u(\cdot,t) : S^1 \to \mathbb{R}^2$, $t \in [0,T)$. The trace of the moving curve is given by

$$\Gamma(t) = u(S^1,t).$$

We shall frequently use the identification $S^1 \cong \mathbb{R}/2\pi$ and (3.1) then reads as follows,

$$u_t - \frac{1}{|u_x|}\left(\frac{u_x}{|u_x|}\right)_x = 0 \quad \text{in } I \times (0,T) \tag{3.2}$$

$$u(\cdot,0) = u_0 \quad \text{in } I,$$

where $I = [0, 2\pi]$. Furthermore, u is assumed to be periodic in x and $|u_x| > 0$. We do not restrict our considerations to curves in $\mathbb{R}^2$ which means that we allow higher codimension for the numerical scheme. Theoretical results for (3.2) can be found in [38], [36] and are to the effect that an embedded curve in $\mathbb{R}^2$ shrinks to round point in finite time, i.e. (3.2) has a smooth solution on some finite time interval $[0, T)$. Curves in $\mathbb{R}^2$ with double points shrink to round points possibly exhibiting cusps during the evolution. In Figure 3.1 we show computationally that an embedded spiral unwinds to a convex curve and then shrinks to a round point. Figure 3.2 shows how cusps appear during curve shortening flow applied to a non-embedded curve, i.e. a curve with double points.

From now on we shall assume that there exists a smooth solution u : $I \times [0,T] \to \mathbb{R}^m$ ($m \geq 2$) of (3.2).

A weak formulation for (3.2) is

$$\int_I u_t \cdot \varphi\,|u_x| + \int_I \frac{u_x \cdot \varphi_x}{|u_x|} = 0 \tag{3.3}$$

for every test function $\varphi \in H^1_{\text{per}}(I; \mathbb{R}^m)$ together with the initial condition $u(\cdot,0) = u_0$.

We use (3.2) in order to discretize in space. For the sake of simplicity let $x_j = jh$ ($j = 0, \ldots, N$) be a uniform grid with gridsize $h = 2\pi/N$ and

$$X_h = \left\{\varphi_h \in C^0(I; \mathbb{R}^m)\,|\; \varphi_h|_{[x_{j-1},x_j]} \in P_1^m, j = 1, \ldots, N,\; \varphi_h(0) = \varphi_h(2\pi)\right\}$$

the space of piecewise linear continuous functions with values in $\mathbb{R}^m$. The spatial discretization of (3.2) is then given by

$$\int_I u_{ht} \cdot \varphi_h |u_{hx}| + \int_I \frac{u_{hx} \cdot \varphi_{hx}}{|u_{hx}|} = 0 \qquad \forall\,\varphi_h \in X_h \tag{3.4}$$

$$u_h(\cdot,0) = u_{h0}$$

where u_{h0} is some approximation of u_0 in X_h. Denoting the usual scalar nodal basis by $\{\phi_1, \ldots, \phi_N\}$ we can expand $u_h(x,t) = \sum_{j=1}^{N} u_j(t)\phi_j(x)$ with vectors $u_j(t) \in \mathbb{R}^m$. This one–dimensional Finite Element formulation can be rewritten as a difference scheme. Namely, if we insert $\varphi_h = \phi_j e^k$, ($k =$

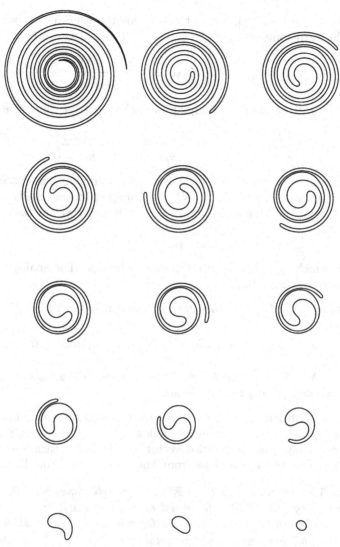

Fig. 3.1. Spiral unwinding under curve shortening flow.

$1, \ldots, m; j = 1, \ldots, N)$ into (3.4) we obtain

$$\int_I u_{ht} \cdot \varphi_h |u_{hx}| \, dx$$

$$= \frac{1}{6} \dot{u}_{j-1}^k |u_j - u_{j-1}| + \frac{1}{3} \dot{u}_j^k \left(|u_j - u_{j-1}| + |u_{j+1} - u_j| \right) + \frac{1}{6} \dot{u}_{j+1}^k |u_{j+1} - u_j|$$

and

$$\int_I \frac{u_{hx} \cdot \varphi_{hx}}{|u_{hx}|} \, dx = -\frac{u_{j+1}^k - u_j^k}{q_{j+1}} + \frac{u_j^k - u_{j-1}^k}{q_j}.$$

Here, $q_j = |u_j - u_{j-1}|$, the dot stands for the time derivative and u_j^k denotes the k–th component of the vector u_j. Thus, (3.4) can be written as

$$\frac{1}{6}\dot{u}_{j-1}q_j + \frac{1}{3}\dot{u}_j(q_j + q_{j+1}) + \frac{1}{6}\dot{u}_{j+1}q_{j+1} = \frac{u_{j+1} - u_j}{q_{j+1}} - \frac{u_j - u_{j-1}}{q_j} \quad (3.5)$$

$(j = 1, \ldots, N)$. If we use mass lumping in (3.5) we get the difference scheme

$$\frac{1}{2}(q_j + q_{j+1})\dot{u}_j = \frac{u_{j+1} - u_j}{q_{j+1}} - \frac{u_j - u_{j-1}}{q_j}, \quad (3.6)$$

with initial values $u_j(0) =: u_{0j}$ $(j = 1, \ldots, N)$. Periodicity in space means that $u_j = u_{j+N}$, $(j = -1, 0, \ldots, N)$. The lumped scheme really shortens length of the curve during evolution. In the continuous case one easily derives the equation

$$|u_x|_t = -|u_t|^2|u_x| \leq 0 \quad (3.7)$$

from (3.2) which says that length decreases pointwise. The analogue is true for the discrete scheme (3.6).

Proposition 3.1. *If $q_j > 0$, then the difference scheme (3.6) implies*

$$\dot{q}_j = -\frac{1}{4}(q_{j-1} + q_j)|\dot{u}_{j-1}|^2 - \frac{1}{4}(q_j + q_{j+1})|\dot{u}_j|^2 \leq 0 \quad (3.8)$$

$(j = 1, \ldots, N)$. That is the faces of the polygon with vertices $u_1, \ldots, u_N$ decrease in length during time evolution.

Under the assumption that a smooth and regular solution of the curve shortening flow (3.2) exists, one obtains the following convergence result together with error estimates for the position vector u and the "curvature vector" u_t. The proof follows as a special case from Theorem 3.2 and from [23].

Theorem 3.1. *Let $u : I \times [0, T] \to \mathbb{R}^m$ be a periodic smooth solution of the curve shortening flow (3.2) with initial data u_0 and $|u_x| \geq c_0 > 0$ where $I = (0, 2\pi)$. Then there exists an $h_0 > 0$ depending on u and T such that for every $0 < h \leq h_0$ there exists a unique solution $u_h(x, t) = \sum_{j=1}^N u_j(t)\phi_j(x)$ of the difference scheme (3.6) with initial data $u_{h0}(x_j) = u_0(x_j)$ $(j = 1, \ldots, N)$ and*

$$\max_{t \in [0,T]} \|u - u_h\|_{L^2(I)} + \left(\int_0^T \|u_x - u_{hx}\|_{L^2(I)}^2 \, dt\right)^{1/2} \leq ch, \quad (3.9)$$

$$\max_{t \in [0,T]} \|u_t - u_{ht}\|_{L^2(I)} + \left(\int_0^T \|u_{tx} - u_{htx}\|_{L^2(I)}^2 \, dt\right)^{1/2} \leq ch,$$

where c depends on u and T.

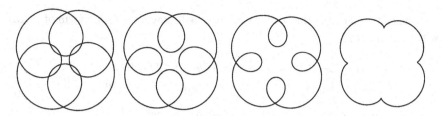

Fig. 3.2. Cusp formation for an initial curve which is not embedded.

3.2 Anisotropic Curve Shortening Flow

By anisotropic interface energy we mean weighted length of the curve Γ

$$E_\gamma = \int_\Gamma \gamma(\nu),$$

where γ is a given admissible anisotropy function as introduced in Section 2.3. The problem of anisotropic curve shortening flow is then defined as the $L^2(\Gamma(t))$–gradient flow for the anisotropic surface energy E_γ.

In what follows we shall restrict ourselves again to curves in the plane. For a given initial curve $\Gamma(0) = u_0(S^1)$ we want to determine a smooth regular parametrization $u = u(x, t)$ of $\Gamma(t) = u(S^1, t)$ on some time interval, such that

$$(u_t, \varphi)_{L^2(\Gamma(t))} = -\langle E'_\gamma(u(\cdot, t)), \varphi \rangle \qquad \forall\, \varphi \in H^1(S^1, \mathbb{R}^2). \qquad (3.10)$$

If Γ is regularly parametrized, then tangent τ and normal ν are given by

$$\tau = \frac{u_x}{|u_x|}, \quad \nu = \tau^\perp, \qquad (3.11)$$

where we use the notation $(a_1, a_2)^\perp = (-a_2, a_1)$. Employing the homogeneity of γ we have

$$E_\gamma(u) = \int_\Gamma \gamma(\nu) = \int_{S^1} \gamma\left(\frac{u_x^\perp}{|u_x|}\right)|u_x|\,dx = \int_{S^1} \gamma(u_x^\perp)\,dx,$$

so that the abstract equation (3.10) for the gradient flow of E_γ can be written in the form

$$\int_{S^1} u_t \cdot \varphi\,|u_x|\,dx + \int_{S^1} \gamma_p(u_x^\perp) \cdot \varphi_x^\perp\,dx = 0 \qquad \forall\, \varphi \in H^1(S^1, \mathbb{R}^2). \ (3.12)$$

This weak form will be the basis for a numerical method for the anisotropic curve shortening flow problem. The classical form of (3.12) is

$$u_t + \frac{1}{|u_x|}\gamma_p(u_x^\perp)_x^\perp = 0 \quad \text{in } S^1 \times (0, T). \qquad (3.13)$$

For the convenience of the reader we note the following detailed form of (3.13) with $u = (u_1, u_2)$:

$$u_{1t}|u_x| - \gamma_{p_2 p_2}(-u_{2x}, u_{1x})u_{1xx} + \gamma_{p_2 p_1}(-u_{2x}, u_{1x})u_{2xx} = 0,$$
$$u_{2t}|u_x| - \gamma_{p_1 p_1}(-u_{2x}, u_{1x})u_{2xx} + \gamma_{p_1 p_2}(-u_{2x}, u_{1x})u_{1xx} = 0.$$

It is easy to see that this system can be written as

$$u_t - \alpha(\frac{u_x^\perp}{|u_x|})\frac{1}{|u_x|}\left(\frac{u_x}{|u_x|}\right)_x = 0$$

where

$$\alpha(p) = \gamma_{pp}(p)\, p^\perp \cdot p^\perp, \quad p \in \mathbb{R}^2 \setminus \{0\}.$$

Analytical results for this problem which generalize the theory for the isotropic case ($\alpha = 1$) have been obtained in [35].

We shall continue to use the form (3.12) because this equation only contains first derivatives of the anisotropy function γ. The following Lemma assures that the length of the curve $\Gamma(t)$ decreases locally during the evolution. For a proof see [25].

Lemma 3.1. *Let u be a solution of (3.13) with initial value $u(\cdot, 0) = u_0$. Then for $t > 0$ we have the energy equation*

$$\int_0^t \int_{S^1} |u_t|^2 |u_x|\, dx\, dt + \int_{S^1} \gamma(u_x^\perp(\cdot, t))\, dx = \int_{S^1} \gamma(u_{0x}^\perp)\, dx. \qquad (3.14)$$

The length element satisfies the equation

$$|u_x|_t = -\frac{|u_x|}{\gamma_{pp}(\nu)\tau \cdot \tau}|u_t|^2. \qquad (3.15)$$

Recall the definition of X_h (with $m = 2$) from Section 3.1. A discrete solution of (3.12) will be a function $u_h : [0, T] \to X_h$, such that

$$u_h(\cdot, 0) = u_{h0} = I_h u_0 = \sum_{j=1}^N u_0(x_j)\phi_j(x),$$

and for all discrete test functions $\varphi_h \in X_h$

$$\int_{S^1} u_{ht} \cdot \varphi_h\, |u_{hx}|\, dx + \int_{S^1} \gamma_p(u_{hx}^\perp) \cdot \varphi_{hx}^\perp\, dx = 0. \qquad (3.16)$$

In the same way as in the isotropic case we can write

$$u_h(x, t) = \sum_{j=1}^N u_j(t)\phi_j(x)$$

with $u_j(t) \in \mathbb{R}^2$ and find that the discrete weak equation (3.16) is equivalent to the following system of $2N$ ordinary differential equations

$$\frac{1}{6}\dot{u}_{j-1}|u_j - u_{j-1}| + \frac{1}{3}\dot{u}_j \left(|u_j - u_{j-1}| + |u_{j+1} - u_j|\right) + \frac{1}{6}\dot{u}_{j+1}|u_{j+1} - u_j|$$
$$+\gamma_p(u_{j+1}^\perp - u_j^\perp)^\perp - \gamma_p(u_j^\perp - u_{j-1}^\perp)^\perp = 0$$

for $j = 1, \ldots, N$, where $u_0 = u_N$, $u_{N+1} = u_1$, and the initial values are given by

$$u_j(0) = u_0(x_j), \ j = 1, \ldots, N.$$

We again use mass lumping which is equivalent to a quadrature formula. And so we replace this system by the lumped scheme

$$\frac{1}{2}\dot{u}_j \left(|u_j - u_{j-1}| + |u_{j+1} - u_j|\right) + \gamma_p(u_{j+1}^\perp - u_j^\perp)^\perp - \gamma_p(u_j^\perp - u_{j-1}^\perp)^\perp = 0,$$
$$u_j(0) = u_0(x_j),$$
(3.17)

for $j = 1, \ldots, N$. We are now ready to say what we mean by a discrete solution of anisotropic curve shortening flow. The system (3.17) is equivalent to the one which we use in the following definition of discrete anisotropic curve shortening flow.

A solution of the discrete anisotropic curve shortening flow for the initial curve $\Gamma_{h0} = u_{h0}(S^1)$ is a polygon $\Gamma_h(t) = u_h(S^1, t)$, which is parametrized by a piecewise linear mapping $u_h(\cdot, t) \in X_h$, $t \in [0, T]$, such that $u_h(\cdot, 0) = u_{h0}$ and for all $\varphi_h \in X_h$

$$\int_{S^1} u_{ht} \cdot \varphi_h |u_{hx}| + \int_{S^1} \gamma_p(u_{hx}^\perp) \cdot \varphi_{hx}^\perp + \frac{1}{6}h^2 \int_{S^1} u_{hxt} \cdot \varphi_{hx} |u_{hx}| = 0. \ (3.18)$$

Here h is the constant grid size of the uniform grid in I. The last term of (3.18) is introduced by mass lumping. One could also define the discrete curve shortening flow without this quantity, but then the geometric property of length shortening would not be true for the discrete problem. In Figure 3.3 we show the results of a computation of anisotropic flow with a special γ. We added the constant right hand side $f = 1$ to the equation (3.13), i.e. we solved

$$u_t + \frac{1}{|u_x|}\gamma_p(u_x^\perp)_x^\perp = f\frac{u_x^\perp}{|u_x|}. \tag{3.19}$$

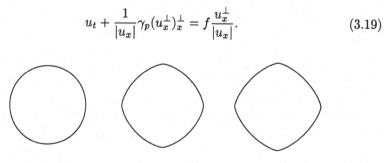

Fig. 3.3. The basic test: evolution of the unit circle into the Wulff shape under anisotropy (2.8) with $A = 0.24$.

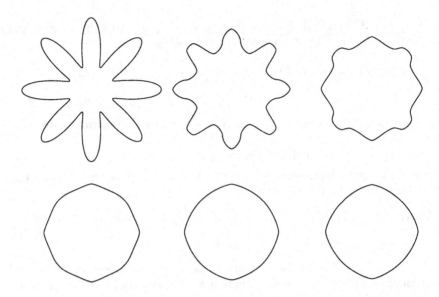

Fig. 3.4. Solution of (3.19) with $f = 1$ under anisotropy (2.8).

Then a stationary solution is a curve of constant anisotropic curvature $H_\gamma = 1$, the Wulff shape. Figure 3.4 contains computations with the same anisotropy for a different initial curve. We add an example of a computation with a "forbidden" anisotropy pointed out to us by M. Paolini. Here γ is not smooth. Smoothness of γ is required for our convergence analysis.

$$\gamma(p) = \begin{cases} \frac{1}{|p_1|} & (|p_1| \geq |p_2|) \\ 2|p_2| & (|p_1| < |p_2|) \end{cases} \tag{3.20}$$

In Figure 3.5 we show the Frank diagram and the Wulff shape for this anisotropy. Figure 3.6 shows computational results of the evolution (3.19) with $f = 1$ for a knotted initial curve.

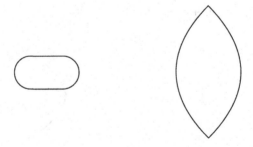

Fig. 3.5. Frank diagram (left) and Wulff shape (right) for the anisotropy function (3.20).

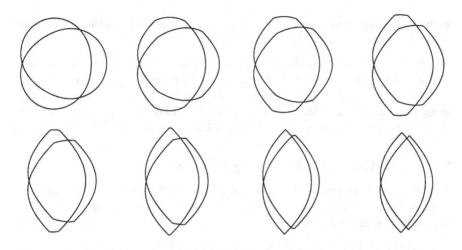

Fig. 3.6. Solution of (3.19) with $f = 1$ under anisotropy (3.20).

In [25] we proved the following convergence result. We formulate the result for the geometric quantities normal, length and normal velocity. The error estimates in standard norms then follow easily.

Theorem 3.2. *Let u be a solution of the anisotropic curve shortening flow (3.12) on the interval $[0, T]$ with $u(\cdot, 0) = u_0$, $\min_{S^1 \times [0,T]} |u_x| \geq c_0 > 0$ and $u_t \in L^2((0,T), H^2(S^1))$. Then there is an $h_0 > 0$ such that for all $0 < h \leq h_0$ there exists a unique solution u_h of the discrete anisotropic curve shortening flow (3.18) on $[0, T]$ with $u_h(\cdot, 0) = u_{h0} = I_h u_0$, and the error between smooth and discrete solution can be estimated as follows:*

$$\int_0^T \int_{S^1} |u_t - u_{ht}|^2 |u_{hx}| \, dx \, dt \leq ch^2, \tag{3.21}$$

$$\sup_{(0,T)} \int_{S^1} |\nu - \nu_h|^2 |u_{hx}| \, dx + \sup_{(0,T)} \int_{S^1} (|u_x| - |u_{hx}|)^2 \, dx \leq ch^2. \tag{3.22}$$

The constants depend on c_0, T and $\|u_t\|_{L^2((0,T), H^2(S^1))}$.

The discretization in space leads to the following algorithm.

Algorithm 3.1. For a given initial discrete curve $\Gamma_{h0} = u_{h0}(S^1)$ solve the system of ordinary differential equations

$$\frac{1}{2} \dot{u}_j \, (q_j + q_{j+1}) - \frac{g_j}{q_j} u_{j-1} + \left(\frac{g_j}{q_j} + \frac{g_{j+1}}{q_{j+1}} \right) u_j - \frac{g_{j+1}}{q_{j+1}} u_{j+1} \tag{3.23}$$

$$+ \frac{g_j'}{q_j} u_{j-1}^\perp - \left(\frac{g_j'}{q_j} + \frac{g_{j+1}'}{q_{j+1}} \right) u_j^\perp + \frac{g_{j+1}'}{q_{j+1}} u_{j+1}^\perp = 0,$$

$$g_j = \gamma\!\left(\frac{(u_j - u_{j-1})^\perp}{q_j} \right), \quad g_j' = \gamma_p\!\left(\frac{(u_j - u_{j-1})^\perp}{q_j} \right) \cdot \frac{u_j - u_{j-1}}{q_j},$$

$$q_j = |u_j - u_{j-1}|,$$

with periodicity conditions $u_0 = u_N$, $u_{N+1} = u_1$ and initial conditions $u_j(0) = u_{h0}(x_j)$, for $j = 1, \ldots, N$.

Again we emphasize that this algorithm does not use the second derivatives of the anisotropy function γ. In the isotropic case one has $g_j = 1$ and $g_j' = 0$.

The system (3.23) can formally be written in complex tridiagonal form using the complex notation $U_j = u_j^1 - iu_j^2 \in \mathbb{C}$, and $G_j = g_j + ig_j' \in \mathbb{C}$. For the details and for a suitable time discretization we refer to [25].

3.3 Mean Curvature Flow of Hypersurfaces

The idea which we presented for curves also works for surfaces. This time let us start with the time discretization of the mean curvature flow problem for hypersurfaces of dimension n.

Algorithm 3.2. Let $\tau > 0$ and the surface Γ_0 be given. For $m = 0, 1, \ldots$ solve the system of $n + 1$ linear partial differential equations

$$\frac{1}{\tau}(u_{m+1} - u_m) - \Delta_{\Gamma_m} u_{m+1} = 0 \quad \text{on} \quad \Gamma_m,$$
$$u_{m+1} = u_m \quad \text{on} \quad \partial\Gamma_m,$$

and set

$$\Gamma_{m+1} = u_{m+1}(\Gamma_m).$$

The key point of this time discretization is that we do not solve a differential equation in the parameter domain, but on the surface which is given by the previous time step. Note that $u^{m+1} : \Gamma_m \to \Gamma_{m+1}$. The differential equations in Algorithm 3.2 are intrinsic and so we should discretise the problem in such a way that the resulting scheme is intrinsic too. This is what we have done in the previous sections for curves. For surfaces we need a Finite Element Method *on* surfaces.

3.4 Finite Elements on Surfaces

In Algorithm 3.2 we have to solve numerically a problem of the form

$$a u - \Delta_\Gamma u = f \quad \text{on} \quad \Gamma$$
$$u = g \quad \text{on} \quad \partial\Gamma$$

with $a > 0$ in each time step and for every component of the map u^{m+1}. The variational form of this equation reads

$$\int_\Gamma au\varphi + \int_\Gamma \nabla_\Gamma u \cdot \nabla_\Gamma \varphi = \int_\Gamma f\varphi \tag{3.24}$$

for every φ which vanishes on $\partial\Gamma$.

For the following we assume that the boundary of Γ is empty. Let Γ be a given smooth hypersurface. We discretise this equation by piecewise linear elements. To do so, we first approximate Γ by a polygonal surface Γ_h, which consists of simplices (triangles for 2d surfaces) and is such that the nodes of Γ_h are points on the smooth surface Γ. This defines an n–dimensional triangulation $\mathcal{T}_h$ in $\mathbb{R}^{n+1}$:

$$\Gamma_h = \bigcup_{T \in \mathcal{T}_h} T.$$

On this discrete surface we define a finite element space by

$$X_h = \left\{ \varphi_h \in C^0(\Gamma_h) \,\middle|\, \varphi_h \text{ is linear affine on each } T \in \mathcal{T}_h \right\}.$$

This leads to the following discretization of (3.24): find $u_h \in X_h$ such that

$$\int_{\Gamma_h} a u_h \varphi_h + \int_{\Gamma_h} \nabla_{\Gamma_h} u_h \cdot \nabla_{\Gamma_h} \varphi_h = \int_{\Gamma_h} f \varphi_h \quad \forall \, \varphi_h \in X_h. \tag{3.25}$$

Let us have a short look at the implementation of this method. For this we need stiffness and mass matrices. The nodal basis function $\phi_j = \phi_j(x_1, \ldots, x_{n+1})$ with respect to the j-th node $a_j \in \Gamma_h$ is defined by

$$\phi_j \in X_h, \qquad \phi_j(a_i) = \delta_{ij}.$$

We observe that the stiffness matrix is then given by

$$S_{ij} = \int_{\Gamma_h} \nabla_{\Gamma_h} \phi_i \cdot \nabla_{\Gamma_h} \phi_j \quad i,j = 1, \ldots, N$$

and is built up from element stiffness matrices.

For the simplex T the element stiffness matrix is, in local indices,

$$S(T)_{ij} = \int_T \nabla_T \phi_i \cdot \nabla_T \phi_j \quad i,j = 1, \ldots, n+1$$

and the mass matrix is

$$M_{ij} = \int_{\Gamma_h} a \phi_i \phi_j \quad i,j = 1, \ldots, N.$$

Writing $u_h = \sum_{j=1}^{N} u_j \phi_j$, the linear system for $U = (u_1, \ldots, u_N)$ then has the usual form

$$MU + SU = b$$

with symmetric and positive definite matrices M and S. To be more precise: S is positive definite only on $\mathbb{R}^N/\mathbb{R}$. The only difference to a "Cartesian" FEM is that the nodes have one more coordinate.

The above ideas can now be used in order to discretise Algorithm 3.2 in space resulting in a fully discrete scheme for the approximation of parametric mean curvature flow. Note that test functions and solutions are now vector-valued. We summarize the method in the following algorithm and refer to [22] for further details.

Algorithm 3.3. Let the surface $\Gamma_{h,0}$ be given (e.g. as an interpolant of the smooth initial surface) and let $X_{h,0}$ be the induced finite element space on $\Gamma_{h,0}$. For $m = 0, 1, \ldots$ solve the system of $n+1$ linear equations for $u_{h,m+1} \in X_{h,m}$ such that for every $\phi_h \in X_{h,m}$

$$\frac{1}{\tau} \int_{\Gamma_{h,m}} (u_{h,m+1} - u_{h,m})\phi_h + \int_{\Gamma_{h,m}} \nabla_{\Gamma_{h,m}} u_{h,m+1} \cdot \nabla_{\Gamma_{h,m}} \phi_h = 0$$

with $u_{h,m+1} = u_{h,m}$ on $\partial\Gamma_{h,m}$ and set $\Gamma_{h,m+1} = u_{h,m+1}(\Gamma_{h,m})$. This defines $X_{h,m+1}$.

Fig. 3.7. Mean curvature flow leads to stable minimal surfaces with prescribed boundary if no singularities develop during the evolution. Initial surface (cylinder), two time steps and the stationary catenoid.

Fig. 3.8. A relatively thin two–dimensional torus shrinking under parametric mean curvature flow to a circle, computed with Algorithm 3.3.

We show the results of computations which were carried out with the help of this algorithm in Figures 3.7, 3.8, 3.9 and 3.10. Figures 3.7 is the standard

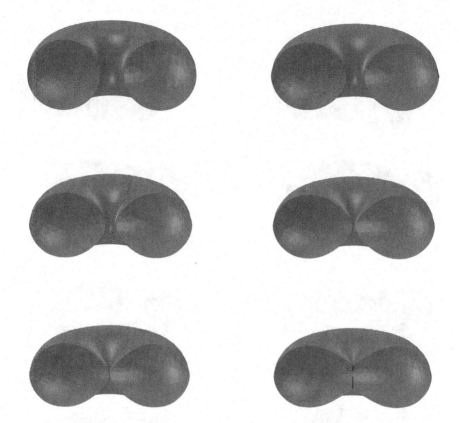

Fig. 3.9. A fat two–dimensional torus (for better visibility cut through along the x_1-x_3-plane) developing a needle singularity while moving under parametric mean curvature flow.

example for mean curvature evolution with fixed boundary. The cylinder evolves into a catenoid. If we start with a torus as an initial surface, then the behaviour of parametric mean curvature flow depends strongly on the initial radii of the torus. In Figure 3.8 a thin torus shrinks down to a circle. In Figure 3.9 a fat torus develops a singularity in its middle and moves towards a sphere. The last image in this figure is beyond reliable numerical methods. Nevertheless it is interesting to note that the algorithm does not break down and produces a "needle" in the center of the torus. If we deform an initial torus by squeezing it together at four positions then mean curvature flow produces four singularities and the torus moves into the direction of four spherical surfaces. Again the last image is beyond reliable numerics, but obviously interesting.

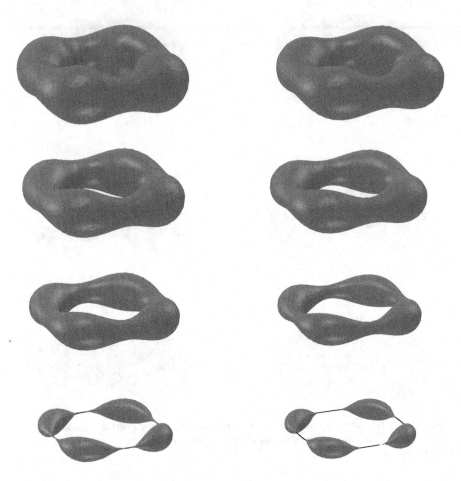

Fig. 3.10. A deformed two–dimensional torus developing four singularities while moving under parametric mean curvature flow.

4   Mean Curvature Flow of Level Sets I

In order to compute topological changes of free boundaries it is necessary to leave the parametric world which fixes the topological type of the interface. One method to do this is to define the interface as the level set of a scalar function:

$$\Gamma(t) = \left\{x \in \mathbb{R}^{n+1} \,|\, u(x,t) = c\right\}.$$

Let us assume for the moment that $\nabla u \neq 0$ in a neighbourhood of Γ. Then a normal to Γ is given by

$$\nu = \frac{\nabla u}{|\nabla u|},$$

its mean curvature is

$$H = \nabla \cdot \frac{\nabla u}{|\nabla u|}$$

and the normal velocity is

$$V = -\frac{u_t}{|\nabla u|}.$$

Thus, $\Gamma(t)$ formally moves by mean curvature if

$$\frac{u_t}{|\nabla u|} - \nabla \cdot \frac{\nabla u}{|\nabla u|} = 0 \quad \text{in } \mathbb{R}^{n+1} \times (0, \infty). \tag{4.1}$$

This partial differential equation is highly nonlinear, degenerate parabolic and not even defined where the gradient of u vanishes. But sets where $\nabla u = 0$ are of interest for us, since typically here the topology of Γ changes. Thus one needs an appropriate notion of solution for (4.1).

4.1   Viscosity Solutions

Starting from (4.1), we are interested in the following problem:

$$u_t = \sum_{i,j=1}^{n+1} \left(\delta_{ij} - \frac{u_{x_i} u_{x_j}}{|\nabla u|^2} \right) u_{x_i x_j} \quad \text{in } \mathbb{R}^{n+1} \times (0, \infty) \tag{4.2}$$

$$u(\cdot, 0) = u_0 \qquad\qquad\qquad \text{in } \mathbb{R}^{n+1}. \tag{4.3}$$

As already mentioned above, the underlying PDE is not strictly parabolic and not defined where ∇u vanishes, so that standard methods for parabolic equations fail. However, an existence and uniqueness theory for (4.2), (4.3) can be carried out within the framework of viscosity solutions.

Definition 4.1. A function $u \in C^0(\mathbb{R}^{n+1} \times [0, \infty))$ is called a viscosity subsolution of (4.2) provided that for each $\phi \in C^\infty(\mathbb{R}^{n+2})$, if $u - \phi$ has a local maximum at $(x_0, t_0) \in \mathbb{R}^{n+1} \times (0, \infty)$, then

$$\phi_t \le \sum_{i,j=1}^{n+1} \left(\delta_{ij} - \frac{\phi_{x_i} \phi_{x_j}}{|\nabla \phi|^2} \right) \phi_{x_i x_j} \quad \text{at } (x_0, t_0), \text{ if } \nabla \phi(x_0, t_0) \neq 0$$

$$\tag{4.4}$$

$$\phi_t \le \sum_{i,j=1}^{n+1} (\delta_{ij} - p_i p_j) \phi_{x_i x_j} \qquad \text{at } (x_0, t_0) \text{ for some } |p| \le 1, \text{ if } \nabla \phi(x_0, t_0) = 0.$$

A viscosity supersolution is defined analogously: maximum is replaced by minimum and $\le$ by $\ge$. A viscosity solution of (4.2) is a function $u \in C^0(\mathbb{R}^n \times [0, \infty))$ that is both a subsolution and a supersolution.

We shall assume that the initial function u_0 is smooth and satisfies

$$u_0(x) = 1, \qquad \text{for } |x| \ge S \tag{4.5}$$

for some $S > 0$. The following existence and uniqueness theorem is a special case of results proved independently by Evans & Spruck [29] and Chen, Giga & Goto [10].

Theorem 4.1. *Assume* $u_0 : \mathbb{R}^{n+1} \to \mathbb{R}$ *satisfies (4.5). Then there exists a unique viscosity solution of (4.2), (4.3), such that*

$$u(x,t) = 1 \qquad \text{for } |x| + t \geq R$$

for some $R > 0$ depending only on S.

The level set approach can now be described as follows: given a compact hypersurface Γ_0, choose a continuous function $u_0 : \mathbb{R}^{n+1} \to \mathbb{R}$ such that $\Gamma_0 = \{x \in \mathbb{R}^{n+1} \mid u_0(x) = 0\}$. If $u : \mathbb{R}^{n+1} \times [0,\infty) \to \mathbb{R}$ is the unique viscosity solution of (4.2, 4.3), we then call

$$\Gamma(t) = \{x \in \mathbb{R}^{n+1} \mid u(x,t) = 0\}, \quad t \geq 0$$

a generalized solution of the mean curvature flow problem. We remark that the authors in [29] and [10] also established that the sets $\Gamma(t) = \{x \in \mathbb{R}^{n+1} \mid u(x,t) = 0\}, t > 0$ are independent of the particular choice of u_0 which has Γ_0 as its zero level set, so that the generalized evolution $(\Gamma(t))_{t\geq 0}$ is well defined for a given Γ_0. As $\Gamma(t)$ exists for all times, it provides a notion of solution beyond singularities in the flow. For this reason, the level set approach has also become very important in the numerical approximation of mean curvature flow and related problems.

The level set solution has been investigated further in several papers, in particular we mention [30], [31], [32] and [52].

4.2   Regularization

Evans & Spruck proved in [29] that the (smooth) solutions u^ϵ of

$$u_t^\epsilon = \sum_{i,j=1}^{n+1} \left(\delta_{ij} - \frac{u_{x_i}^\epsilon u_{x_j}^\epsilon}{\epsilon^2 + |\nabla u^\epsilon|^2} \right) u_{x_i x_j}^\epsilon \quad \text{in } \mathbb{R}^{n+1} \times (0,\infty) \qquad (4.6)$$

$$u^\epsilon(\cdot,0) = u_0 \qquad \qquad \text{in } \mathbb{R}^{n+1} \qquad (4.7)$$

converge locally uniformly as $\epsilon \to 0$ to the unique viscosity solution of (4.2), (4.3). For numerical purposes it is important to know the asymptotic error between the viscosity solution and the solution of the regularised problem quantitatively as $\epsilon \to 0$. In [14] there is proof of the following theorem together with several *a priori* estimates and their dependence on ϵ.

Theorem 4.2. *For every $\alpha \in (0, \frac{1}{2}), 0 < T < \infty$ there is a constant $C = C(u_0, T, \alpha)$ such that*

$$\sup_{0 \leq t \leq T} \|u - u^\epsilon\|_{L^\infty(\mathbb{R}^{n+1})} \leq C\epsilon^\alpha \qquad \text{for all } \epsilon > 0.$$

If one wants to calculate approximations to the viscosity solution u of (4.2, 4.3) then, according to Theorem 4.2, it is enough to solve the regularised problem (4.6, 4.7). But one has to restrict the computations to some finite domain. This means that one has to solve the problem

$$u_{\epsilon t} = \sum_{i,j=1}^{n+1} \left(\delta_{ij} - \frac{u_{\epsilon x_i} u_{\epsilon x_j}}{\epsilon^2 + |\nabla u_\epsilon|^2} \right) u_{\epsilon x_i x_j} \quad \text{in } \Omega \times (0,\infty) \qquad (4.8)$$

$$u_\epsilon(\cdot, 0) = u_0 \qquad\qquad\qquad \text{in } \Omega \qquad (4.9)$$

on a domain $\Omega \subset \mathbb{R}^{n+1}$ which is large enough. One easily proves the following Corollary of Theorem 4.2 with the use of the parabolic comparison theorem.

Corollary 4.1. *Let $\Omega = B_{\tilde{S}}(0)$ with $\tilde{S} > R = R(S)$, where R is the radius from Theorem 4.1. Then for every $\alpha \in (0, \frac{1}{2}), 0 < T < \infty$ there is a constant $C = C(u_0, T, \alpha)$ such that*

$$\|u - u_\epsilon\|_{L^\infty(\Omega \times (0,T))} \le C\epsilon^\alpha. \qquad (4.10)$$

We are now in position to look at the regularised level set mean curvature flow problem as a problem for graphs. If we scale

$$U := \frac{u_\epsilon}{\epsilon} \qquad (4.11)$$

then U is a solution of the mean curvature flow problem for graphs, see (5.4),

$$U_t - \sqrt{1 + |\nabla U|^2} \, \nabla \cdot \frac{\nabla U}{\sqrt{1 + |\nabla U|^2}} = 0 \text{ in } \Omega \times (0, T) \qquad (4.12)$$

$$U = \frac{u_0}{\epsilon} \text{ on } \partial\Omega \times (0,T) \cup \Omega \times \{0\}.$$

This is a theoretical observation and implies that we can apply the techniques developed for the mean curvature flow of graphs to the mean curvature flow of level sets. But for computations we shall not use (4.12) but the unscaled version for u_ϵ.

5   Mean Curvature Flow of Graphs

5.1   The Differential Equation

We study the numerical computation of the isotropic evolution of n-dimensional surfaces which are graphs in $\mathbb{R}^{n+1}$. The law of motion is given by

$$V = -H, \qquad (5.1)$$

where V represents the normal velocity and H is the mean curvature of the interface

$$\Gamma(t) = \{(x, u(x,t)) \mid x \in \Omega\}$$

above a bounded domain $\Omega \subset \mathbb{R}^n$. The surface element is given by

$$Q(u) = \sqrt{1 + |\nabla u|^2}$$

while the normal to Γ is chosen to be

$$\nu(u) = \frac{(\nabla u, -1)}{\sqrt{1 + |\nabla u|^2}} = \frac{(\nabla u, -1)}{Q(u)}.$$

Thus the normal velocity can be written as

$$V(u) = -\frac{u_t}{Q(u)} \tag{5.2}$$

and the mean curvature of Γ is given by

$$H(u) = \nabla \cdot \frac{\nabla u}{\sqrt{1 + |\nabla u|^2}}.$$

Equation (5.1) then leads to the differential equation

$$-\frac{u_t}{\sqrt{1 + |\nabla u|^2}} = -\nabla \cdot \frac{\nabla u}{\sqrt{1 + |\nabla u|^2}} \tag{5.3}$$

or to the initial boundary value problem

$$u_t - \sqrt{1 + |\nabla u|^2} \, \nabla \cdot \frac{\nabla u}{\sqrt{1 + |\nabla u|^2}} = 0 \text{ in } \Omega \times (0, T) \tag{5.4}$$

$$u = u_0 \text{ on } \partial\Omega \times (0, T) \cup \Omega \times \{0\}.$$

We have chosen the boundary data to be independent of time. A generalization to time dependent boundary values is possible and only adds some technical problems. In more explicit form the differential equation reads

$$u_t - \sum_{i,j=1}^{n} \left(\delta_{ij} - \frac{u_{x_i} u_{x_j}}{1 + |\nabla u|^2} \right) u_{x_i x_j} = 0. \tag{5.5}$$

The equation is parabolic but uniformly parabolic only if $|\nabla u|$ is uniformly bounded. With

$$a_{ij} = \delta_{ij} - \frac{u_{x_i} u_{x_j}}{1 + |\nabla u|^2}$$

one has for arbitrary $\xi \in \mathbb{R}^n$

$$\sum_{i,j=1}^{n} a_{ij} \xi_i \xi_j = |\xi|^2 - \frac{(\nabla u \cdot \xi)^2}{1 + |\nabla u|^2} \geq |\xi|^2 \left(1 - \frac{|\nabla u|^2}{1 + |\nabla u|^2} \right) = |\xi|^2 \frac{1}{1 + |\nabla u|^2}.$$

We also observe that the equation is *not* in divergence form.

5.2   Analytical Results

Analytical results for the problem (5.4) are due to Lieberman [44] and Huisken [40].

Theorem 5.1. *Let Ω be a bounded domain in $\mathbb{R}^n$ with $\partial\Omega \in C^{2+\alpha}$ and suppose that the mean curvature of $\partial\Omega$ is nonnegative everywhere. Let $u_0 \in C^{2+\alpha}(\overline{\Omega})$ satisfy the compatibility condition $H(u_0) = 0$ on $\partial\Omega$. Then there exists a unique solution $u \in C^{2+\alpha,1+\frac{\alpha}{2}}(\overline{\Omega} \times [0,\infty))$ of problem (5.4).*

The assumption that the boundary of the domain has nonnegative mean curvature is a necessary condition. If it is dropped, the gradient of the solution will become infinite on the boundary, see [47]; however the discrete solution exists for all times. The main tool in the proof of the previous theorem was the derivation of an evolution equation for the surface element.

Lemma 5.1. *Let u be a smooth solution of (5.4). Then $Q = \sqrt{1 + |\nabla u|^2}$ satisfies the differential equation*

$$
Q_t - \sum_{i,j=1}^{n} (\delta_{ij} - \nu_i\nu_j)Q_{x_i x_j} + \frac{1}{Q}\sum_{i,j=1}^{n}(\delta_{ij} - \nu_i\nu_j)Q_{x_i}Q_{x_j}
$$

$$
+ \frac{1}{Q}\sum_{i,j,m=1}^{n}(\delta_{ij} - \nu_i\nu_j)u_{x_m x_i}u_{x_m x_j} = 0 \quad in \ \Omega \times (0,T),
$$

where $\nu_i = u_{x_i}/Q$ $(i = 1,\ldots,n)$.

Our numerical algorithms will be based on a variational formulation of the problem, i.e. on an $L^2(\Omega)$-formulation. For that purpose, integral estimates for the solution are important.

Lemma 5.2. *The following energy equation holds for the solution of problem (5.4)*

$$
\int_\Omega \frac{u_t^2}{Q(u)} + \frac{d}{dt}\int_\Omega Q(u) = 0. \tag{5.6}
$$

If $u_0 = 0$ on $\partial\Omega$, then

$$
\frac{d}{dt}\int_\Omega u^2 Q(u) + 2\int_\Omega \frac{|\nabla u|^2}{Q(u)} + \int_\Omega \frac{u_t^2 u^2}{Q(u)} = 0. \tag{5.7}
$$

For later use it is necessary to identify the important geometric quantities in the energy equations. The equations (5.6) and (5.7) can be written as

$$
\int_\Gamma (V(u))^2 + \frac{d}{dt}|\Gamma| = 0,
$$

$$
\frac{d}{dt}\int_\Gamma u^2 + 2\int_\Gamma |\underline{\nabla}u|^2 + \int_\Gamma (V(u))^2 u^2 = 0,
$$

where $\underline{\nabla}$ denotes the tangential gradient on Γ and $|\Gamma|$ is the measure of Γ.

Proof. Proposition 2.1. □

5.3   Spatial Discretization

Let $\mathcal{T}_h$ be an admissible nondegenerate triangulation of the domain Ω with mesh size bounded by h, simplices S and Ω_h be the corresponding discrete domain. We assume that vertices on $\partial\Omega_h$ are contained in $\partial\Omega$. The discrete space is chosen to be

$$X_h = \{v_h \in C^0(\overline{\Omega}_h) \mid v_h \text{ is a linear polynomial on each } S \in \mathcal{T}_h\}.$$

The subspace containing functions with zero boundary values will be denoted by X_{h0}. We assume that there exists an interpolation operator $I_h : H^2(\Omega) \to X_h$ which satisfies

$$\|v - I_h v\|_{L^2(\Omega \cap \Omega_h)} + h\|\nabla(v - I_h v)\|_{L^2(\Omega \cap \Omega_h)} \leq ch^2\|v\|_{H^2(\Omega)} \quad \forall\, v \in H^2(\Omega). \tag{5.8}$$

For dimensions $n \leq 3$, we can e.g. choose the usual Lagrange interpolation operator, in higher dimensions a possible choice is the Clément operator.

Although the differential equation is not in divergence form, one can easily derive a variational form of (5.4), namely

$$\int_\Omega \frac{u_t \phi}{Q(u)} + \int_\Omega \frac{\nabla u \cdot \nabla \phi}{Q(u)} = 0, \qquad \phi \in H_0^1(\Omega),\ 0 < t < T. \tag{5.9}$$

The semidiscrete approximation of (5.9) is to find $u_h(\cdot, t) \in X_h$ with $u_h(\cdot, t) - I_h u_0 \in X_{h0}$ and $u_h(\cdot, 0) = u_{h0} = I_h u_0$ such that

$$\int_{\Omega_h} \frac{u_{ht}\phi_h}{Q(u_h)} + \int_{\Omega_h} \frac{\nabla u_h \cdot \nabla \phi_h}{Q(u_h)} = 0 \tag{5.10}$$

for all $t \in (0, T)$ and for all discrete test functions $\phi_h \in X_{h0}$.

5.4   Estimate of the Spatial Error

In order prove error estimates for the semidiscrete problem we need to make regularity assumptions on the solution of the continuous problem. Let us suppose that u satisfies

$$\sup_{t \in (0,T)} \left(\|u(\cdot, t)\|_{H^{2,\infty}(\Omega)} + \|u_t(\cdot, t)\|_{H^{1,\infty}(\Omega)} \right)$$

$$+ \int_0^T \left(\|u_t\|_{H^2(\Omega)}^2 + \|u_{tt}\|^2 \right) ds \leq N \tag{5.11}$$

for some $N > 0$ (see [16] for sufficient conditions which imply (5.11)). In the following we shall assume that we have a solution of this kind until the time T. The error estimate is then valid as long as this solution exists. There is no assumption on the geometric form of the boundary $\partial\Omega$ here.

Theorem 5.2. *Let u be a solution of the continuous problem (5.4) which satisfies (5.11). Then there exists a unique solution $u_h \in C^1((0,T), X_h)$ of the semidiscrete problem (5.10) and*

$$\int_0^T \int_{\Omega \cap \Omega_h} (V(u) - V(u_h))^2 \, Q(u_h) + \sup_{(0,T)} \int_{\Omega \cap \Omega_h} |\nu(u) - \nu(u_h)|^2 \, Q(u_h) \le ch^2.$$

The constant c depends on N.

The *proof* will follow as a special case from the anisotropic error estimates in Theorem 6.1.

5.5 Time Discretization

The derivation of fully discrete schemes for the mean curvature flow problem is an important task for both stability and efficiency of the algorithm. There are now several approaches to the time discretization of curvature flow problems (for example see [24], [16], [54]). In the following we list three different time discretizations of (5.10). We shall skip the index h at the function u and at the domain Ω because the estimates are based on the weak formulations. We only use these for the proof, so that the estimates remain valid both for continuous and spatially discrete solutions.

We use the notation

$$v^m(x) = v(x, m\tau)$$

for $m = 0, \ldots, M$ with time step $\tau > 0$ and $M \le [T/\tau]$.

The *fully implicit* scheme is given by

$$\frac{1}{\tau} \int_\Omega \frac{(u^{m+1} - u^m)\phi}{Q(u^{m+1})} + \int_\Omega \frac{\nabla u^{m+1} \cdot \nabla \phi}{Q(u^{m+1})} = 0 \quad (m = 0, \ldots, M - 1).$$

Thus, in every time step one has to solve a highly nonlinear elliptic problem and therefore has to use some solver for the nonlinear problem. The simplest time discretization is the *explicit* one,

$$\frac{1}{\tau} \int_\Omega \frac{(u^{m+1} - u^m)\phi}{Q(u^m)} + \int_\Omega \frac{\nabla u^m \cdot \nabla \phi}{Q(u^m)} = 0 \quad (m = 0, \ldots, M - 1),$$

for which numerically one only has to invert the mass matrix (weighted by $Q(u^m)$). If one uses mass lumping, every time step only requires one matrix multiplication. But similar to the heat equation, a restriction on the smallness of the time step size with respect to the grid size h will be necessary. It is certainly worth trying some *semi–implicit* scheme for time discretization. We shall discuss the scheme

$$\frac{1}{\tau} \int_\Omega \frac{(u^{m+1} - u^m)\phi}{Q(u^m)} + \int_\Omega \frac{\nabla u^{m+1} \cdot \nabla \phi}{Q(u^m)} = 0 \quad (m = 0, \ldots, M - 1). \quad (5.12)$$

92 Klaus Deckelnick, Gerhard Dziuk

Here at every time step one has to invert a Laplace–type equation with stiffness matrix weighted by $Q(u^m)$.

As a stability criterion we use the basic energy norms introduced in (5.6).

Theorem 5.3. *The solution $u^m, 0 \le m \le M$ of (5.12) satisfies for every $m \in \{1,\dots,M\}$*

$$\tau \sum_{k=0}^{m-1} \int_\Omega |V^k|^2 Q(u^k) + \sum_{k=0}^{m-1} \int_\Omega (Q(u^{k+1}) - Q(u^k))^2 \frac{1}{Q(u^k)}$$

$$+\frac{1}{2}\sum_{k=0}^{m-1} \int_\Omega |\nu(u^{k+1}) - \nu(u^k)|^2 Q(u^{k+1}) + \int_\Omega Q(u^m) = \int_\Omega Q(u^0) \quad (5.13)$$

where

$$V^k = -\frac{(u^{k+1} - u^k)/\tau}{Q(u^k)}$$

is the discrete normal velocity.

Proof. We choose $\phi = u^{k+1} - u^k$ as a test function in (5.12) for $m = k$ and get

$$\frac{1}{\tau}\int_\Omega \frac{(u^{k+1}-u^k)^2}{Q(u^k)} + \int_\Omega \frac{\nabla u^{k+1} \cdot \nabla(u^{k+1}-u^k)}{Q(u^k)} = 0. \quad (5.14)$$

Let us use the notation $\underline{\nu}(v) = \frac{\nabla v}{Q(v)}$. Then

$$\frac{\nabla u^{k+1} \cdot \nabla(u^{k+1}-u^k)}{Q(u^k)} = \frac{Q(u^{k+1})^2 - 1}{Q(u^k)} - \underline{\nu}(u^{k+1}) \cdot \underline{\nu}(u^k) Q(u^{k+1})$$

$$= \frac{Q(u^{k+1})^2}{Q(u^k)} + \frac{1}{2}|\nu(u^{k+1}) - \nu(u^k)|^2 Q(u^{k+1}) - Q(u^{k+1})$$

$$= \frac{1}{2}|\nu(u^{k+1}) - \nu(u^k)|^2 + Q(u^{k+1}) - Q(u^k) + \frac{(Q(u^{k+1}) - Q(u^k))^2}{Q(u^k)}.$$

We insert this result into (5.14) and sum over $k = 0,\dots,m-1$. $\square$

Let us emphasize once more that the semi–implicit scheme is unconditionally stable even though the nonlinear expressions are treated explicitly.

We now use the finite element scheme (5.10) together with the semi-implicit time discretization (5.15) for the numerical solution of problem (5.4). The scheme can be summarized in the following algorithm.

Algorithm 5.1. Let $u_h^0 = I_h u_0$. For $m = 0,\dots,M-1$ compute $u_h^{m+1} \in X_h$ such that $u_h^{m+1} - u_h^0 \in X_{h0}$ and for every $\phi_h \in X_{h0}$

$$\frac{1}{\tau}\int_{\Omega_h} \frac{u_h^{m+1}\phi_h}{Q(u_h^m)} + \int_{\Omega_h} \frac{\nabla u_h^{m+1} \cdot \nabla \phi_h}{Q(u_h^m)} = \frac{1}{\tau}\int_{\Omega_h} \frac{u_h^m \phi_h}{Q(u_h^m)}. \quad (5.15)$$

Choose on open set $\Omega' \subset \mathbb{R}^n$ which contains $\overline{\Omega} \cup \Omega_h$ for all $h \leq 1$. In view of the regularity (5.11) of u and since $\partial\Omega$ is smooth, there exists an extension $\overline{u} : \Omega' \times [0,T] \to \mathbb{R}$ such that $\overline{u}_{|\Omega \times [0,T]} = u$ and

$$\sup_{(0,T)} \|\overline{u}\|_{H^{2,\infty}(\Omega')} + \sup_{(0,T)} \|\overline{u}_t\|_{H^{1,\infty}(\Omega')} + \int_0^T \|\overline{u}_t\|_{H^2(\Omega')}^2 + \int_0^T \|\overline{u}_{tt}\|^2 \leq cN,$$

(5.16)

where N is that in (5.11).

Theorem 5.4. *Assume that there exists a solution of (5.4) on $\Omega \times (0,T)$ which satisfies (5.11) and let u_h^m, $(m = 1, \ldots, M)$ be the solution of Algorithm 5.1. Then there exists a $\tau_0 > 0$ such that for all $0 < \tau \leq \tau_0$*

$$\tau \sum_{m=0}^{M-1} \int_{\Omega \cap \Omega_h} (V(u^m) - V_h^m)^2 Q(u_h^m) \leq c(\tau^2 + h^2),$$

(5.17)

$$\sup_{m=0,\ldots,M} \int_{\Omega \cap \Omega_h} |\nu(u^m) - \nu(u_h^m)|^2 Q(u_h^m) \leq c(\tau^2 + h^2),$$

(5.18)

where $M = [\frac{T}{\tau}]$ and

$$V_h^m = -\frac{u_h^{m+1} - u_h^m}{\tau} \frac{1}{Q(u_h^m)}$$

is the discrete normal velocity.

Proof. This is a special case of the results obtained in [18]. □

6   Anisotropic Curvature Flow of Graphs

For a graph of a height function u over some base domain $\Omega \subset \mathbb{R}^n$, i.e. $\Gamma = \{(x, u(x)) \,|\, x \in \Omega\}$ we have that (isotropic) mean curvature H is given as minus the first variation of area. Similarly, we define anisotropic mean curvature as minus the first variation of anisotropic area E_γ. Just as above, the area element and a unit normal are then given by

$$Q(u) = \sqrt{1 + |\nabla u|^2} \quad \text{and} \quad \nu(u) = \frac{(\nabla u, -1)}{\sqrt{1 + |\nabla u|^2}} = \frac{(\nabla u, -1)}{Q(u)}.$$

(6.1)

Let γ be an admissible weight function. We calculate the weighted area for a graph Γ given by the height function u as

$$E_\gamma(\Gamma) = E_\gamma(u) = \int_\Omega \gamma(\nu(u)) Q(u) = \int_\Omega \gamma(\nabla u, -1)$$

because of the homogeneity of γ. The first variation of E_γ in the direction of a function $\phi \in C_0^\infty(\Omega)$ is then

$$\frac{d}{d\epsilon} E_\gamma(u + \epsilon\phi)\Big|_{\epsilon=0} = \sum_{i=1}^n \int_\Omega \gamma_{p_i}(\nabla u, -1)\phi_{x_i} = -\sum_{i,j=1}^n \int_\Omega \gamma_{p_i p_j}(\nabla u, -1)u_{x_i x_j}\phi$$

$$= -\int_\Omega H_\gamma \phi,$$

(6.2)

94   Klaus Deckelnick, Gerhard Dziuk

where we have set
$$H_\gamma = \sum_{i,j=1}^{n} \gamma_{p_i p_j}(\nabla u, -1) u_{x_i x_j}.$$

We call H_γ the γ-mean curvature or anisotropic mean curvature. The isotropic case is recovered by setting $\gamma(p) = |p|$:

$$H = \frac{1}{Q(u)} \sum_{i,j=1}^{n} \left(\delta_{ij} - \frac{u_{x_i} u_{x_j}}{Q(u)^2} \right) u_{x_i x_j} = \nabla \cdot \frac{\nabla u}{\sqrt{1 + |\nabla u|^2}}.$$

Let us now consider the law of motion
$$V = -H_\gamma, \tag{6.3}$$

which says that the surface is moved in normal direction with velocity given by the anisotropic mean curvature. In general one uses laws which include a mobility term $\beta = \beta(\nu)$, so that $\beta V = -H_\gamma$. Such a mobility can be included in all our considerations under suitable assumptions. But for the sake of simplicity we do not include it in what follows.

For graphs the equation (6.3) leads to the initial boundary value problem

$$u_t - \sqrt{1 + |\nabla u|^2} \sum_{i=1}^{n} \frac{\partial}{\partial x_i}(\gamma_{p_i}(\nabla u, -1)) = 0 \text{ in } \Omega \times (0,T) \tag{6.4}$$
$$u = u_0 \text{ on } \partial\Omega \times (0,T)$$
$$u(\cdot,0) = u_0 \text{ in } \Omega.$$

In the sequel we shall again assume that this problem has a solution u which satisfies (5.11) and refer to [16] for a corresponding existence and uniqueness result.

6.1 Discretization in Space and Estimate of the Error

As in the isotropic case we may use a variational approach even though the differential equation is not in divergence form. Starting from (6.4) we obtain

$$\int_\Omega \frac{u_t \varphi}{\sqrt{1 + |\nabla u|^2}} + \sum_{i=1}^{n} \int_\Omega \gamma_{p_i}(\nabla u, -1)\varphi_{x_i} = 0 \tag{6.5}$$

for all $\varphi \in H_0^1(\Omega)$, $t \in (0,T)$ together with the above initial and boundary conditions. We now consider a semidiscrete approximation of (6.5): find $u_h(\cdot,t) \in X_h$ with $u_h(\cdot,t) - u_h^0 \in X_{h0}$ such that

$$\int_{\Omega_h} \frac{u_{h,t}\varphi_h}{\sqrt{1 + |\nabla u_h|^2}} + \sum_{i=1}^{n} \int_{\Omega_h} \gamma_{p_i}(\nabla u_h, -1)\varphi_{h,x_i} = 0 \quad \forall \, \varphi_h \in X_{h0}, t \in [0,T]$$
$$u_h(\cdot,0) = u_h^0 \tag{6.6}$$

where $u_h^0 = I_h u_0 \in X_h$ is the Lagrange interpolant of u_0.

Our main result gives an error bound for the important geometric quantities V and ν.

Theorem 6.1. *Suppose that (6.4) has a solution u that satisfies (5.11). Then (6.6) has a unique solution u_h and*

$$\int_0^T \|V(u) - V(u_h)\|_{L^2(\Gamma_h(t))}^2 \, dt + \sup_{t \in (0,T)} \|(\nu(u) - \nu(u_h))(\cdot, t)\|_{L^2(\Gamma_h(t))}^2 \le Ch^2.$$

Here, $\Gamma_h(t) = \{(x, u_h(x,t)) \mid x \in \Omega_h \cap \Omega\}$.
Proof. We prove the result for the case of a convex domain Ω. The general case only adds technical difficulties which can be treated with standard numerical analysis. Therefore from now on that we assume that $\Omega_h \subset \Omega$.

Choosing $\varphi = \varphi_h \in X_{h0}$ in (6.5) and taking the difference between the resulting equation and (6.6) yields

$$-\int_{\Omega_h} (V(u) - V(u_h))\,\varphi_h + \sum_{j=1}^n \int_{\Omega_h} \left(\gamma_{p_j}(\nu(u)) - \gamma_{p_j}(\nu(u_h))\right)\varphi_{hx_j} = 0$$

for every $\varphi_h \in X_{h0}$. Since the boundary values are time independent, we use

$$\varphi_h = I_h u_t - u_{ht} = (u_t - u_{ht}) - (u_t - I_h u_t)$$

as a test function and obtain

$$-\int_{\Omega_h} (V(u) - V(u_h))\,(u_t - u_{ht}) \tag{6.7}$$

$$+ \sum_{j=1}^n \int_{\Omega_h} \left(\gamma_{p_j}(\nu(u)) - \gamma_{p_j}(\nu(u_h))\right)(u_{tx_j} - u_{htx_j})$$

$$= -\int_{\Omega_h} (V(u) - V(u_h))\,(u_t - I_h u_t)$$

$$+ \sum_{j=1}^n \int_{\Omega_h} \left(\gamma_{p_j}(\nu(u)) - \gamma_{p_j}(\nu(u_h))\right)(u_{tx_j} - (I_h u_t)_{x_j}).$$

Let us estimate the terms in this equation separately. We begin with the two estimates for the terms on the left hand side of (6.7). Firstly,

$$-\int_{\Omega_h} (V(u) - V(u_h))\,(u_t - u_{ht}) \tag{6.8}$$

$$= \int_{\Omega_h} (V(u) - V(u_h))^2 Q(u_h) + \int_{\Omega_h} V(u)(Q(u) - Q(u_h))(V(u) - V(u_h))$$

$$\ge \int_{\Omega_h} (V(u) - V(u_h))^2 Q(u_h)$$

$$-\|u_t\|_{L^\infty(\Omega)} \int_{\Omega_h} |\nu(u) - \nu(u_h)||V(u) - V(u_h)|Q(u_h)$$

$$\ge \frac{3}{4} \int_{\Omega_h} (V(u) - V(u_h))^2 Q(u_h) - \|u_t\|_{L^\infty(\Omega)}^2 \int_{\Omega_h} |\nu(u) - \nu(u_h)|^2 Q(u_h).$$

Here we have used Young's inequality and the fact that

$$|Q(u) - Q(u_h)| \leq |\nu(u) - \nu(u_h)|Q(u)Q(u_h). \tag{6.9}$$

The second term on the left hand side of (6.7) is the one which contains the anisotropy. Observing that the $(n+1)$–st component of the vector

$$\nu(u)Q(u) - \nu(u_h)Q(u_h)$$

equals zero we conclude that

$$\sum_{j=1}^{n} \int_{\Omega_h} \left(\gamma_{p_j}(\nu(u)) - \gamma_{p_j}(\nu(u_h))\right) \partial_t(u_{x_j} - u_{hx_j}) \tag{6.10}$$

$$= \int_{\Omega_h} \left(\gamma_p(\nu(u)) - \gamma_p(\nu(u_h))\right) \cdot \partial_t \left(\nu(u)Q(u) - \nu(u_h)Q(u_h)\right)$$

$$= -\int_{\Omega_h} \left(\gamma_p(\nu(u)) - \gamma_p(\nu(u_h))\right) \cdot \Big\{\nu(u_h)\partial_t Q(u_h) + \partial_t(\nu(u_h))Q(u_h)$$

$$-\partial_t(\nu(u))Q(u) - \nu(u)\partial_t Q(u)\Big\}$$

$$= I_1 + I_2 + I_3 + I_4. \tag{6.11}$$

We use the homogeneity properties (2.4) of γ. For $I_1, \ldots, I_4$ this gives

$$I_1 = -\int_{\Omega_h} \left(\gamma_p(\nu(u)) - \gamma_p(\nu(u_h))\right) \cdot \nu(u_h)\partial_t Q(u_h) \tag{6.12}$$

$$= \int_{\Omega_h} \left(\gamma(\nu(u_h)) - \gamma_p(\nu(u)) \cdot \nu(u_h)\right) \partial_t Q(u_h)$$

$$= \frac{d}{dt} \int_{\Omega_h} \left(\gamma(\nu(u_h)) - \gamma_p(\nu(u)) \cdot \nu(u_h)\right) Q(u_h)$$

$$+ \int_{\Omega_h} \gamma_{pp}(\nu(u))\nu(u_h) \cdot \partial_t \nu(u) \, Q(u_h)$$

$$+ \int_{\Omega_h} \gamma_p(\nu(u)) \cdot \partial_t \nu(u_h) \, Q(u_h) - \int_{\Omega_h} \partial_t \gamma(\nu(u_h)) \, Q(u_h).$$

For the second integral we have

$$I_2 = -\int_{\Omega_h} \left(\gamma_p(\nu(u)) - \gamma_p(\nu(u_h))\right) \cdot \partial_t(\nu(u_h)) \, Q(u_h) \tag{6.13}$$

$$= -\int_{\Omega_h} \gamma_p(\nu(u)) \cdot \partial_t(\nu(u_h))Q(u_h) + \int_{\Omega_h} \partial_t \gamma(\nu(u_h))Q(u_h),$$

and I_4 can be written as

$$I_4 = \int_{\Omega_h} \left(\gamma_p(\nu(u)) - \gamma_p(\nu(u_h))\right) \cdot \nu(u)\partial_t Q(u) \tag{6.14}$$

$$= \int_{\Omega_h} \gamma(\nu(u))\partial_t Q(u) - \int_{\Omega_h} \gamma_p(\nu(u_h)) \cdot \nu(u)\partial_t Q(u).$$

We add (6.12), (6.13), (6.14) and I_3,

$$I_3 = \int_{\Omega_h} (\gamma_p(\nu(u)) - \gamma_p(\nu(u_h))) \cdot \partial_t \nu(u) \, Q(u),$$

and arrive at the following expression for (6.10):

$$\sum_{j=1}^{n} \int_{\Omega_h} (\gamma_{p_j}(\nu(u)) - \gamma_{p_j}(\nu(u_h))) \, \partial_t(u_{x_j} - u_{hx_j}) \qquad (6.15)$$

$$= \frac{d}{dt} \int_{\Omega_h} (\gamma(\nu(u_h)) - \gamma_p(\nu(u)) \cdot \nu(u_h)) \, Q(u_h)$$

$$+ \int_{\Omega_h} (\gamma(\nu(u)) - \gamma_p(\nu(u_h)) \cdot \nu(u)) \, \partial_t Q(u)$$

$$+ \int_{\Omega_h} (\gamma_p(\nu(u)) - \gamma_p(\nu(u_h))) \cdot \partial_t \nu(u) \, Q(u)$$

$$+ \int_{\Omega_h} \gamma_{pp}(\nu(u)) \nu(u_h) \cdot \partial_t \nu(u) \, Q(u_h)$$

$$= \frac{d}{dt} \int_{\Omega_h} (\gamma(\nu(u_h)) - \gamma_p(\nu(u)) \cdot \nu(u_h)) \, Q(u_h)$$

$$+ \int_{\Omega_h} \{\gamma(\nu(u)) - [\gamma(\nu(u_h)) + \gamma_p(\nu(u_h)) \cdot (\nu(u) - \nu(u_h))]\} \, \partial_t Q(u)$$

$$- \int_{\Omega_h} \{\gamma_p(\nu(u_h)) - [\gamma_p(\nu(u)) + \gamma_{pp}(\nu(u))(\nu(u_h) - \nu(u))]\} \cdot \partial_t \nu(u) \, Q(u)$$

$$+ \int_{\Omega_h} \gamma_{pp}(\nu(u))(\nu(u_h) - \nu(u)) \cdot \partial_t \nu(u)(Q(u_h) - Q(u)).$$

Two terms in the above formula contain Taylor expansions of the anisotropy function γ. If we let $\nu_s = s\nu(u) + (1-s)\nu(u_h)$, $\tilde{\nu}_s = s\nu(u_h) + (1-s)\nu(u)$ and observe that $\nu_s, \tilde{\nu}_s \neq 0$ for all $s \in [0,1]$ we can continue

$$= \frac{d}{dt} \int_{\Omega_h} (\gamma(\nu(u_h)) - \gamma_p(\nu(u)) \cdot \nu(u_h)) \, Q(u_h)$$

$$+ \int_{\Omega_h} \int_0^1 (1-s)\gamma_{pp}(\nu_s)(\nu(u) - \nu(u_h)) \cdot (\nu(u) - \nu(u_h)) \partial_t Q(u)$$

$$- \sum_{k=1}^{n+1} \int_{\Omega_h} \int_0^1 (1-s)\gamma_{p_k pp}(\tilde{\nu}_s)(\nu(u) - \nu(u_h)) \cdot (\nu(u) - \nu(u_h)) \partial_t \nu(u)_k \, Q(u)$$

$$+ \int_{\Omega_h} \gamma_{pp}(\nu(u))(\nu(u_h) - \nu(u)) \cdot \partial_t \nu(u)(Q(u_h) - Q(u)).$$

In order to estimate the second and the third integral we require a lower bound on $|\nu_s|$ and $|\tilde{\nu}_s|$ respectively. To this end we calculate for $s \in [0,1]$

$$|\nu_s|^2 = s^2 + (1-s)^2 + 2s(1-s) \frac{\nabla u \cdot \nabla u_h + 1}{Q(u)Q(u_h)}$$

$$\geq 1 - 2s(1-s)\left(1 - \frac{\nabla u}{Q(u)} \cdot \frac{\nabla u_h}{Q(u_h)}\right) \geq 1 - \frac{1}{2}\left(1 + \frac{|\nabla u|}{Q(u)}\right).$$

98   Klaus Deckelnick, Gerhard Dziuk

Since (5.11) implies that $|\nabla u| \leq M$ we obtain

$$|\nu_s|^2 \geq 1 - \frac{1}{2}\left(1 + \frac{M}{\sqrt{1+M^2}}\right)$$

so that the homogeneity of γ yields

$$|\gamma_{pp}(\nu_s)| = \left|\gamma_{pp}\left(\frac{\nu_s}{|\nu_s|}\right)\right|\frac{1}{|\nu_s|} \leq c(M)\sup_{p\in S^n}|\gamma_{pp}(p)|$$

uniformly in $s \in [0,1]$. Thus, we can estimate

$$\int_{\Omega_h}\int_0^1 (1-s)\gamma_{pp}(\nu_s)(\nu(u)-\nu(u_h))\cdot(\nu(u)-\nu(u_h))\partial_t Q(u)$$

$$\leq \int_{\Omega_h}\int_0^1 (1-s)|\gamma_{pp}(\nu_s)||\nu(u)-\nu(u_h)|^2|\nabla u_t|$$

$$\leq c(M)\|\nabla u_t\|_{L^\infty(\Omega)}\int_{\Omega_h}|\nu(u)-\nu(u_h)|^2 Q(u_h).$$

If we apply a similar argument to the third integral above and also use the simple inequality $|Q(u)\partial_t\nu(u)| \leq c|\nabla u_t|$ we finally obtain from (6.15)

$$\sum_{j=1}^n \int_{\Omega_h}\left(\gamma_{p_j}(\nu(u)) - \gamma_{p_j}(\nu(u_h))\right)\partial_t(u_{x_j}-u_{hx_j}) \tag{6.16}$$

$$\geq \frac{d}{dt}\int_{\Omega_h}\left(\gamma(\nu(u_h)) - \gamma_p(\nu(u))\cdot\nu(u_h)\right)Q(u_h)$$

$$-c(M)\|\nabla u_t\|_{L^\infty(\Omega)}\int_{\Omega_h}|\nu(u)-\nu(u_h)|^2 Q(u_h).$$

Let us now estimate the right hand side of (6.7).

$$-\int_{\Omega_h}(V(u)-V(u_h))(u_t-I_h u_t) \tag{6.17}$$

$$\leq \left(\int_{\Omega_h}(V(u)-V(u_h))^2 Q(u_h)\right)^{\frac{1}{2}}\left(\int_{\Omega_h}\frac{(u_t-I_h u_t)^2}{Q(u_h)}\right)^{\frac{1}{2}}$$

$$\leq \left(\int_{\Omega_h}(V(u)-V(u_h))^2 Q(u_h)\right)^{\frac{1}{2}}\|u_t-I_h u_t\|_{L^2(\Omega_h)}$$

and using (5.8)

$$\leq ch^2\|u_t\|_{H^2(\Omega)}\left(\int_{\Omega_h}(V(u)-V(u_h))^2 Q(u_h)\right)^{\frac{1}{2}}$$

$$\leq \frac{1}{4}\int_{\Omega_h}(V(u)-V(u_h))^2 Q(u_h) + ch^4\|u_t\|_{H^2(\Omega)}^2.$$

For the second term on the right hand side of (6.7) we have

$$\sum_{j=1}^{n} \int_{\Omega_h} \left(\gamma_{p_j}(\nu(u)) - \gamma_{p_j}(\nu(u_h)) \right) \partial_t (u_{x_j} - (I_h u)_{x_j}) \qquad (6.18)$$

$$\leq c \left(\int_{\Omega_h} |\nu(u) - \nu(u_h)|^2 Q(u_h) \right)^{\frac{1}{2}} \left(\int_{\Omega_h} \frac{|\nabla(u_t - I_h u_t)|^2}{Q(u_h)} \right)^{\frac{1}{2}}$$

$$\leq c \int_{\Omega_h} |\nu(u) - \nu(u_h)|^2 Q(u_h) + c h^2 \|u_t\|_{H^2(\Omega)}^2.$$

We collect the estimates (6.8), (6.16), (6.17), (6.18) and obtain

$$\frac{1}{2} \int_{\Omega_h} (V(u) - V(u_h))^2 Q(u_h) \qquad (6.19)$$

$$+ \frac{d}{dt} \int_{\Omega_h} (\gamma(\nu(u_h)) - \gamma_p(\nu(u)) \cdot \nu(u_h)) Q(u_h)$$

$$\leq c(M) \left(1 + \|\nabla u_t\|_{L^\infty(\Omega)} + \|u_t\|_{L^\infty(\Omega)}^2 \right) \int_{\Omega_h} |\nu(u) - \nu(u_h)|^2 Q(u_h)$$

$$+ c h^2 \|u_t\|_{H^2(\Omega)}^2.$$

In view of Lemma 2.2 we may estimate

$$\gamma(\nu(u_h)) - \gamma_p(\nu(u)) \cdot \nu(u_h) \geq c_1 |\nu(u) - \nu(u_h)|^2. \qquad (6.20)$$

Integrating with respect to time and using (6.20) we finally arrive at the estimate

$$\int_0^t \int_{\Omega_h} (V(u) - V(u_h))^2 Q(u_h) \, dt + \int_{\Omega_h} |\nu(u) - \nu(u_h)|^2 Q(u_h) \qquad (6.21)$$

$$\leq c(M) \int_0^t \left(\|\nabla u_t\|_{L^\infty(\Omega)} + \|u_t\|_{L^\infty(\Omega)}^2 \right) \int_{\Omega_h} |\nu(u) - \nu(u_h)|^2 Q(u_h) \, dt$$

$$+ c h^2 \int_0^t \|u_t\|_{H^2(\Omega)}^2 \, dt + c h^2 \|u_0\|_{H^2(\Omega)}^2.$$

A Gronwall argument completes the proof of Theorem 6.1. □

6.2   Fully Discrete Scheme, Stability and Error Estimate

We denote by $\tau > 0$ the time step. Our numerical method is based on the variational formulation (6.5), which we can write with the help of (2.5) as

$$\int_\Omega \frac{u_t(\cdot, m\tau)}{Q^m} \varphi + \sum_{i=1}^n \int_\Omega \gamma_{p_i}(\nu^m)\varphi_{x_i} = 0 \qquad \forall\, \varphi \in H_0^1(\Omega). \qquad (6.22)$$

Here we have used the abbreviations

$$u^m = u(\cdot, m\tau), \quad Q^m = Q(u^m) \quad \text{and} \quad \nu^m = \nu(u^m). \qquad (6.23)$$

Defining in an analogous way $Q_h^m = Q(u_h^m)$ and $\nu_h^m = \nu(u_h^m)$ our scheme reads as follows:

Algorithm 6.1. Given u_h^m, find $u_h^{m+1} \in X_h$ such that $u_h^{m+1} - I_h u_0 \in X_{h0}$ and

$$\frac{1}{\tau} \int_{\Omega_h} \frac{u_h^{m+1} - u_h^m}{Q_h^m} \varphi_h + \sum_{i=1}^n \int_{\Omega_h} \gamma_{p_i}(\nu_h^m)\varphi_{hx_i}$$

$$+ \lambda \int_{\Omega_h} \frac{\gamma(\nu_h^m)}{Q_h^m} \nabla\left(u_h^{m+1} - u_h^m\right) \cdot \nabla\varphi_h = 0 \quad (6.24)$$

for all $\varphi_h \in X_{h0}$. Here we have set $u_h^0 = I_h u_0$.

The above scheme is semi–implicit and requires the solution of a linear system in each time step. We shall see that it is unconditionally stable provided the parameter λ is chosen appropriately.

Theorem 6.2. *Let* $\bar\gamma = \frac{1}{\sqrt5-1} \max\left\{ \sup_{|p|=1} |\gamma_p(p)|, \sup_{|p|=1} |\gamma_{pp}(p)| \right\}$. *Then we have for* $0 \le M \le [\frac{T}{\tau}]$

$$\tau \sum_{m=1}^{M-1} \int_{\Omega_h} \frac{1}{Q_h^m} \left|\frac{u_h^{m+1} - u_h^m}{\tau}\right|^2|$$

$$+ \left(\lambda \inf_{|p|=1} \gamma(p) - \bar\gamma\right)\tau \sum_{m=1}^{M-1} \int_{\Omega_h} \left|\frac{\nu_h^{m+1} - \nu_h^m}{\sqrt\tau}\right|^2 Q_h^{m+1}$$

$$+ \lambda\tau \sum_{m=1}^{M-1} \int_{\Omega_h} \frac{\gamma(\nu_h^m)}{Q_h^m}\left(\frac{Q_h^{m+1} - Q_h^m}{\sqrt\tau}\right)^2 + \int_{\Omega_h} \gamma(\nu_h^M)Q_h^M \le \int_{\Omega_h} \gamma(\nu_h^0)Q_h^0.$$

In particular, if λ *is chosen in such a way that* $\rho := \lambda \inf_{|p|=1} \gamma(p) - \bar\gamma > 0$, *then*

$$\sup_{0 \le m \le [\frac{T}{\tau}]} \int_{\Omega_h} Q_h^m \le C(u_0, \gamma). \qquad (6.25)$$

Thus we have proved stability for the semi–implicit scheme without any restriction on the time step size.

In [18] we have proved convergence together with suitable error estimates for the fully discrete Algorithm 6.1.

Theorem 6.3. *Suppose that*

$$\lambda \inf_{|p|=1} \gamma(p) > \bar{\gamma} := \frac{1}{\sqrt{5}-1} \max \left\{ \sup_{|p|=1} |\gamma_p(p)|, \sup_{|p|=1} |\gamma_{pp}(p)| \right\}$$

holds. Then there exists $\tau_0 > 0$ such that for all $0 < \tau \leq \tau_0$

$$\sum_{m=0}^{[\frac{T}{\tau}]-1} \tau \int_{\Omega \cap \Omega_h} (V^m - V_h^m)^2 Q_h^m + \max_{0 \leq m \leq [\frac{T}{\tau}]} \int_{\Omega \cap \Omega_h} |\nu^m - \nu_h^m|^2 Q_h^m \leq c(\tau^2 + h^2).$$

Consideration of the isotropic case gives the following result. In Theorem 5.4 we analyzed the following scheme for the isotropic case $\gamma(p) = |p|$:

$$\frac{1}{\tau} \int_{\Omega} \frac{(u_h^{m+1} - u_h^m)\varphi_h}{Q_h^m} + \int_{\Omega} \frac{\nabla u_h^{m+1} \cdot \nabla \varphi_h}{Q_h^m} = 0$$

which is (6.24) for the choice $\lambda = 1$. Since $\inf_{|p|=1} \gamma(p) = 1$, $\sup_{|p|=1} |\gamma_p(p)| = \sup_{|p|=1} |\gamma_{pp}(p)| = 1$ and $1 > \frac{1}{\sqrt{5}-1}$ we recover the unconditional convergence of this scheme (see also [24]). We include Table 6.1 with computational results

h	$err(\nu)$	eoc	$err(V)$	eoc	$L^\infty(H^1)$	eoc	$L^\infty(L^2)$	eoc
7.0711e-2	9.7027e-2	-	2.3278e-2	-	1.3366e-1	-	3.1700e-3	-
3.5355e-2	2.3213e-2	2.06	6.0827e-3	1.94	4.4935e-2	1.57	7.3639e-4	2.11
2.6050e-2	2.4818e-2	-0.22	7.5203e-3	-0.70	4.5372e-2	-0.03	3.0728e-4	2.86
1.4861e-2	1.3868e-2	1.04	4.1117e-3	1.08	2.4163e-2	1.12	8.8373e-5	2.22
7.8462e-3	7.0232e-3	1.07	1.9806e-3	1.14	1.2256e-2	1.06	2.7864e-5	1.81
4.0210e-3	3.5368e-3	1.03	1.0176e-3	1.00	6.1725e-3	1.03	1.2378e-5	1.22
2.0342e-3	1.8103e-3	0.98	5.2675e-4	0.97	3.1225e-3	1.00	6.3199e-6	0.99
1.0229e-3	9.2938e-4	0.97	2.6988e-4	0.97	1.5799e-3	0.99	3.2715e-6	0.96

Table 6.1. Absolute errors for the test problem with $\tau = 0.01h$.

for fully discrete anisotropic mean curvature flow (with mobility $1/\gamma$) of a graph from [18]. There we computed a known exact solution $u = u(x,t)$ defined on

$$\Omega = \{x \in \mathbb{R}^2 \mid |x| < R = 0.035355\}$$

for $t \in (0, 0.125)$ by $\gamma^*(x, u(x,t)) = \sqrt{1-4t}$ with the anisotropy $\gamma(p) = \sqrt{0.01p_1^2 + p_2^2 + p_3^2}$. The table shows the error and the experimental order of convergence for the quantities from Theorem 6.3, namely

$$err(V)^2 = \sum_{m=0}^{M} \tau \int_{\Omega_h} |V^m - V_h^m|^2 Q_h^m, \quad err(\nu)^2 = \max_{0 \leq m \leq M} \int_{\Omega_h} |\nu^m - \nu_h^m|^2 Q_h^m,$$

and also for the norms $L^\infty((0,T), H^1(\Omega))$, $L^\infty((0,T), L^2(\Omega))$. The computations confirm the theoretical results. Note that the convergence in L^2 is linear because of the linear coupling between h and τ.

7 Mean Curvature Flow of Level Sets II

7.1 The Approximation of Viscosity Solutions

In the following we use the abbreviations

$$\nu_\epsilon(v) = \frac{(\nabla v, -\epsilon)}{Q_\epsilon(v)}, \quad Q_\epsilon(v) = \sqrt{\epsilon^2 + |\nabla v|^2}, \quad V_\epsilon(v) = -\frac{v_t}{Q_\epsilon(v)}.$$

Our results for the mean curvature flow of a graph can directly be transformed into a convergence result for the regularised level set problem.

Theorem 7.1. *Let u_ϵ be the solution of (4.8), (4.9) and $u_{\epsilon h}$ be the solution of the semidiscrete problem $u_{\epsilon h}(\cdot, t) \in X_h$ with $u_{\epsilon h}(\cdot, t) - I_h u_0 \in X_{h0}$, $u_{\epsilon h}(\cdot, 0) = u_{h0} = I_h u_0$ and*

$$\int_{\Omega_h} \frac{u_{\epsilon h t} \phi_h}{Q_\epsilon(u_{\epsilon h})} + \int_{\Omega_h} \frac{\nabla u_{\epsilon h} \cdot \nabla \phi_h}{Q_\epsilon(u_{\epsilon h})} = 0 \qquad (7.1)$$

for all $t \in (0, T)$ and all discrete test functions $\phi_h \in X_{h0}$. Then

$$\int_0^T \int_{\Omega \cap \Omega_h} (V_\epsilon(u_\epsilon) - V_\epsilon(u_{\epsilon h}))^2 Q_\epsilon(u_{\epsilon h}) \le c_\epsilon h^2,$$

$$\sup_{(0,T)} \int_{\Omega \cap \Omega_h} |\nu_\epsilon(u_\epsilon) - \nu_\epsilon(u_{\epsilon h})|^2 Q_\epsilon(u_{\epsilon h}) \le c_\epsilon h^2.$$

We omit the *proof* as it is based on the scaling argument which we shall use for the fully discrete scheme below.

In two space dimensions we can prove that the computed solutions $u_{\epsilon h}$ converge in L^∞ to the viscosity solution. The proof is contained in [19].

Theorem 7.2. *Let u be the viscosity solution of (4.2), (4.3) and let $u_{\epsilon h}$ be the solution of the problem (7.1) with $\Omega \subset \mathbb{R}^2$ as in Corollary 4.1. Then there exists a function $h = h(\epsilon) \to 0$ as $\epsilon \to 0$ such that*

$$\lim_{\epsilon \to 0} \|u - u_{\epsilon h(\epsilon)}\|_{L^\infty(\Omega \times (0,T))} = 0.$$

A similar transformation of the graph problem into the level set problem is available for the fully discrete scheme. We now have the following convergence theorem for the fully discrete regularised level set problem. The required regularity of the continuous solution is available because the domain is a ball and thus mean convex.

The fully discrete numerical scheme for (regularised) isotropic mean curvature flow of level sets reads as follows:

Algorithm 7.1. Let $u_{\epsilon h}^0 = I_h u_0$. For $m = 0, \ldots, M - 1$ compute $u_{\epsilon h}^{m+1} \in X_h$ such that $u_{\epsilon h}^{m+1} - u_h^0 \in X_{h0}$ and for every $\phi_h \in X_{h0}$

$$\frac{1}{\tau} \int_{\Omega_h} \frac{u_{\epsilon h}^{m+1} \phi_h}{Q_\epsilon(u_{\epsilon h}^m)} + \int_{\Omega_h} \frac{\nabla u_{\epsilon h}^{m+1} \cdot \nabla \phi_h}{Q_\epsilon(u_{\epsilon h}^m)} = \frac{1}{\tau} \int_{\Omega_h} \frac{u_{\epsilon h}^m \phi_h}{Q_\epsilon(u_{\epsilon h}^m)}. \qquad (7.2)$$

For this scheme we have the following convergence result.

Theorem 7.3. *Let u_ϵ be the solution of (4.8), (4.9) and let $u_{\epsilon h}^m$, ($m = 1, \ldots, M$) be the solution from Algorithm 7.1. Then there exists a $\tau_0 > 0$ such that for all $0 < \tau \leq \tau_0$*

$$\tau \sum_{m=0}^{M-1} \int_{\Omega \cap \Omega_h} (V_\epsilon(u^m) - V_{\epsilon h}^m)^2 Q_\epsilon(u_{\epsilon h}^m) \leq c_\epsilon(\tau^2 + h^2), \qquad (7.3)$$

$$\sup_{m=0,\ldots,M} \int_{\Omega \cap \Omega_h} |\nu_\epsilon(u_\epsilon^m) - \nu_\epsilon(u_{\epsilon h}^m)|^2 Q_\epsilon(u_{\epsilon h}^m) \leq c_\epsilon(\tau^2 + h^2), \qquad (7.4)$$

with $M = [\frac{T}{\tau}]$. Here $V_{\epsilon h}^m = -(u_{\epsilon h}^{m+1} - u_{\epsilon h}^m)/(\tau\, Q_\epsilon(u_{\epsilon h}^m))$ is the regularised discrete normal velocity.

This result implies the convergence of the fully discrete regularized solution to the viscosity solution.

Theorem 7.4. *Let u be the viscosity solution from Theorem 4.1 and let Ω be the domain from Corollary 4.1 in $\mathbb{R}^2$. Denote by $u_{\epsilon h \tau}$ the time interpolated solution of the fully discrete scheme (7.2). Then there exist functions $h = h(\epsilon) \to 0$ and $\tau = \tau(\epsilon) \to 0$ as $\epsilon \to 0$ such that*

$$\|u - u_{\epsilon h(\epsilon)\tau(\epsilon)}\|_{L^\infty(\Omega_h \times (0,T))} \to 0.$$

We illustrate our algorithm by computing the viscosity solution for a torus. Because of the rotational symmetry (see Figures 3.8 and 3.9) we can avoid working in three space dimensions. It is an easy task to generalize the Cartesian Algorithm 7.1 to the axially symmetric case. Let

$$\Omega = \{(r, x_3)| -4 < r, x_3 < 4\} \subset \mathbb{R}^2.$$

On this domain we solve the axially symmetric mean curvature flow problem in the level set formulation with the initial function

$$u_0(r, x_3) = \min\left\{\sqrt{(r_0 - r)^2 + x_3^2} - r_1, \sqrt{(r_0 + r)^2 + x_3^2} - r_1\right\}. \qquad (7.5)$$

Here r_0 is the distance of the center of the torus to the origin and r_1 is the inner radius of the torus. To give an idea of how the solution surface u_ϵ looks, in Figure 7.1 we show the initial value and the solution at a later time. Mean curvature flow in the level set form moves *all* level sets by their mean curvature. And this is shown in the series of level set plots in Figure 7.2. We have chosen $r_0 = 1$ and $r_1 = 0.75$. Picking the level line which corresponds to $u_\epsilon = 0$, one observes the merging of the torus thus continuing the evolution from Figure 3.9. If one follows the motion of a thin torus, then obviously the torus shrinks and vanishes, thus continuing the motion from Figure 3.8.

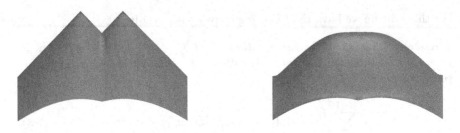

Fig. 7.1. Initial surface u_0 and solution u_ϵ at the 100-th time step (upside down) for Example 7.5.

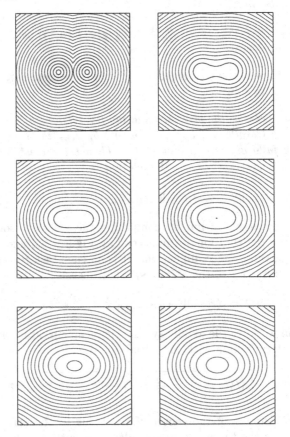

Fig. 7.2. Level sets $u_\epsilon = const$ for the time steps $0, 50, 100, 150, 200, 250$ for Example 7.5.

7.2 Anisotropic Mean Curvature Flow of Level Sets

The equation for anisotropic mean curvature flow of all level sets

$$\Gamma(t) = \left\{ x \in \mathbb{R}^{n+1} \,|\, u(x,t) = c \right\}$$

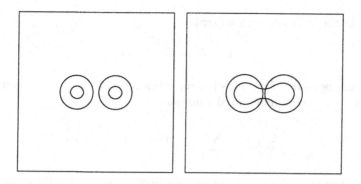

Fig. 7.3. Shrinking torus (left) and merging torus (right) for Example 7.5.

of the scalar function u is

$$u_t - |\nabla u| \nabla \cdot \gamma_p(\nabla u) = 0 \tag{7.6}$$

or equivalently,

$$\frac{u_t}{|\nabla u|} - \sum_{j,k=1}^{n} \gamma_{p_j p_k}(\nabla u) u_{x_j x_k} = 0 \quad \text{in } \mathbb{R}^{n+1} \times (0, \infty). \tag{7.7}$$

We only give the idea of the regularization of this degenerate partial differential equation. The numerical treatment of the regularised problem is then similar to that of the isotropic problem. We regularise the equation by using an extension of the anisotropy to $n + 2$ space dimensions. We assume that there exists an admissible weight function $\bar{\gamma}$,

$$\bar{\gamma} = \bar{\gamma}(p_1, \ldots, p_{n+1}, p_{n+2})$$

such that

$$\bar{\gamma}(p_1, \ldots, p_{n+1}, 0) = \gamma(p_1, \ldots, p_{n+1}).$$

In the following we denote this extension again by γ. Rather than treating (7.6) we introduce for a (small) positive parameter ϵ the regularised problem

$$\frac{u_{\epsilon t}}{\sqrt{\epsilon^2 + |\nabla u_\epsilon|^2}} - \nabla \cdot \gamma_p(\nabla u_\epsilon, -\epsilon) = 0 \tag{7.8}$$

or written explicitly,

$$\frac{u_{\epsilon t}}{\sqrt{\epsilon^2 + |\nabla u_\epsilon|^2}} - \sum_{j,k=1}^{n+1} \gamma_{p_j p_k}(\nabla u_\epsilon, -\epsilon) u_{\epsilon x_j x_k} = 0.$$

We consider this differential equation on $\Omega \times (0, T)$, where $\Omega \subset \mathbb{R}^{n+1}$ is a bounded smooth domain and $T > 0$ is some final time. Furthermore, we prescribe the initial and boundary conditions

$$u_\epsilon(x, t) = u_0(x), \quad (x, t) \in \overline{\Omega} \times \{0\} \cup \partial\Omega \times [0, T].$$

Similar to the isotropic case we scale

$$U = \frac{u_\epsilon}{\epsilon},$$

use the homogeneity of degree -1 of the second derivatives of γ and transform (7.8) into the partial differential equation

$$\frac{U_t}{\sqrt{1 + |\nabla U|^2}} - \sum_{j,k=1}^{n} \gamma_{p_j p_k}(\nabla U, -1) U_{x_j x_k} = 0.$$

This is the equation for the anisotropic mean curvature flow of the graph given by U. Thus we may use our results for the approximation of this problem in a similar way as in the isotropic case.

References

1. ALMGREN, R. Variational algorithms and pattern formation in dendritic solidification. *J. Comput. Phys.* **106**, 337–354 (1993).
2. ANGENENT, S., GURTIN, M. Multiphase thermomechanics with interfacial structure, 2. Evolution of an isothermal interface. *Arch. Rat. Mech. Anal.* **108** (1989), 323–391.
3. AUBERT, G., KORNPROBST, P. *Mathematical Problems in Image Processing.* Applied Mathematical Sciences **147**. Springer (2002).
4. BELLETTINI, G., PAOLINI, M. Anisotropic motion by mean curvature in the context of Finsler geometry. *Hokkaido Math. J.* **25**, 537–566, 1996.
5. BLOWEY, J. F., ELLIOTT, C. M. The Cahn–Hilliard gradient theory for phase separation with non–smooth free energy Part I: Mathematical Analysis. *European J. Applied Mathematics* **2**, 233–280 (1991).
6. BLOWEY, J. F., ELLIOTT, C. M. The Cahn–Hilliard gradient theory for phase separation with non–smooth free energy Part II: Numerical Analysis. *European J. Applied Mathematics* **3**, 147–179 (1993).
7. BRAKKE, K. A. The motion of a surface by its mean curvature. *Princeton University Press* (1978).
8. CAHN, J. W., TAYLOR, J. E. Surface motion by surface diffusion. *Acta Metall. Mater.* **42**, 1045–1063 (1994).
9. CASELLES, V., KIMMEL, R., SAPIRO, G., SBERT, C. Minimal surfaces: a geometric three dimensional segmentation approach. *Numer. Math.* **77**, (1997) 423–451.
10. CHEN, Y–G., GIGA, Y., GOTO, S. Uniqueness and existence of viscosity solutions of generalized mean curvature flow equations. *J. Diff. Geom* **33**, (1991) 749–786.
11. CHEN, Y.–G., GIGA, Y., HITAKA, Y.T., HONMA, M. A stable difference scheme for computing motion of level surfaces by the mean curvature. In KIM, D. *et al.* (eds), *Proceedings of the Global Analysis Research Center Symposium,* Seoul, Korea, (1994), pp 1–19.
12. CHOPP, D. L. Computing minimal surfaces via level set curvature flow. *J. Comp. Phys.* **106**, 77–91 (1993).

13. CRANDALL, M.G., LIONS, P.L. Convergent difference schemes for nonlinear parabolic equations and mean curvature motion. *Numer. Math.* **75**, (1996) 17–41.

14. DECKELNICK, K. Error analysis for a difference scheme approximating mean curvature flow. *Interfaces and Free Boundaries* **2**, (2000) 117–142.

15. DECKELNICK, K., DZIUK, G. Convergence of a finite element method for non-parametric mean curvature flow. *Numer. Math.* **72**, (1995) 197–222.

16. DECKELNICK, K., DZIUK, G. Discrete anisotropic curvature flow of graphs. *Math. Modelling Numer. Anal.* **33**, (1999) 1203–1222.

17. DECKELNICK, K., DZIUK, G. Error estimates for a semi implicit fully discrete finite element scheme for the mean curvature flow of graphs. *Interfaces and Free Boundaries* **2**, (2000) 341-359.

18. DECKELNICK, K., DZIUK, G. A fully discrete numerical scheme for weighted mean curvature flow. *Numer. Math.* **91**, (2002) 423–452.

19. DECKELNICK, K., DZIUK, G. Convergence of numerical schemes for the approximation of level set solutions to mean curvature flow. *M. Falcone, C. Makridakis (eds.), Numerical Methods for Viscosity Solutions and Applications, Series Adv. Math. Appl. Sciences* **59**, 77-94 (2001).

20. DECKELNICK, K., DZIUK, G. A finite element level set method for anisotropic mean curvature flow with space dependent weight. *Abschlußband SFB 256*, Bonn (2002).

21. DECKELNICK, K., ELLIOTT, C. M. Finite element error bounds for curve shrinking with prescribed normal contact to a fixed boundary. *IMA J. Numer. Anal.* **18**, 635-654 (1998).

22. DZIUK, G. An algorithm for evolutionary surfaces. *Numer. Math.* **58**, 603-611, (1991).

23. DZIUK, G. Convergence of a semi–discrete scheme for the curve shortening flow. *Math. Mod. Meth. Appl. Sc.* **4**, 589–606 (1994).

24. DZIUK, G. Numerical schemes for the mean curvature flow of graphs. *in P. Argoul, M. Frémond, Q. S. Nguyen (Eds.): IUTAM Symposium on Variations of Domains and Free-Boundary Problems in Solid Mechanics.* 63–70, Kluwer Academic Publishers, Dordrecht-Boston-London 1999

25. DZIUK, G. Discrete anisotropic curve shortening flow. *SIAM J. Numer. Anal.* **36**, 1808–1830 (1999).

26. ECKER, K., HUISKEN, G. Mean curvature evolution of entire graphs. *Ann. of Math.* **130**, (1989) 453–471.

27. ELLIOTT, C. M. Approximation of curvature dependent interface motion. *in Duff, I. S. (ed.) et al, The state of the art in numerical analysis.*, Oxford: Clarendon Press. Inst. Math. Appl. Conf. Ser., New Ser. 63, 407–440 (1997).

28. EVANS, L.C. Convergence of an algorithm for mean curvature motion. *Indiana Univ. Math. J.* **42**, (1993) 533–557.

29. EVANS, L.C., SPRUCK, J. Motion of level sets by mean curvature I. *J. Diff. Geom.* **33**, (1991) 636–681.

30. EVANS, L.C., SPRUCK, J. Motion of level sets by mean curvature II. *Trans. Am. Math. Soc.* **330**, (1992) 321–332.

31. EVANS, L.C., SPRUCK, J. Motion of level sets by mean curvature III. *J. Geom. Anal.* **2**, (1992) 121–150.

32. EVANS, L.C., SPRUCK, J. Motion of level sets by mean curvature IV. *J. Geom. Anal.* **5**, (1995) 77–114.

33. FIERRO, F., GOGLIONE, R., PAOLINI, M. Finite element minimization of curvature functionals with anisotropy. *Calcolo* 3–4, 191–210 (1994).

34. FRIED, M. Niveauflächen zur Berechnung zweidimensionaler Dendrite. *Dissertation Freiburg* 1999

35. GAGE, M. Evolving plane curves by curvature in relative geometries. *Duke Math. J.* **72**, (1993) 441–466.

36. GAGE, M., HAMILTON, R. S. The heat equation shrinking convex plane curves. *J. Diff. Geom.* **23** 69–96 (1986)

37. GIRAO, P. M. Convergence of a crystalline algorithm for the motion of a simple closed convex curve by weighted curvature. *SIAM J. Numer. Anal.* **32**, 886–899 (1995).

38. GRAYSON, M. A. The heat equation shrinks embedded plane curves to round points. *J. Diff. Geom.* **26**, 285 - 314 (1987).

39. HUISKEN, G. Flow by mean curvature of convex surfaces into spheres. *J. Diff. Geom.* **20**, 237–266 (1984).

40. HUISKEN, G. Non–parametric mean curvature evolution with boundary conditions. *J. Differential Equations* **77**, 369–378 (1989).

41. JOHNSON, C., THOMÉE, V. Error estimates for a finite element approximation of a minimal surface. *Math. Comp.* **29**, 343-349 (1975).

42. KIMURA, M. Numerical analysis of moving boundary problems using the boundary tracking method. *Japan J. Industr. Appl. Math.* **14**, 373–398 (1997).

43. LADYZHENSKAYA, O. A., SOLONNIKOV, V. A., URAL'TSEVA, N. N. Linear and quasilinear equations of parabolic type. *Amer. Math. Soc., Providence, R. I.* (1968).

44. LIEBERMAN, G. The first initial–boundary value problem for quasilinear second order parabolic equations. *Ann. Scuola Norm. Sup. Pisa* **13**, 347–387 (1986).

45. MERRIMAN, B., BENCE, J.K., OSHER, S. Motion of multiple junctions: A level set approach. *J. Comput. Phys.* **112**, (1994) 343–363.

46. NOCHETTO, R.H., VERDI, C. Convergence past singularities for a fully discrete approximation of curvature driven interfaces. *SIAM J. Numer. Anal.* **34**, 490–512 (1997).

47. OLIKER, V.I., URALTSEVA, N.N.: Evolution of nonparametric surfaces with speed depending on curvature II. The mean curvature case. *Commun. Pure Appl. Math.* **46**, No 1, 97–135 (1993).

48. OSHER, S., SETHIAN, J.A. Fronts propagating with curvature dependent speed: Algorithms based on Hamilton–Jacobi formulations. *J. Comp. Phys.* **79**, (1988) 12–49.

49. SCHMIDT, A. Computation of three dimensional dendrites with finite elements. *J. Comp. Phys.* **125**, 293-312, 1996.

50. PREUSSER, T. RUMPF, M. A level set method for anisotropic geometric diffusion in 3D image processing. *Report SFB 256 Bonn*, **37**, 2000.

51. SETHIAN, J.A. *Level set methods*. Cambridge Monographs on Applied and Computational Mathematics **3**. Cambridge University Press (1996).

52. SONER, H.M. Motion of a set by the curvature of its boundary. *J. Differ. Equations* **101**, (1993) 313–372.

53. TAYLOR, J. E. Mean curvature and weighted mean curvature. *Acta Metall. Mater.* **40**, 1475–1485 (1992)

54. WALKINGTON, N.J. Algorithms for computing motion by mean curvature. *SIAM J. Numer. Anal.* **33**, (1996) 2215–2238.

An Introduction to Algorithms for Nonlinear Optimization

Nicholas I. M. Gould[1]* and Sven Leyffer[2]

[1] Computational Science and Engineering Department, Rutherford Appleton Laboratory, Chilton, Oxfordshire, OX11 0QX, England
e-mail: n.gould@rl.ac.uk
[2] Mathematics and Computer Science Division, Argonne National Laboratory, 9700 S. Cass Avenue, Argonne, IL 60439, USA. e-mail: leyffer@mcs.anl.gov

Abstract. We provide a concise introduction to modern methods for solving nonlinear optimization problems. We consider both linesearch and trust-region methods for unconstrained minimization, interior-point methods for problems involving inequality constraints, and SQP methods for those involving equality constraints. Theoretical as well as practical aspects are emphasised. We conclude by giving a personal view of some of the most significant papers in the area, and a brief guide to on-line resources.

Introduction

The solution of nonlinear optimization problems–that is the minimization or maximization of an objective function involving unknown parameters/variables in which the variables may be restricted by constraints–is one of the core components of computational mathematics. Nature (and man) loves to optimize, and the world is far from linear. In his book on Applied Mathematics, the eminent mathematician Gil Strang opines that optimization, along with the solution of systems of linear equations, and of (ordinary and partial) differential equations, is one of the three cornerstones of modern applied mathematics. It is strange, then, that despite fathering many of the pioneers in the area, the United Kingdom (in particular, higher education) has turned away from the subject in favour of Strang's other key areas. Witness the numbers of lectures in the EPSRC summer school over the past ten years (broadly 3 series of lectures on numerical linear algebra, 4 on ODEs, 17 on PDEs, 4 in other areas, but *none* in optimization in Summer Schools V-XI) and the relative paucity of undergraduate or postgraduate courses in the area.

It is timely, therefore, to be given the opportunity to be able to review developments in nonlinear optimization. The past 10 years have shown an incredible growth in the power and applicability of optimization techniques, fueled in part by the "interior-point revolution" of the late 1980s. We have purposely chosen not to consider linear or discrete problems, nor to describe important applications such as optimal control, partially because there are

* This work was supported in part by the EPSRC grant GR/R46641

other experts better able to review these fields, and partly because they would all make exciting courses on their own. Indeed, the prospect of simply describing nonlinear optimization in five lectures is extremely daunting, and each of the subjects we shall describe could easily fill the whole course! Of course, there is a strong cross-fertilisation of ideas from discrete to linear to nonlinear optimization, just as there are strong influences both to and from other branches of numerical analysis and computational mathematics.

This article is partitioned in broadly the same way as the course on which it is based. Optimality conditions play a vital role in optimization, both in the identification of optima, and in the design of algorithms to find them. We consider these in Section 1. Sections 2 and 3 are concerned with the two main techniques for solving unconstrained optimization problems. Although it can be argued that such problems arise relatively infrequently in practice (nonlinear fitting being a vital exception), the underlying linesearch and trust-region ideas are so important that it is best to understand them first in their simplest setting. The remaining two sections cover the problems we really wish to solve, those involving constraints. We purposely consider inequality constraints (alone) in one and equality constraints (alone) in the other, since then the key ideas may be developed without the complication of treating both kinds of constraints at once. Of course, real methods cope with both, and suitable algorithms will be hybrids of the methods we have considered.

We make no apologies for mixing theory in with algorithms, since (most) good algorithms have good theoretical underpinnings. So as not to disturb the development in the main text, the proofs of stated theorems have been relegated to Appendix C. In addition, we do not provide citations in the main text, but have devoted Appendix A to an annotated bibliography of what we consider to be essential references in nonlinear optimization. Such a list is, by its nature, selective, but we believe that the given references form a corpus of seminal work in the area, which should be read by any student interested in pursuing a career in optimization.

Before we start, we feel that one key development during the last five years has done more to promote the use of optimization than possibly any other. This is NEOS, the Network Enabled Optimization Server, at Argonne National Laboratory and Northwestern University in Chicago, see http://www-neos.mcs.anl.gov/neos . Here, users are able to submit problems for remote solution, without charge, by a large (and expanding) collection of the world's best optimization solvers, many of them being only available otherwise commercially. Further details of what may be found on the World-Wide-Web are given in Appendix B.

1 Optimality Conditions and Why They Are Important

1.1 Optimization Problems

As we have said optimization is concerned with the minimization or maximization of an objective function, say, $f(x)$. Since

$$\text{maximum } f(x) = - \text{ minimum } (-f(x))$$

there is no loss in generality in concentrating in this article on minimization–throughout, minimization will take place with respect to an n-vector, x, of real unknowns. A bit of terminology here: the smallest value of f gives its *minimum*, while any (there may be more than one) corresponding values of x are a *minimizer*.

There are a number of important subclasses of optimization problems. The simplest is *unconstrained minimization*, where we aim to

$$\underset{x \in \mathbb{R}^n}{\text{minimize}} \, f(x)$$

where the *objective function* $f \colon \mathbb{R}^n \longrightarrow \mathbb{R}$. One level up is *equality constrained minimization*, where now we try to

$$\underset{x \in \mathbb{R}^n}{\text{minimize}} \, f(x) \text{ subject to } c(x) = 0$$

where the *constraints* $c \colon \mathbb{R}^n \longrightarrow \mathbb{R}^m$. For consistency we shall assume that $m \le n$, for otherwise it is unlikely (but not impossible) that there is an x that satisfies all of the equality constraints. Another important problem is *inequality constrained minimization*, in which we aim to

$$\underset{x \in \mathbb{R}^n}{\text{minimize}} \, f(x) \text{ subject to } c(x) \ge 0$$

where $c \colon \mathbb{R}^n \longrightarrow \mathbb{R}^m$ and now m may be larger than n. The most general problem involves both equality and inequality constraints–some inequalities may have upper as well as lower bounds–and may be further sub-classified depending on the nature of the constraints. For instance, some of the $c_i(x)$ may be linear (that is $c_i(x) = a_i^T x - b_i$ for some vector a_i and scalar b_i), some may be simple bounds on individual components of x (for example, $c_i(x) = x_i$), or some may result from a network ("flow in = flow out").

1.2 Notation

It is convenient to introduce our most common notation and terminology at the outset. Suppose that $f(x)$ is at least twice continuously differentiable ($f \in \mathcal{C}^2$). We let $\nabla_x f(x)$ denote the vector of first partial derivatives, whose i-th component is $\partial f(x)/\partial x_i$. Similarly, the i,j-th component of the (symmetric) matrix $\nabla_{xx} f(x)$ is the second partial derivative $\partial^2 f(x)/\partial x_i \partial x_j$. We also write the usual Euclidean inner product between two p-vectors u and v as $\langle u, v \rangle \overset{\text{def}}{=}$ $\sum_{i=1}^{p} u_i v_i$ (and mention, for those who care, that some but not all of what we have to say remains true in more general Hilbert spaces!). We denote the set of points for which all the constraints are satisfied as $\mathcal{C}$, and say that any $x \in \mathcal{C}$ (resp. $x \notin \mathcal{C}$) is *feasible* (resp. *infeasible*).

With this in mind we define the *gradient* and *Hessian* (matrix) of the objective function f to be $g(x) \overset{\text{def}}{=} \nabla_x f(x)$ and $H(x) \overset{\text{def}}{=} \nabla_{xx} f(x)$, respectively. Likewise, the gradient and Hessian of the i-th constraint are $a_i(x) \overset{\text{def}}{=} \nabla_x c_i(x)$ and $H_i(x) \overset{\text{def}}{=} \nabla_{xx} c_i(x)$. The *Jacobian* (matrix) is

$$A(x) \overset{\text{def}}{=} \nabla_x c(x) \equiv \begin{pmatrix} a_1^T(x) \\ \cdots \\ a_m^T(x) \end{pmatrix}.$$

Finally, if y is a vector (of so-called *Lagrange multipliers*), the *Lagrangian* (function) is

$$\ell(x, y) \overset{\text{def}}{=} f(x) - \langle y, c(x) \rangle,$$

while its gradient and Hessian with respect to x are, respectively,

$$g(x, y) \overset{\text{def}}{=} \nabla_x \ell(x, y) \equiv g(x) - \sum_{i=1}^{m} y_i a_i(x) \equiv g(x) - A^T(x)y \text{ and}$$

$$H(x, y) \overset{\text{def}}{=} \nabla_{xx} \ell(x, y) \equiv H(x) - \sum_{i=1}^{m} y_i H_i(x).$$

One last piece of notation: e_i is the i-th unit vector, while e is the vector of ones, and I is the (appropriately dimensioned) identity matrix.

1.3 Lipschitz Continuity and Taylor's Theorem

It might be argued that those who understand Taylor's theorem and have a basic grasp of linear algebra have all the tools they need to study continuous optimization–of course, this leaves aside all the beautiful mathematics needed to fully appreciate optimization in abstract settings, yet another future EP-SRC summer school course, we hope!

Taylor's theorem(s) can most easily be stated for functions with Lipschitz continuous derivatives. Let $\mathcal{X}$ and $\mathcal{Y}$ be open sets, let $F : \mathcal{X} \to \mathcal{Y}$, and let $\| \cdot \|_{\mathcal{X}}$ and $\| \cdot \|_{\mathcal{Y}}$ be norms on $\mathcal{X}$ and $\mathcal{Y}$ respectively. Then F is *Lipschitz continuous at* $x \in \mathcal{X}$ if there exists a function $\gamma(x)$ such that

$$\|F(z) - F(x)\|_{\mathcal{Y}} \leq \gamma(x)\|z - x\|_{\mathcal{X}}$$

for all $z \in \mathcal{X}$. Moreover F is *Lipschitz continuous throughout/in* $\mathcal{X}$ if there exists a constant γ such that

$$\|F(z) - F(x)\|_{\mathcal{Y}} \leq \gamma\|z - x\|_{\mathcal{X}}$$

for all x and $z \in \mathcal{X}$. Lipschitz continuity relates (either locally or globally) the changes that occur in F to those that are permitted in x.

Armed with this, we have the following *Taylor* approximation results. The first suggests how good (or bad) a first-order (linear) or second-order (quadratic) Taylor series approximation to a scalar-valued function may be.

Theorem 1.1. *Let S be an open subset of $\mathbb{R}^n$, and suppose $f : S \to \mathbb{R}$ is continuously differentiable throughout S. Suppose further that $g(x)$ is Lipschitz continuous at x, with Lipschitz constant $\gamma^L(x)$ in some appropriate vector norm. Then, if the segment $x + \theta s \in S$ for all $\theta \in [0, 1]$,*

$$|f(x + s) - m^L(x + s)| \leq \tfrac{1}{2}\gamma^L(x)\|s\|^2, \quad \text{where}$$

$$m^L(x + s) = f(x) + \langle g(x), s \rangle.$$

If f is twice continuously differentiable throughout S and $H(x)$ is Lipschitz continuous at x, with Lipschitz constant $\gamma^Q(x)$,

$$|f(x + s) - m^Q(x + s)| \leq \tfrac{1}{6}\gamma^Q(x)\|s\|^3, \quad \text{where}$$

$$m^Q(x + s) = f(x) + \langle g(x), s \rangle + \tfrac{1}{2}\langle s, H(x)s \rangle.$$

The second result compares how bad a first-order Taylor series approximation to a vector valued function might be.

Theorem 1.2. *Let S be an open subset of $\mathbb{R}^n$, and suppose $F : S \to \mathbb{R}^m$ is continuously differentiable throughout S. Suppose further that $\nabla_x F(x)$ is Lipschitz continuous at x, with Lipschitz constant $\gamma^L(x)$ in some appropriate vector norm and its induced matrix norm. Then, if the segment $x + \theta s \in S$ for all $\theta \in [0, 1]$,*

$$\|F(x + s) - M^L(x + s)\| \leq \tfrac{1}{2}\gamma^L(x)\|s\|^2,$$

where

$$M^L(x + s) = F(x) + \nabla_x F(x)s.$$

1.4 Optimality Conditions

Now is the time to come clean. It is very, very difficult to say anything about the solutions to the optimization problems given in Section 1.1. This is almost entirely because we are considering very general problems, for which there may be many local, often non-global, minimizers. There are two possible ways around this. We might choose to restrict the class of problems we allow, so that all local minimizers are global. But since this would rule out the vast majority of nonlinear problems that arise in practice, we instead choose to lower our sights, and only aim for local minimizers–there are methods that offer some guarantee of global optimality, but to date they are really restricted to small or very specially structured problems.

Formally, we still need to define what we mean by a local minimizer. A feasible point x_* is a *local* minimizer of $f(x)$ if there is an open neighbourhood $\mathcal{N}$ of x_* such that $f(x_*) \leq f(x)$ for all $x \in \mathcal{C} \bigcap \mathcal{N}$. If there is an open neighbourhood $\mathcal{N}$ of x_* such that $f(x_*) < f(x)$ for all $x \neq x_* \in \mathcal{C} \bigcap \mathcal{N}$, it is *isolated.*

While such definitions agree with our intuition, they are of very little use in themselves. What we really need are optimality conditions. Optimality conditions are useful for three reasons. Firstly, the provide a means of guaranteeing that a candidate solution is indeed (locally) optimal–these are the so-called *sufficient conditions*. Secondly, they indicate when a point is not optimal–these are the *necessary conditions*. Finally they guide us in the design of algorithms, since lack of optimality indicates when we may improve our objective. We now give details.

1.5   Optimality Conditions for Unconstrained Minimization

We first consider what we might deduce if we were fortunate enough to have found a local minimizer of $f(x)$. The following two results provide first- and second-order necessary optimality conditions (respectively).

Theorem 1.3. *Suppose that $f \in C^1$, and that x_* is a local minimizer of $f(x)$. Then*

$$g(x_*) = 0.$$

Theorem 1.4. *Suppose that $f \in C^2$, and that x_* is a local minimizer of $f(x)$. Then $g(x_*) = 0$ and $H(x_*)$ is positive semi-definite, that is*

$$\langle s, H(x_*)s \rangle \geq 0 \ \text{ for all } \ s \in \mathbb{R}^n.$$

But what if we have found a point that satisfies the above conditions? Is it a local minimizer? Yes, an isolated one, provided the following second-order sufficient optimality conditions are satisfied.

Theorem 1.5. *Suppose that $f \in C^2$, that x_* satisfies the condition $g(x_*) = 0$, and that additionally $H(x_*)$ is positive definite, that is*

$$\langle s, H(x_*)s \rangle > 0 \ \text{ for all } \ s \neq 0 \in \mathbb{R}^n.$$

Then x_* is an isolated local minimizer of f.

Notice how slim is the difference between these necessary and sufficient conditions.

1.6   Optimality Conditions for Constrained Minimization

When constraints are present, things get more complicated. In particular, the geometry of the feasible region at (or near) to a minimizer plays a very subtle role. Consider a suspected minimizer x_*. We shall say that a constraint is *active* at x_* if and only if $c_i(x_*) = 0$. By necessity, equality constraints will be active, while determining which (if any) of the inequalities is active is probably the overriding concern in constrained optimization.

In order to say anything about optimality, it is unfortunately necessary to rule out "nasty" local minimizers such as cusps on the constraint boundary. This requires that we have to ask that so-called *constraint qualifications* hold–essentially these say that linear approximations to the constraints characterize all feasible perturbations about x_* and that perturbations which keep strongly active constraints strongly active (a *strongly* active constraint is one that will still be active if the data, and hence minimizer, is slightly perturbed) are completely characterized by their corresponding linearizations being forced to be active. Fortunately, such assumptions are automatically satisfied if the constraints are linear, or if the constraints that are active have independent gradients, and may actually be guaranteed in far weaker circumstances than these.

1.6.1   Optimality Conditions for Equality-Constrained Minimization Given constraint qualifications, first- and second-order necessary optimality conditions for problems involving equality constraints are (respectively) as follows.

Theorem 1.6. *Suppose that f, $c \in C^1$, and that x_* is a local minimizer of $f(x)$ subject to $c(x) = 0$. Then, so long as a first-order constraint qualification holds, there exists a vector of Lagrange multipliers y_* such that*

$$c(x_*) = 0 \ (primal\ feasibility)\ and$$
$$g(x_*) - A^T(x_*)y_* = 0 \ (dual\ feasibility).$$

Theorem 1.7. *Suppose that f, $c \in C^2$, and that x_* is a local minimizer of $f(x)$ subject to $c(x) = 0$. Then, provided that first- and second-order constraint qualifications hold, there exists a vector of Lagrange multipliers y_* such that*

$$\langle s, H(x_*, y_*)s \rangle \geq 0 \ for\ all\ s \in \mathcal{N} \tag{1.1}$$

where

$$\mathcal{N} = \{s \in \mathbb{R}^n \mid A(x_*)s = 0\}.$$

Notice that there are two first-order optimality requirements: primal feasibility (the constraints are satisfied), and dual feasibility (the gradient of the objective function is expressible as a linear combination of the gradients of the constraints). It is not hard to anticipate that, just as in the unconstrained case, sufficient conditions occur when the requirement (1.1) is strengthened to $\langle s, H(x_*, y_*)s \rangle > 0$ for all $s \in \mathcal{N}$.

**1.6.2   Optimality Conditions for Inequality-Constrained Minimiz-
ation** Finally, when the problem involves inequality constraints, it is easy
to imagine that only the constraints that are active at x_* play a role–the
inactive constraints play no part in defining the minimizer–and indeed this is
so. First- and second-order necessary optimality conditions are (respectively)
as follows.

Theorem 1.8. *Suppose that f, $c \in C^1$, and that x_* is a local minimizer of
$f(x)$ subject to $c(x) \geq 0$. Then, provided that a first-order constraint qualification holds, there exists a vector of Lagrange multipliers y_* such that*

$$c(x_*) \geq 0 \text{ (primal feasibility)},$$
$$g(x_*) - A^T(x_*)y_* = 0 \text{ and } y_* \geq 0 \text{ (dual feasibility) and} \qquad (1.2)$$
$$c_i(x_*)[y_*]_i = 0 \text{ (complementary slackness)}.$$

Theorem 1.9. *Suppose that f, $c \in C^2$, and that x_* is a local minimizer of
$f(x)$ subject to $c(x) \geq 0$. Then, provided that first- and second-order constraint qualifications hold, there exists a vector of Lagrange multipliers y_* for
which primal/dual feasibility and complementary slackness requirements hold
as well as*

$$\langle s, H(x_*, y_*)s \rangle \geq 0 \text{ for all } s \in \mathcal{N}_+$$

where

$$\mathcal{N}_+ = \left\{ s \in \mathbb{R}^n \;\middle|\; \begin{array}{l} \langle s, a_i(x_*) \rangle = 0 \text{ if } c_i(x_*) = 0 \text{ \& } [y_*]_i > 0 \text{ and} \\ \langle s, a_i(x_*) \rangle \geq 0 \text{ if } c_i(x_*) = 0 \text{ \& } [y_*]_i = 0 \end{array} \right\}. \qquad (1.3)$$

See how dual feasibility now imposes an extra requirement, that the Lagrange
multipliers be non-negative, while as expected there is an additional (complementary slackness) assumption that inactive constraints necessarily have
zero Lagrange multipliers. Also notice that $\mathcal{N}_+$, the set over which the Hessian of the Lagrangian is required to be positive semi-definite, may now be
the intersection of a linear manifold and a cone, a particularly unpleasant set
to work with.

The by-now obvious sufficient conditions also hold:

Theorem 1.10. *Suppose that f, $c \in C^2$, and that x_* and a vector of Lagrange multipliers y_* satisfy (1.2) and*

$$\langle s, H(x_*, y_*)s \rangle > 0$$

*for all s in the set $\mathcal{N}_+$ given in (1.3). Then x_* is an isolated local minimizer
of $f(x)$ subject to $c(x) \geq 0$.*

2   Linesearch Methods for Unconstrained Optimization

In this and the next sections, we shall concentrate on the unconstrained minimization problem,

$$\underset{x \in \mathbb{R}^n}{\text{minimize}} \, f(x),$$

where the objective function $f \colon \mathbb{R}^n \longrightarrow \mathbb{R}$. We shall assume that $f \in C^1$ (sometimes C^2) with Lipschitz continuous derivatives. Often in practice this assumption is violated, but nonetheless the methods converge (perhaps by good fortune) regardless.

Despite knowing how to characterise local minimizers of our problem, in practice it is rather unusual for us to be able to provide or compute an explicit minimizer. Instead, we would normally expect to fall back on a suitable iterative process. An *iteration* is simply a procedure whereby a sequence of points

$$\{x_k\}, \quad k = 1, 2, \ldots$$

is generated, starting from some initial "guess" x_0, with the overall aim of ensuring that (a subsequence) of the $\{x_k\}$ has favourable limiting properties. These might include that any limit generated satisfies first-order or, even better, second-order necessary optimality conditions.

Notice that we will not be able to guarantee that our iteration will converge to a global minimizer unless we know that f obeys very strong conditions, nor regrettably in general that any limit point is even a local minimizer (unless by chance it happens to satisfy second-order sufficiency conditions). What we normally do try to ensure is that, at the very least, the iteration is *globally* convergent, that is that (for at least) a subsequence of iterates $\{g(x_k)\}$ converges to zero. And our hope is that such a sequence converges at a reasonably fast asymptotic rate. These two preoccupations lie at the heart of computational optimization.

For brevity, in what follows, we shall write $f_k = f(x_k)$, $g_k = g(x_k)$ and $H_k = H(x_k)$.

2.1   Linesearch Methods

Generically, linesearch methods work as follows. Firstly, a *search direction* p_k is calculated from x_k. This direction is required to be a *descent direction*, i.e.,

$$\langle p_k, g_k \rangle < 0 \ \text{ if } \ g_k \neq 0,$$

so that, for small steps along p_k, Taylor's theorem (Theorem 1.1) guarantees that the objective function may be reduced. Secondly, a suitable *steplength* $\alpha_k > 0$ is calculated so that

$$f(x_k + \alpha_k p_k) < f_k.$$

The computation of α_k is the *linesearch*, and may itself be an iteration. Finally, given both search direction and steplength, the iteration concludes by setting

$$x_{k+1} = x_k + \alpha_k p_k.$$

Such a scheme sounds both natural and simple. But as with most simple ideas, it needs to be refined somewhat in order to become a viable technique. What might go wrong? Firstly, consider the example in Figure 2.1.

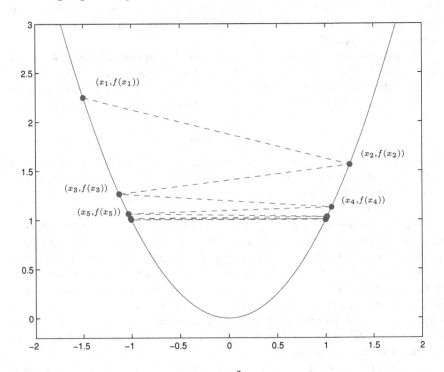

Fig. 2.1. The objective function $f(x) = x^2$ and the iterates $x_{k+1} = x_k + \alpha_k p_k$ generated by the descent directions $p_k = (-1)^{k+1}$ and steps $\alpha_k = 2 + 3/2^{k+1}$ from $x_0 = 2$.

Here the search direction gives a descent direction, and the iterates oscillate from one side of the minimizer to the other. Unfortunately, the decrease per iteration is ultimately so small that the iterates converge to the pair ± 1, neither of which is a stationary point. What has gone wrong? Simply the steps are too long relative to the amount of objective-function decrease that they provide.

Is this the only kind of failure? Unfortunately, no. For consider the example in Figure 2.2.

Now the iterates approach the minimizer from one side, but the stepsizes are so small that each iterate falls woefully short of the minimizer, and ultimately converge to the non-stationary value 1.

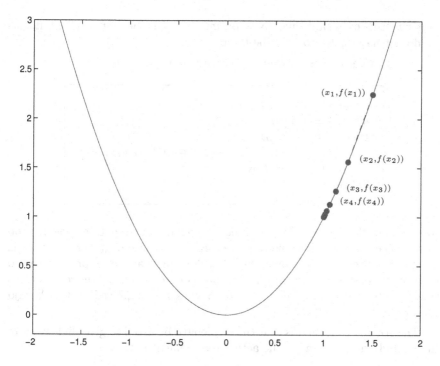

Fig. 2.2. The objective function $f(x) = x^2$ and the iterates $x_{k+1} = x_k + \alpha_k p_k$ generated by the descent directions $p_k = -1$ and steps $\alpha_k = 1/2^{k+1}$ from $x_0 = 2$.

So now we can see that a simple-minded linesearch method can fail if the linesearch allows steps that are either too long or too short relative to the amount of decrease that might be obtained with a well-chosen step.

2.2   Practical Linesearch Methods

In the early days, it was often suggested that α_k should be chosen to minimize $f(x_k + \alpha p_k)$. This is known as an *exact* linesearch. In most cases, exact linesearches prove to be both very expensive–they are essentially univariate minimizations–and most definitely not cost effective, and are consequently rarely used nowadays.

Modern linesearch methods prefer to use *inexact* linesearches, which are guaranteed to pick steps that are neither too long nor too short. In addition, they aim to pick a "useful" initial "guess" for each stepsize so as to ensure fast asymptotic convergence–we will return to this when we discuss Newton's method. The main contenders amongst the many possible inexact linesearches are the so-called "backtracking- Armijo" and the "Armijo-Goldstein" varieties. The former are extremely easy to implement, and form the backbone of most Newton-like linesearch methods. The latter are particularly impor-

tant when using secant quasi-Newton methods (see Section 2.5.3), but alas
we do not have space to describe them here.

Here is a basic *backtracking* linesearch to find α_k:

Given $\alpha_{\text{init}} > 0$ (e.g., $\alpha_{\text{init}} = 1$),
let $\alpha^{(0)} = \alpha_{\text{init}}$ and $l = 0$.
Until $f(x_k + \alpha^{(l)}p_k) < f_k$
   set $\alpha^{(l+1)} = \tau\alpha^{(l)}$, where $\tau \in (0,1)$ (e.g., $\tau = \frac{1}{2}$)
   and increase l by 1.
Set $\alpha_k = \alpha^{(l)}$.

Notice that the backtracking strategy prevents the step from getting too
small, since the first allowable value stepsize of the form $\alpha_{\text{init}}\tau^i$, $i = 0, 1, \ldots$ is
accepted. However, as it stands, there is still no mechanism for preventing too
large steps relative to decrease in f. What is needed is a tighter requirement
than simply that $f(x_k + \alpha^{(l)}p_k) < f_k$. Such a role is played by the Armijo
condition.

The *Armijo condition* is that the steplength be asked to give slightly more
than simply decrease in f. The actual requirement is that

$$f(x_k + \alpha_k p_k) \le f(x_k) + \alpha_k\beta\langle p_k, g_k\rangle$$

for some $\beta \in (0,1)$ (e.g., $\beta = 0.1$ or even $\beta = 0.0001$)–this requirement is
often said to give *sufficient decrease*. Observe that, since $\langle p_k, g_k\rangle < 0$, the
longer the step, the larger the required decrease in f. The range of permitted
values for the stepsize is illustrated in Figure 2.3.

The Armijo condition may then be inserted into our previous backtracking
scheme to give the aptly-named *Backtracking-Armijo* linesearch:

Given $\alpha_{\text{init}} > 0$ (e.g., $\alpha_{\text{init}} = 1$),
let $\alpha^{(0)} = \alpha_{\text{init}}$ and $l = 0$.
Until $f(x_k + \alpha^{(l)}p_k) \le f(x_k) + \alpha^{(l)}\beta\langle p_k, g_k\rangle$
   set $\alpha^{(l+1)} = \tau\alpha^{(l)}$, where $\tau \in (0,1)$ (e.g., $\tau = \frac{1}{2}$)
   and increase l by 1.
Set $\alpha_k = \alpha^{(l)}$.

Of course, it is one thing to provide likely-sounding rules to control stepsize
selection, but another to be sure that they have the desired effect. Indeed,
can we even be sure that there are points which satisfy the Armijo condition?
Yes, for we have

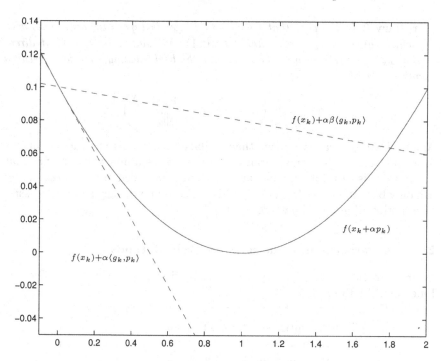

Fig. 2.3. A steplength of anything up to 1.8 is permitted for this example, in the case where $\beta = 0.2$.

Theorem 2.1. *Suppose that $f \in C^1$, that $g(x)$ is Lipschitz continuous with Lipschitz constant $\gamma(x)$, that $\beta \in (0,1)$ and that p is a descent direction at x. Then the Armijo condition*

$$f(x + \alpha p) \le f(x) + \alpha \beta \langle p, g(x) \rangle$$

is satisfied for all $\alpha \in [0, \alpha_{\max(x,p)}]$, where

$$\alpha_{\max}(x,p) = \frac{2(\beta - 1)\langle p, g(x) \rangle}{\gamma(x)\|p\|_2^2}.$$

Note that since $\gamma(x)$ is rarely known, the theorem does not provide a recipe for computing $\alpha_{\max}(x,p)$, merely a guarantee that there is such a suitable value. The numerator in $\alpha_{\max}(x,p)$ corresponds to the slope and the denominator to the curvature term. It can be interpreted as follows: If the curvature term is large, then the admissible range of α is small. Similarly, if the projected gradient along the search direction is large, then the range of admissible α is larger.

It then follows that the Backtracking-Armijo linesearch can be guaranteed to terminate with a suitably modest stepsize.

Corollary 2.1. *Suppose that $f \in C^1$, that $g(x)$ is Lipschitz continuous with Lipschitz constant γ_k at x_k, that $\beta \in (0,1)$ and that p_k is a descent direction at x_k. Then the stepsize generated by the backtracking-Armijo linesearch terminates with*

$$\alpha_k \geq \min \left(\alpha_{\text{init}}, \frac{2\tau(\beta - 1)\langle p_k, g_k \rangle}{\gamma_k \|p_k\|_2^2} \right).$$

Again, since γ_k is rarely known, the corollary does not give a practical means for computing α_k, just an assurance that there is a suitable value. Notice that the stepsize is certainly not too large, since it is bounded above by $\alpha_{\max}$, and can only be small when $\langle p, g(x) \rangle / \|p\|_2^2$ is. This will be the key to the successful termination of generic linesearch methods.

2.3   Convergence of Generic Linesearch Methods

In order to tie all of the above together, we first need to state our Generic Linesearch Method:

Given an initial guess x_0, let $k = 0$
Until convergence:
    Find a descent direction p_k at x_k.
    Compute a stepsize α_k using a
        backtracking-Armijo linesearch along p_k.
    Set $x_{k+1} = x_k + \alpha_k p_k$, and increase k by 1.

It is then quite straightforward to apply Corollary 2.1 to deduce the following very general convergence result.

Theorem 2.2. *Suppose that $f \in C^1$ and that g is Lipschitz continuous on $\mathbb{R}^n$. Then, for the iterates generated by the Generic Linesearch Method,*

either

$$g_l = 0 \text{ for some } l \geq 0$$

or

$$\lim_{k \to \infty} f_k = -\infty$$

or

$$\lim_{k \to \infty} \min \left(|\langle p_k, g_k \rangle|, \frac{|\langle p_k, g_k \rangle|}{\|p_k\|_2} \right) = 0.$$

In words, either we find a first-order stationary point in a finite number of iterations, or we encounter a sequence of iterates for which the objective function is unbounded from below, or the slope (or a normalized slope) along the search direction converges to zero. While the first two of these possibilities are straightforward and acceptable consequences, the latter is perhaps not.

For one thing, it certainly does not say that the gradient converges to zero, that is the iterates may not ultimately be first-order critical, since it might equally occur if the search direction and gradient tend to be mutually orthogonal. Thus we see that simply requiring that p_k be a descent direction is not a sufficiently demanding requirement. We will return to this shortly, but first we consider *the* archetypical globally convergent algorithm, the method of steepest descent.

2.4   Method of Steepest Descent

We have just seen that the Generic Linesearch Method may not succeed if the search direction becomes orthogonal to the gradient. Is there a direction for which this is impossible? Yes, when the search direction is the descent direction

$$p_k = -g_k,$$

the so-called *steepest-descent* direction–the epithet is appropriate since this direction solves the problem

$$\underset{p \in \mathbb{R}^n}{\text{minimize}}\ m_k^L(x_k + p) \overset{\text{def}}{=} f_k + \langle p, g_k \rangle \text{ subject to } \|p\|_2 = \|g_k\|_2,$$

and thus gives the greatest possible reduction in a first-order model of the objective function for a step whose length is specified. Global convergence follows immediately from Theorem 2.2.

Theorem 2.3. *Suppose that $f \in C^1$ and that g is Lipschitz continuous on $\mathbb{R}^n$. Then, for the iterates generated by the Generic Linesearch Method using the steepest-descent direction,*

either

$$g_l = 0 \text{ for some } l \geq 0$$

or

$$\lim_{k \to \infty} f_k = -\infty$$

or

$$\lim_{k \to \infty} g_k = 0.$$

As we mentioned above, this theorem suggests that steepest descent really is the archetypical globally convergent method, and in practice many other methods resort to steepest descent when they run into trouble. However, the method is not scale invariant, as re-scaling variables can lead to widely different "steepest-descent" directions. Even worse, as we can see in Figure 2.4, convergence may be (and actually almost always is) very slow in theory, while numerically convergence sometimes does not occur at all as the iteration stagnates. In practice, steepest-descent is all but worthless in most

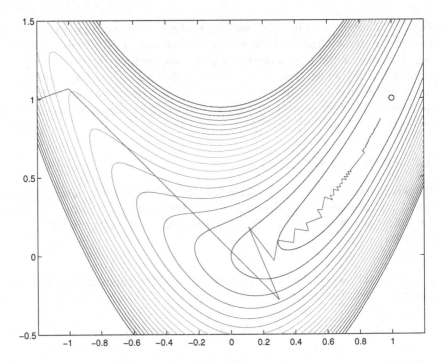

Fig. 2.4. Contours for the objective function $f(x,y) = 10(y - x^2)^2 + (x - 1)^2$, and the iterates generated by the Generic Linesearch steepest-descent method.

cases. The figure exhibits quite typical behaviour in which the iterates repeatedly oscillate from one side of a objective function "valley" to the other. All of these phenomena may be attributed to a lack of attention to problem curvature when building the search direction. We now turn to methods that try to avoid this defect.

2.5   More General Descent Methods

2.5.1   Newton and Newton-Like Methods Let B_k be a symmetric, positive definite matrix. Then it is trivial to show that the search direction p_k for which

$$B_k p_k = -g_k$$

is a descent direction. In fact, this direction solves the direction-finding problem

$$\underset{p \in \mathbb{R}^n}{\text{minimize}}\ m_k^Q(x_k + p) \overset{\text{def}}{=} f_k + \langle p, g_k \rangle + \tfrac{1}{2}\langle p, B_k p \rangle, \qquad (2.1)$$

where $m_k^Q(x_k + p)$ is a quadratic approximation to the objective function at x_k.

Of particular interest is the possibility that $B_k = H_k$, for in this case $m_k^Q(x_k + p)$ gives a second-order Taylor's approximation to $f(x_k + p)$. The resulting direction for which

$$H_k p_k = -g_k$$

is known as the *Newton* direction, and any method which uses it is a Newton method. But notice that the Newton direction is only guaranteed to be useful in a linesearch context if the Hessian H_k is positive definite, for otherwise p_k might turn out to be an ascent direction.

It is also worth saying that while one can motivate such Newton-like methods from the prospective of minimizing a local second-order model of the objective function, one could equally argue that they aim to find a zero of a local first-order model

$$g(x_k + p) \approx g_k + B_k p_k$$

of its gradient. So long as B_k remains "sufficiently" positive definite, we can make precisely the same claims for these second-order methods as for those based on steepest descent.

Indeed, one can regard such methods as "scaled" steepest descent, but they have the advantage that they can be made scale invariant for suitable B_k, and crucially, as we see in Figure 2.5, their convergence is often significantly faster than steepest descent. In particular, in the case of the Newton direction, the Generic Linesearch method will usually converge very rapidly indeed.

Theorem 2.4. *Suppose that $f \in C^1$ and that g is Lipschitz continuous on $\mathbb{R}^n$. Then, for the iterates generated by the Generic Linesearch Method using the Newton or Newton-like direction,*

either

$$g_l = 0 \text{ for some } l \geq 0$$

or

$$\lim_{k \to \infty} f_k = -\infty$$

or

$$\lim_{k \to \infty} g_k = 0$$

provided that the eigenvalues of B_k are uniformly bounded and bounded away from zero.

Theorem 2.5. *Suppose that $f \in C^2$ and that H is Lipschitz continuous on $\mathbb{R}^n$. Then suppose that the iterates generated by the Generic Linesearch Method with $\alpha_{\text{init}} = 1$ and $\beta < \frac{1}{2}$, in which the search direction is chosen to be the Newton direction $p_k = -H_k^{-1} g_k$ whenever H_k is positive definite, has a limit point x_* for which $H(x_*)$ is positive definite. Then*

(i) $\alpha_k = 1$ for all sufficiently large k,
(ii) the entire sequence $\{x_k\}$ converges to x_*, and
(iii) the rate is Q-quadratic, i.e, there is a constant $\kappa \geq 0$.

$$\lim_{k \to \infty} \frac{\|x_{k+1} - x_*\|_2}{\|x_k - x_*\|_2^2} \leq \kappa.$$

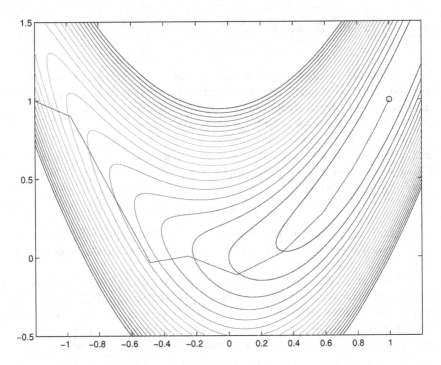

Fig. 2.5. Contours for the objective function $f(x,y) = 10(y - x^2)^2 + (x - 1)^2$, and the iterates generated by the Generic Linesearch Newton method.

2.5.2 Modified-Newton Methods Of course, away from a local minimizer there is no reason to believe that H_k will be positive definite, so precautions need to be taken to ensure that Newton and Newton-like linesearch methods, for which B_k is (or is close to) H_k, satisfy the assumptions of the global convergence Theorem 2.4. If H_k is indefinite, it is usual to solve instead

$$(H_k + M_k)p_k = -g_k,$$

where M_k is chosen so that $H_k + M_k$ is "sufficiently" positive definite and $M_k = 0$ when H_k is itself "sufficiently" positive definite. This may be achieved in a number of ways.

Firstly, if H_k has the spectral (that is eigenvector-eigenvalue) decomposition $H_k = Q_k D_k Q_k^T$, then M_k may be chosen so that

$$H_k + M_k = Q_k \max(\epsilon I, |D_k|)Q_k^T$$

for some "small" ϵ". This will shift all the insufficiently positive eigenvalues by as little as possible as is needed to make the overall matrix positive definite. While such a decomposition may be too expensive to compute for larger problems, a second, cheaper alternative is to find (or estimate) the smallest (necessarily real!) eigenvalue, $\lambda_{\min}(H_k)$, of H_k, and to set

$$M_k = \max(0, \epsilon - \lambda_{\min}(H_k))I$$

so as to shift *all* the eigenvalues by just enough as to make the smallest "sufficiently" positive. While this is often tried in practice, in the worst case it may have the effect of over-emphasising one large, negative eigenvalue at the expense of the remaining small, positive ones, and in producing a direction which is essentially steepest descent. Finally, a good compromise is instead to attempt a Cholesky factorization of H_k, and to alter the generated factors if there is evidence that the factorization will otherwise fail. There are a number of so-called *Modified Cholesky* factorizations, each of which will obtain

$$H_k + M_k = L_k L_k^T,$$

where M_k is zero for sufficiently positive-definite H_k, and "not-unreasonably large" in all other cases.

2.5.3   Quasi-Newton Methods

It was fashionable in the 1960s and 1970s to attempts to build suitable approximations B_k to the Hessian, H_k. Activity in this area has subsequently died down, possibly because people started to realize that computing exact second derivatives was not as onerous as they had previously contended, but these techniques are still of interest particularly when gradients are awkward to obtain (such as when the function values are simply given as the result of some other, perhaps hidden, computation). There are broadly two classes of what may be called quasi-Newton methods.

The first are simply based on estimating columns of H_k by *finite differences*. For example, we might use the approximation

$$(H_k)e_i \approx h^{-1}(g(x_k + he_i) - g_k) \stackrel{\text{def}}{=} (B_k)e_i$$

for some "small" scalar $h > 0$. The difficulty here is in choosing an appropriate value for h: too large a value gives inaccurate approximations, while a too small one leads to large numerical cancellation errors.

The second sort of quasi-Newton methods are known as *secant approximations*, and try to ensure the *secant condition*

$$B_{k+1}s_k = y_k, \text{ where } s_k = x_{k+1} - x_k \text{ and } y_k = g_{k+1} - g_k,$$

that would be true if $H(x)$ were constant, is satisfied. The secant condition gives a lot of flexibility, and among the many methods that have been discovered, the *Symmetric Rank-1* method, for which

$$B_{k+1} = B_k + \frac{(y_k - B_k s_k)(y_k - B_k s_k)^T}{\langle s_k, y_k - B_k s_k \rangle},$$

and the *BFGS* method, for which

$$B_{k+1} = B_k + \frac{y_k y_k^T}{\langle s_k, y_k \rangle} - \frac{B_k s_k s_k^T B_k}{\langle s_k, B_k s_k \rangle}$$

are the best known (and generally the best). Note that the former may give
indefinite approximations (or even fail), while the latter is guaranteed to
generate symmetric and positive definite matrices so long as B_0 is positive
definite and $\langle s_k, y_k \rangle > 0$ (the last condition may be ensured by an appropriate "Goldstein" linesearch). Since both of these secant methods are based on
low-rank updates, it is possible to keep the per-iteration linear algebraic requirements at a more modest level for such methods than is generally possible
with Newton or finite-difference methods.

2.5.4 Conjugate-Gradient and Truncated-Newton Methods

And
what if the problem is large and matrix factorization is out of the question?
We have already considered (and rejected) steepest-descent methods. Is there
something between the simplicity of steepest descent and the power (but
expense) of Newton-like methods? Fortunately, the answer is yes.

Suppose that instead of solving (2.1), we instead find our search direction
as

$$p_k = \text{(approximate)} \arg \min_{p \in \mathbb{R}^n} q(p) = f_k + \langle p, g_k \rangle + \tfrac{1}{2}\langle p, B_k p \rangle,$$

where we assume that B_k is positive definite–the key word here is *approximate*. Suppose that instead of minimizing q over all $p \in \mathbb{R}^n$, we restrict
p to lie in a (much) smaller subspace–of course if we do this we will not
(likely) obtain the optimal value of q, but we might hope to obtain a good
approximation with considerably less effort.

Let $D^i = (d^0 : \cdots : d^{i-1})$ be any collection of i vectors, let

$$\mathcal{D}^i = \{p \mid p = D^i p_d \text{ for some } p_d \in \mathbb{R}^i\}$$

be the subspace spanned by D^i, and suppose that we choose to pick

$$p^i = \arg \min_{p \in \mathcal{D}^i} q(p).$$

Then immediately $D^{i\,T} g^i = 0$, where $g^i = B_k p^i + g_k$ is the gradient of q at
p^i. More revealingly, since $p^{i-1} \in \mathcal{D}^i$, it follows that $p^i = p^{i-1} + D^i p_d^i$, where

$$p_d^i = \arg \min_{p_d \in \mathbb{R}^i}\langle p_d, D^{i\,T} g^{i-1}\rangle + \tfrac{1}{2}\langle p_d, D^{i\,T} B_k D^i p_d\rangle$$
$$= -(D^{i\,T} B_k D^i)^{-1} D^{i\,T} g^{i-1} = -\langle d^{i-1}, g^{i-1}\rangle(D^{i\,T} B_k D^i)^{-1} e^i.$$

Hence

$$p^i = p^{i-1} - \langle d^{i-1}, g^{i-1}\rangle D^i (D^{i\,T} B_k D^i)^{-1} e^i. \qquad (2.2)$$

All of this is true regardless of D^i. But now suppose that the members of $\mathcal{D}^i$
are B_k-*conjugate*, that is to say that $\langle d_i, B_k d_j\rangle = 0$ for all $i \neq j$. If this is so
(2.2) becomes

$$p^i = p^{i-1} + \alpha^{i-1} d^{i-1}, \quad \text{where } \alpha^{i-1} = -\frac{\langle d^{i-1}, g^{i-1}\rangle}{\langle d^{i-1}, B_k d^{i-1}\rangle}. \qquad (2.3)$$

Thus so long as we can generate B_k-conjugate vectors, we can build up successively improving approximations to the minimize of q by solving a sequence of *one-dimensional* minimization problems–the relationship (2.3) may be interpreted as finding α^{i-1} to minimize $q(p^{i-1} + \alpha d^{i-1})$. But can we find suitable B_k-conjugate vectors?

Surprisingly perhaps, yes, it is easy. Since g^i is independent of $\mathcal{D}^i$, let

$$d^i = -g^i + \sum_{j=0}^{i-1} \beta^{ij} d^j$$

for some unknown β^{ij}. Then elementary manipulation (and a cool head) shows that if we choose β^{ij} so that d^i is B-conjugate to $\mathcal{D}^i$, we obtain the wonderful result that

$$\beta^{ij} = 0 \text{ for } j < i - 1, \text{ and } \beta^{i\,i-1} \equiv \beta^{i-1} = \frac{\|g_i\|_2^2}{\|g_{i-1}\|_2^2}.$$

That is, almost all of the β^{ij} are zero! Summing all of this up, we arrive at the method of *conjugate gradients* (CG):

> Given $p^0 = 0$, set $g^0 = g_k$, $d^0 = -g_k$ and $i = 0$.
> Until g^i is "small", iterate:
> $\quad \alpha^i = \|g^i\|_2^2/\langle d^i, Bd^i \rangle$
> $\quad p^{i+1} = p^i + \alpha^i d^i$
> $\quad g^{i+1} = g^i + \alpha^i B_k d^i$
> $\quad \beta^i = \|g^{i+1}\|_2^2/\|g^i\|_2^2$
> $\quad d^{i+1} = -g^{i+1} + \beta^i d^i$
> and increase i by 1.

Important features are that $\langle d^j, g^{i+1} \rangle = 0$ and $\langle g^j, g^{i+1} \rangle = 0$ for all $j = 0, \ldots, i$, and most particularly that $\langle p^i, g_k \rangle \leq \langle p^{i-1}, g_k \rangle < 0$ for $i = 1, \ldots, n$, from which we see that *any* $p_k = p^i$ is a descent direction.

In practice the above conjugate gradient iteration may be seen to offer a compromise between the steepest-descent direction (stopping when $i = 1$) and a Newton (-like) direction (stopping when $i = n$). For this reason, using such a curtailed conjugate gradient step within a linesearch (or trust-region) framework is often known as a *truncated*-Newton method. Frequently the size of g^i relative to g_k is used as a stopping criteria, a particularly popular rule being to stop the conjugate-gradient iteration when

$$\|g^i\| \leq \min(\|g_k\|^\omega, \eta)\|g_k\|,$$

where η and $\omega \in (0, 1)$, since then a faster-than-linear asymptotic convergence rate may be achieved if $B_k = H_k$.

3 Trust-Region Methods for Unconstrained Optimization

In this section, we continue to concentrate on the unconstrained minimization problem, and shall as before assume that the objective function is C^1 (sometimes C^2) with Lipschitz continuous derivatives.

3.1 Linesearch Versus Trust-Region Methods

One might view linesearch methods as naturally "optimistic". Fairly arbitrary search directions are permitted–essentially 50% of all possible directions give descent from a given point–while unruly behaviour is held in check via the linesearch. There is, however, another possibility, that more control is taken when choosing the search direction, with the hope that this will then lead to a higher probability that the (full) step really is useful for reducing the objective. This naturally "conservative" approach is the basis of trust-region methods.

As we have seen, linesearch methods pick a descent direction p_k, then pick a stepsize α_k to "reduce" $f(x_k + \alpha p_k)$ and finally accept $x_{k+1} = x_k + \alpha_k p_k$. *Trust-region* methods, by contrast, pick the overall step s_k to reduce a "model" of $f(x_k + s)$, and accept $x_{k+1} = x_k + s_k$ if the decrease predicted by the model is realised by $f(x_k + s_k)$. Since there is no guarantee that this will always be so, the fall-back mechanism is to set $x_{k+1} = x_k$, and to "refine" the model when the existing model produces a poor step. Thus, while a linesearch method recovers from a poor step by retreating along a parametric (usually linear) curve, a trust-region method recovers by reconsidering the whole step-finding procedure.

3.2 Trust-Region Models

It is natural to build a model of $f(x_k + s)$ by considering Taylor series approximations. Of particular interest are the *linear* model

$$m_k^L(s) = f_k + \langle s, g_k \rangle,$$

and the *quadratic* model

$$m_k^Q(s) = f_k + \langle s, g_k \rangle + \tfrac{1}{2} \langle s, B_k s \rangle,$$

where B_k is a symmetric approximation to the local Hessian matrix H_k. However, such models are far from perfect. In particular, the models are unlikely to resemble $f(x_k + s)$ if s is large. More seriously, the models may themselves be unbounded from below so that any attempts to minimize them may result in a large step. This defect will always occur for the linear model (unless $g_k = 0$), and also for the quadratic model if B_k is indefinite (and possibly if B_k is only positive semi-definite). Thus simply using a Taylor-series model is fraught with danger.

There is, fortunately, a simple and effective way around this conundrum. The idea is to prevent the model $m_k(s)$ from being unboundedness by imposing a *trust-region* constraint

$$\|s\| \leq \Delta_k,$$

for some "suitable" scalar *radius* $\Delta_k > 0$, on the step. This is a natural idea, since we know from Theorem 1.1 that we can improve the approximation error $|f(x_k + s) - m_k(s)|$ by restricting the allowable step. Thus our *trust-region subproblem* is to

$$\text{approximately} \underset{s\in\mathbb{R}^n}{\text{minimize}} \; m_k(s) \text{ subject to } \|s\| \leq \Delta_k,$$

and we shall choose s_k as approximate solution of this problem. In theory, it does not depend on which norm $\| \cdot \|$ we use (at least, in finite-dimensional spaces), but in practice it might!

For simplicity, we shall concentrate on the second-order (Newton-like) model

$$m_k(s) = m_k^Q(s) = f_k + \langle s, g_k \rangle + \tfrac{1}{2}\langle s, B_k s \rangle$$

and any (consistent) trust-region norm $\| \cdot \|$ for which

$$\kappa_s \| \cdot \| \leq \| \cdot \|_2 \leq \kappa_l \| \cdot \|$$

for some $\kappa_l \geq \kappa_s > 0$. Notice that the gradient of $m_k(s)$ at $s = 0$ coincides with the gradient of f at x_k, and also, unlike for linesearch methods, $B_k = H_k$ is always allowed. The vast majority of models use the ℓ_1, ℓ_2 or ℓ_∞ norms on $\mathbb{R}^n$, and for these we have $\| \cdot \|_2 \leq \| \cdot \|_2 \leq \| \cdot \|_2$ (obviously!!), $n^{-\frac{1}{2}} \| \cdot \|_1 \leq \| \cdot \|_2 \leq \| \cdot \|_1$ and $\| \cdot \|_\infty \leq \| \cdot \|_2 \leq n \| \cdot \|_\infty$.

3.3 Basic Trust-Region Method

Having decided upon a suitable model, we now turn to the trust-region algorithm itself. As we have suggested, we shall choose to "accept" $x_{k+1} = x_k + s_k$ whenever (a reasonable fraction of) the predicted model decrease $f_k - m_k(s_k)$ is realized by the actual decrease $f_k - f(x_k + s_k)$. We measure this by computing the ratio

$$\rho_k = \frac{f_k - f(x_k + s_k)}{f_k - m_k(s_k)}$$

of actual to predicted decrease, and accepting the trust-region step when ρ_k is not unacceptably smaller than 1.0. If the ratio is close to (or larger than) 1.0, there is good reason to believe that future step computations may well benefit from an increase in the trust-region radius, so we allow a radius increase in this case. If, by contrast, there is poor agreement between the actual and predicted decrease (and particularly, if f actually increases), the current step is poor and should be rejected. In this case, we reduce the trust-region radius to encourage a more suitable step at the next iteration.

We may summarize the basic trust-region method as follows:

Given $k = 0$, $\Delta_0 > 0$ and x_0, until "convergence" do:
Build the second-order model $m(s)$ of $f(x_k + s)$.
"Solve" the trust-region subproblem to find s_k
for which $m(s_k)$ "$<$" f_k and $\|s_k\| \le \Delta_k$, and define

$$\rho_k = \frac{f_k - f(x_k + s_k)}{f_k - m_k(s_k)}.$$

If $\rho_k \ge \eta_v$ [*very successful*] $\boxed{0 < \eta_v < 1}$

    set $x_{k+1} = x_k + s_k$ and $\Delta_{k+1} = \gamma_i \Delta_k$. $\boxed{\gamma_i \ge 1}$

Otherwise if $\rho_k \ge \eta_s$ then [*successful*] $\boxed{0 < \eta_s \le \eta_v < 1}$

    set $x_{k+1} = x_k + s_k$ and $\Delta_{k+1} = \Delta_k$.

Otherwise [*unsuccessful*]

    set $x_{k+1} = x_k$ and $\Delta_{k+1} = \gamma_d \Delta_k$. $\boxed{0 < \gamma_d < 1}$

Increase k by 1.

Reasonable values might be $\eta_v = 0.9$ or 0.99, $\eta_s = 0.1$ or 0.01, $\gamma_i = 2$, and $\gamma_d = 0.5$. In practice, these parameters might even be allowed to vary (within reasonable limits) from iteration to iteration. In particular, there would seem to be little justification in increasing the trust region radius following a very successful iteration unless $\|s_k\| \approx \Delta_k$, or in decreasing the radius by less than is required to "cut off" an unsuccessful s_k.

In practice, the trust-region radius is *not* increased for a very successful iterations, if the step is much shorter, say less than half the trust-region radius. There exist various schemes for choosing an initial trust-region radius. However, if the problem is well scaled, then $\Delta_0 = O(1)$ is reasonable. Poor scaling can affect the performance of trust-region methods. In practice it often suffices that the variables of the (scaled) problem have roughly the same order of magnitude.

It remains for us to decide what we mean by "solving" the trust-region subproblem. We shall see in Section 3.5 that (at least in the ℓ_2-trust-region norm case) it is possible to find the (global) solution to the subproblem. However, since this may result in a considerable amount of work, we first seek "minimal" conditions under which we can guarantee convergence of the above algorithm to a first-order critical point.

We have already seen that steepest-descent linesearch methods have very powerful (theoretical) convergence properties. The same is true in the trust-region framework. Formally, at the very least, we shall require that we achieve as much reduction in the model as we would from an iteration of steepest descent. That is, if we define the *Cauchy* point as $s_k^C = -\alpha_k^C g_k$, where

$$\alpha_k^C = \arg\min_{\alpha>0} m_k(-\alpha g_k) \text{ subject to } \alpha\|g_k\| \le \Delta_k$$
$$= \arg\min_{0<\alpha\le\Delta_k/\|g_k\|} m_k(-\alpha g_k),$$

we shall require that our step s_k satisfies

$$m_k(s_k) \le m_k(s_k^C) \text{ and } \|s_k\| \le \Delta_k. \tag{3.1}$$

Notice that the Cauchy point is extremely easy to find, since it merely requires that we minimize the quadratic model along a line segment. In practice, we shall hope to–and can–do far better than this, but for now (3.1) suffices.

Figure 3.1 illustrates the trust-region problem in four different situations. The contours of the original function are shown as dotted lines, while the contours of the trust-region model appear as solid lines with the ℓ_2 trust-region ball in bold. Clockwise from top left, the plots depict the following situations: first, a quadratic model with positive definite Hessian, next a linear model about the same point, the third plot shows a quadratic model with indefinite Hessian and the final plot is a quadratic model with positive definite Hessian whose minimizers lies outside the trust-region.

We now examine the convergence of this trust-region method.

3.4   Basic Convergence of Trust-Region Methods

The first thing to note is that we can guarantee a reasonable reduction in the model at the Cauchy point.

Theorem 3.1. *If $m_k(s)$ is the second-order model and s_k^C is its Cauchy point within the trust-region $\|s\| \le \Delta_k$, then*

$$f_k - m_k(s_k^C) \ge \tfrac{1}{2}\|g_k\|_2 \min\left[\frac{\|g_k\|_2}{1+\|B_k\|_2}, \kappa_s\Delta_k\right].$$

Observe that the guaranteed reduction depends on how large the current gradient is, and is also affected by the size of both the trust-region radius and the (inverse) of the Hessian.

Since our algorithm requires that the step does at least as well as the Cauchy point, we then have the following immediate corollary.

Corollary 3.1. *If $m_k(s)$ is the second-order model, and s_k is an improvement on the Cauchy point within the trust-region $\|s\| \le \Delta_k$,*

$$f_k - m_k(s_k) \ge \tfrac{1}{2}\|g_k\|_2 \min\left[\frac{\|g_k\|_2}{1+\|B_k\|_2}, \kappa_s\Delta_k\right].$$

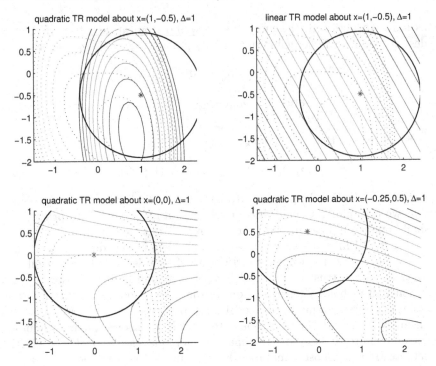

Fig. 3.1. Trust-region models of $f(x) = x_1^4 + x_1 x_2 + (1+x_2)^2$ about different points.

This is a typical trust-region result, in that it relates the model reduction to a measure of the distance to optimality, in this case measured in terms of the norm of the gradient.

It is also necessary to say something about how much the model and the objective can vary. Since we are using a second-order model for which the first-two terms are exactly those from the Taylor's approximation, it is not difficult to believe that the difference between model and function will vary like the square of the norm of s_k, and indeed this is so.

Lemma 3.1. *Suppose that $f \in C^2$, and that the true and model Hessians satisfy the bounds $\|H_k\|_2 \leq \kappa_h$ and $\|B_k\|_2 \leq \kappa_b$ for all k and some $\kappa_h \geq 1$ and $\kappa_b \geq 0$. Then*
$$|f(x_k + s_k) - m_k(s_k)| \leq \kappa_d \Delta_k^2,$$
where $\kappa_d = \frac{1}{2} \kappa_l^2 (\kappa_h + \kappa_b)$, for all k.

Actually the result is slightly weaker than necessary since, for our purposes, we have chosen to replace $\|s_k\|$ by its (trust-region) bound Δ_k.

Armed with these bounds, we now arrive at a crucial result, namely that it will always be possible to make progress from a non-optimal point ($g_k \neq 0$).

Lemma 3.2. *Suppose that $f \in C^2$, that the true and model Hessians satisfy the bounds $\|H_k\|_2 \leq \kappa_h$ and $\|B_k\|_2 \leq \kappa_b$ for all k and some $\kappa_h \geq 1$ and $\kappa_b \geq 0$, and that $\kappa_d = \frac{1}{2}\kappa_l^2(\kappa_h + \kappa_b)$. Suppose furthermore that $g_k \neq 0$ and that*

$$\Delta_k \leq \|g_k\|_2 \min\left(\frac{1}{\kappa_s(\kappa_h + \kappa_b)}, \frac{\kappa_s(1 - \eta_v)}{2\kappa_d}\right).$$

Then iteration k is very successful and

$$\Delta_{k+1} \geq \Delta_k.$$

This result is fairly intuitive, since when the radius shrinks the model looks more and more like its first-order Taylor expansion (provided B_k is bounded) and thus ultimately there must be good local agreement between the model and objective functions.

The next result is a variation on its predecessor, and says that the radius is uniformly bounded away from zero if the same is true of the sequence of gradients, that is the radius will not shrink to zero at non-optimal points.

Lemma 3.3. *Suppose that $f \in C^2$, that the true and model Hessians satisfy the bounds $\|H_k\|_2 \leq \kappa_h$ and $\|B_k\|_2 \leq \kappa_b$ for all k and some $\kappa_h \geq 1$ and $\kappa_b \geq 0$, and that $\kappa_d = \frac{1}{2}\kappa_l^2(\kappa_h + \kappa_b)$. Suppose furthermore that there exists a constant $\epsilon > 0$ such that $\|g_k\|_2 \geq \epsilon$ for all k. Then*

$$\Delta_k \geq \kappa_\epsilon \stackrel{\text{def}}{=} \epsilon \gamma_d \min\left(\frac{1}{\kappa_s(\kappa_h + \kappa_b)}, \frac{\kappa_s(1 - \eta_v)}{2\kappa_d}\right)$$

for all k.

We may then deduce that if there are only a finite number of successful iterations, the iterates must be first-order optimal after the last of these.

Lemma 3.4. *Suppose that $f \in C^2$, and that both the true and model Hessians remain bounded for all k. Suppose furthermore that there are only finitely many successful iterations. Then $x_k = x_*$ for all sufficiently large k and $g(x_*) = 0$.*

Having ruled out this special (and highly unlikely) case, we then have our first global convergence result, namely that otherwise there is at least one sequence of gradients that converge to zero.

Theorem 3.2. *Suppose that $f \in C^2$, and that both the true and model Hessians remain bounded for all k. Then either*

$$g_l = 0 \text{ for some } l \geq 0$$

or

$$\lim_{k \to \infty} f_k = -\infty$$

or

$$\liminf_{k \to \infty} \|g_k\| = 0.$$

Is this all we can show? Is it possible for a second subsequence of gradients to stay bounded away from zero? Fortunately, no.

Corollary 3.2. *Suppose that $f \in C^2$, and that both the true and model Hessians remain bounded for all k. Then either*

$$g_l = 0 \text{ for some } l \geq 0$$

or

$$\lim_{k \to \infty} f_k = -\infty$$

or

$$\lim_{k \to \infty} g_k = 0.$$

Thus we have the highly-satisfying result that the gradients of the sequence $\{x_k\}$ generated by our algorithm converge to, or are all ultimately, zero. This does not mean that a subsequence of $\{x_k\}$ itself converges, but if it does, the limit is first-order critical.

It is also possible to show that an enhanced version of our basic algorithm converges to second-order critical points. To do so, we need to ensure that the Hessian of the model converges to that of the objective (as would obviously be the case if $B_k = H_k$), and that the step s_k has a significant component along the eigenvector corresponding to the most negative eigenvalue of B_k (if any). It is also possible to show that if $B_k = H_k$, if $\{x_k\}$ has a limit x_* for which $H(x_*)$ is positive definite, and if s_k is chosen to

$$\operatorname*{minimize}_{s \in \mathbb{R}^n} m_k(s) \text{ subject to } \|s\| \leq \Delta_k, \tag{3.2}$$

the step Δ_k stays bounded away from zero, and thus the iteration ultimately becomes Newton's method (c.f. (2.1)).

In conclusion, we have seen that trust-region methods have a very rich underlying convergence theory. But so much for theory. We now turn to the outstanding practical issue, namely how one might hope to find a suitable step s_k. We will consider two possibilities, one that aims to get a very good approximation to (3.2), and a second, perhaps less ambitious method that is more geared towards large-scale computation.

3.5   Solving the Trust-Region Subproblem

For brevity, we will temporarily drop the iteration subscript, and consider the problem of

$$(\text{approximately}) \operatorname*{minimize}_{s \in \mathbb{R}^n} q(s) \equiv \langle s, g \rangle + \tfrac{1}{2} \langle s, Bs \rangle \text{ subject to } \|s\| \leq \Delta.$$
$$\tag{3.3}$$

As we have already mentioned, our aim is to find s_* so that

$$q(s_*) \leq q(s^c) \text{ and } \|s_*\| \leq \Delta,$$

where s^c is the Cauchy point. We shall consider two approaches in this section. The first aims to solve (3.3) exactly, in which case our trust-region method will be akin to a Newton-like method. The second aims for an approximate solution using a conjugate-gradient like method. For simplicity, we shall only consider the ℓ_2-trust region $\|s\| \leq \Delta$, mainly because there are very powerful methods in this case, but of course other norms are possible and are sometimes preferred in practice.

3.5.1 Solving the ℓ_2-Norm Trust-Region Subproblem
There is a really powerful solution characterisation result for the ℓ_2-norm trust-region subproblem.

Theorem 3.3. *Any* global *minimizer s_* of $q(s)$ subject to $\|s\|_2 \leq \Delta$ satisfies the equation*

$$(B + \lambda_* I)s_* = -g,$$

where $B + \lambda_* I$ is positive semi-definite, $\lambda_* \geq 0$ and $\lambda_*(\|s_*\|_2 - \Delta) = 0$. If $B + \lambda_* I$ is positive definite, s_* is unique.

This result is extraordinary as it is very unusual to be able to give necessary and sufficient *global* optimality conditions for a nonconvex optimization problem (that is, a problem which might have a number of local minimizers). Even more extraordinary is the fact that the necessary and sufficient conditions are identical. But most crucially, these optimality conditions also suggest how we might solve the problem.

There are two cases to consider. If B is positive definite and the solution s to

$$Bs = -g \qquad (3.4)$$

satisfies $\|s\|_2 \leq \Delta$, then it immediately follows that $s_* = s$ ($\lambda_* = 0$ in Theorem 3.3)–this potential solution may simply be checked by seeing if B has Cholesky factors and, if so, using these factors to solve (3.4) $Bs = -g$ and subsequently evaluate $\|s\|_2$. Otherwise, either B is positive definite but the solution to (3.4) satisfies $\|s\|_2 > \Delta$ or B is singular or indefinite. In these cases, Theorem 3.3 then says that s_* satisfies

$$(B + \lambda I)s = -g \text{ and } \langle s, s \rangle = \Delta^2, \qquad (3.5)$$

which is a *nonlinear* (quadratic) system of algebraic equations in the $n + 1$ unknowns s and λ. Thus, we now concentrate on methods for solving this system.

Suppose B has the spectral decomposition

$$B = U^T \Lambda U;$$

here U is a matrix of (orthonormal) eigenvectors while the diagonal matrix Λ is made up of eigenvalues $\lambda_1 \leq \lambda_2 \leq \ldots \leq \lambda_n$. Theorem 3.3 requires that

$B + \lambda I$ be positive semi-definite, and so the solution (s, λ) to (3.5) that we seek necessarily satisfies $\lambda \geq -\lambda_1$. The first part of (3.5) enables us to write s explicitly in terms of λ, that is

$$s(\lambda) = -(B + \lambda I)^{-1} g;$$

we will temporarily disregard the possibility that the theorem permits a singular $B + \lambda I$. Notice that once we have found λ,

$$(B + \lambda I)s = -g \tag{3.6}$$

is a linear system. In this case, we may substitute $s(\lambda)$ into the second part of (3.5) to reveal that

$$\psi(\lambda) \overset{\text{def}}{=} \|s(\lambda)\|_2^2 = \|U^T(\Lambda + \lambda I)^{-1} U g\|_2^2 = \sum_{i=1}^{n} \frac{\gamma_i^2}{(\lambda_i + \lambda)^2} = \Delta^2, \tag{3.7}$$

where $\gamma_i = \langle e_i, Ug \rangle = \langle U^T e_i, g \rangle$. Thus to solve the trust-region subproblem, it appears that all we have to do is find a particular root of a univariate nonlinear equation.

We illustrate this in Figures 3.2–3.4.

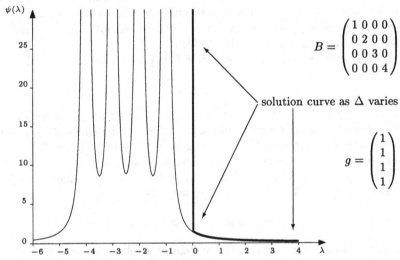

Fig. 3.2. A plot of $\psi(\lambda)$ as λ varies from -6 to 4. Note the poles at the negatives of the eigenvalues of H. The heavy curve plots λ against Δ; the vertical component corresponds to interior solutions while the remaining segment indicates boundary solutions.

The first shows a convex example (B positive definite). For Δ^2 larger than roughly 1.5, the solution to the problem lies in the interior of the trust region, and may be found directly from (3.4). When Δ is smaller than this,

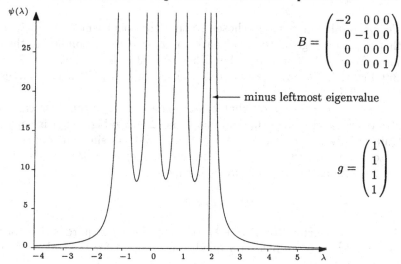

Fig. 3.3. A plot of $\psi(\lambda)$ as λ varies from -4 to 5. Again, note the poles at the negatives of the eigenvalues of H.

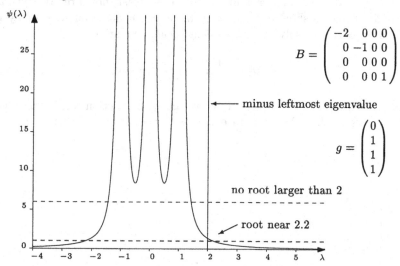

Fig. 3.4. A plot of $\psi(\lambda)$ for the modified model as λ varies from -4 to 5. Note that there is no solution with to the equation $\psi(\lambda) = \Delta^2$ with $\lambda \geq 2$ for Δ larger than roughly 1.2.

the solution lies on the boundary of the trust region, and can be found as the right-most root of (3.7). The second example is nonconvex (B indefinite). Now the solution must lie on the boundary of the trust region for all values of Δ, and again can be found as the right-most root of (3.7), to the right of $-\lambda_1$.

In both Figures 3.2 and 3.3 everything seems easy, and at least a semblance of an algorithm is obvious. But now consider the example in Figure 3.4.

This example is especially chosen so that the coefficient γ_1 in (3.7) is zero, that is g is orthogonal to the eigenvector u_1 of B corresponding to the eigenvalue $\lambda_1 = -2$. Remember that Theorem 3.3 tells us that $\lambda \geq 2 = -\lambda_1$. But Figure 3.4 shows that there is no such root of (3.7) if Δ is larger than (roughly) 1.2.

This is an example of what has become known as the *hard* case, which always arises when $\lambda_1 < 0$, $\langle u_1, g \rangle = 0$ and Δ is too big. What is happening? Quite simply, in the hard case $\lambda = -\lambda_1$ and (3.6) is a singular (but consistent) system–it is consistent precisely because $\langle u_1, g \rangle = 0$. But this system has other solutions $s + \alpha u_1$ for any α, because

$$(B + \lambda I)(s + \alpha u_1) = -g,$$

and u_1 is an eigenvector of $B + \lambda I$. The solution we require is that for which $\|s + \alpha u_1\|_2^2 = \Delta^2$, which is a quadratic equation for the unknown α, and either root suffices.

In the easy (that is not "hard") case, it remains to see how best to solve $\|s(\lambda)\|_2 = \Delta$. The answer is blunt. Don't! At least, not directly, since as the previous figures showed, $\psi(\lambda)$ is an unappealing function with many poles. It is far better to solve the equivalent *secular* equation

$$\phi(\lambda) \overset{\text{def}}{=} \frac{1}{\|s(\lambda)\|_2} - \frac{1}{\Delta} = 0,$$

as this has no poles, indeed it is an analytic function, and thus ideal for Newton's method. We illustrate the secular equation in Figure 3.5.

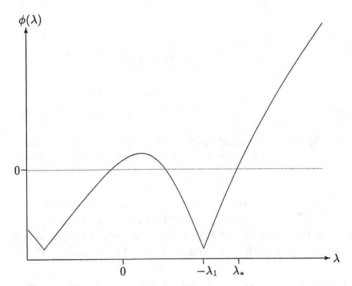

Fig. 3.5. A plot of $\phi(\lambda)$ against λ for the problem of minimizing $-\frac{1}{4}s_1^2 + \frac{1}{4}s_2^2 + \frac{1}{2}s_1 + s_2$ subject to $\|s\|_2 \leq 0.4$.

Without giving details (for these, see the appendix, page 193), Newton's method for the secular equation is as follows

---

Let $\lambda > -\lambda_1$ and $\Delta > 0$ be given.
Until "convergence" do:
    Factorize $B + \lambda I = LL^T$.
    Solve $LL^T s = -g$.
    Solve $Lw = s$.
    Replace λ by

$$\lambda + \left(\frac{\|s\|_2 - \Delta}{\Delta}\right)\left(\frac{\|s\|_2^2}{\|w\|_2^2}\right).$$

This is globally and ultimately quadratically convergent when started in the interval $[-\lambda_1, \lambda_*]$ except in the hard case, but needs to be safeguarded to make it robust for the hard and interior solution cases. Notice that the main computational cost per iteration is a Cholesky factorization of $B + \lambda I$, and while this may be reasonable for small problems, it may prove unacceptably expensive when the number of variables is large. We consider an alternative for this case next.

3.6 Solving the Large-Scale Problem

Solving the large-scale trust-region subproblem using the above method is likely out of the question in all but very special cases. The obvious alternative is to use an iterative method to approximate its solution. The simplest approximation that is consistent with our fundamental requirement that we do as least as well as we would at the Cauchy point is to use the Cauchy point itself. Of course, this is simply the steepest descent method, and thus unlikely to be a practical method. The obvious generalization is the conjugate-gradient method, since the first step of CG is in the steepest-descent direction and, as subsequent CG steps further reduce the model, any step generated by the method is allowed by our theory. However, there are a number of other issues we need to address first. In particular, what about the interaction between conjugate gradients and the trust region? And what if B is indefinite?

The conjugate-gradient method to find an approximation to a minimizer of $q(s)$ may be summarised as follows.

Given $s^0 = 0$, set $g^0 = g$, $d^0 = -g$ and $i = 0$.
Until "breakdown" or g^i "small", iterate:
    $\alpha^i = \|g^i\|_2^2 / \langle d^i, Bd^i \rangle$
    $s^{i+1} = s^i + \alpha^i d^i$
    $g^{i+1} = g^i + \alpha^i Bd^i$
    $\beta^i = \|g^{i+1}\|_2^2 / \|g^i\|_2^2$
    $d^{i+1} = -g^{i+1} + \beta^i d^i$
    and increase i by 1.

Notice that we have inserted a termination statement concerning "break-down". This is intended to cover the fatal case when $\langle d^i, Bd^i \rangle = 0$ (or, in practice, is close to zero), for which the iteration is undefined, and the non-fatal case when $\langle d^i, Bd^i \rangle < 0$ for which $q(s)$ is unbounded from below along the so-called *direction of negative curvature* d_i.

But what of the trust-region constraint? Here we have a crucial result.

Theorem 3.4. *Suppose that the conjugate gradient method is applied to min-imize $q(s)$ starting from $s^0 = 0$, and that $\langle d^i, Bd^i \rangle > 0$ for $0 \le i \le k$. Then the iterates s^j satisfy the inequalities*

$$\|s^j\|_2 < \|s^{j+1}\|_2$$

for $0 \le j \le k - 1$.

Simply put, since the norm of the approximate solution generated by the conjugate gradients increases in norm at each iteration, if there is an iteration for which $\|s^j\|_2 > \Delta$, it must be that the solution to the trust-region subproblem lies on the trust-region boundary. That is $\|s_*\|_2 = \Delta$. This then suggests that we should apply the basic conjugate-gradient method above but terminate at iteration i if either (a) $\langle d^i, Bd^i \rangle \le 0$, since this implies that $q(s)$ is unbounded along d^i, or (b) $\|s^i + \alpha^i d^i\|_2 > \Delta$, since this implies that the solution must lie on the trust-region boundary. In both cases, the simplest strategy is to stop on the boundary at $s = s^i + \alpha^\mathrm{B} d^i$, where α^B chosen as positive root of the quadratic equation

$$\|s^i + \alpha^\mathrm{B} d^i\|_2^2 = \Delta^2.$$

Crucially this s satisfies

$$q(s) \le q(s^\mathrm{C}) \quad \text{and} \quad \|s\|_2 \le \Delta$$

and thus Corollary 3.2 shows that the overall trust-region algorithm converges to a first-order critical point.

How good is this truncated conjugate-gradient strategy? In the convex case, it turns out to be very good. Indeed, no worse than half optimal!

Theorem 3.5. *Suppose that the truncated conjugate gradient method is ap-plied to approximately minimize $q(s)$ within $\|s\|_2 \le \Delta$, and that B is positive definite. Then the computed and actual solutions to the problem, s and s_*, satisfy the bound $q(s) \le \frac{1}{2} q(s_*)$.*

In the nonconvex (B_k indefinite) case, however, the strategy may be rather poor. For example, if $g = 0$ and B is indefinite, the above truncated conjugate-gradient method will terminate at $s = 0$, while the true solution lies on the trust-region boundary.

What can we do in the nonconvex case? The answer is quite involved, but one possibility is to recall that conjugate-gradients is trying to solve the overall problem by successively solving the problem over a sequence of nested subspaces. As we saw, the CG method uses B-conjugate subspaces. But there is an equivalent method, the *Lanczos* method, that uses instead orthonormal bases. Essentially this may be achieved by applying the Gram-Schmidt procedure to the CG basis $\mathcal{D}^i$ to build the equivalent basis $\mathcal{Q}^i = \{s \mid s = Q^i s_q \text{ for some } s_q \in \mathbb{R}^i\}$. It is easy to show that for this Q^i,

$$Q^{i\,T}Q^i = I \quad \text{and} \quad Q^{i\,T}BQ^i = T^i,$$

where T^i is tridiagonal, and $Q^{i\,T}g = \|g\|_2\, e_1$, and it is trivial to generate Q^i from the CG $\mathcal{D}^i$. In this case the trust-region subproblem (3.3) may be rewritten as

$$s_q^i = \arg\min_{s_q \in \mathcal{R}^i} \|g\|_2 \langle e_1, s_q\rangle + \tfrac{1}{2}\langle s_q, T^i s_q\rangle \text{ subject to } \|s_q\|_2 \leq \Delta,$$

where $s^i = Q^i s_q^i$. Since T^i is tridiagonal, $T^i + \lambda I$ has very sparse Cholesky factors, and thus we can afford to solve this problem using the earlier secular equation approach. Moreover, since we will need to solve a sequence of related problems over nested subspaces, it is easy to imagine that one can use the solution for one problem to initialize the next. In practice, since the approach is equivalent to conjugate gradients, it is best to use CG until the trust-region boundary is reached and then to switch to the Lanczos method at that stage. Such a method has turned out to be most effective in practice.

4   Interior-Point Methods for Inequality Constrained Optimization

Having given a break-neck description of methods for unconstrained minimization, we now turn our attention to the real problems of interest, namely those involving constraints. This section will focus on problems involving inequality constraints, while its successor will be concerned with equality constraints. But before we start, we need to discuss the conflicting nature of constrained optimization problems, and how we might deal with them.

Unconstrained minimization is "simple" because there is but one goal, namely to minimize the objective. This is not so for constrained minimization because there is now a conflict of requirements, the aforementioned objective minimization but at the same time a requirement of feasibility of the solution. While in some instances (such as for linear equality constraints and, to a certain extent, all inequality constraints) it may be possible to generate feasible iterates, and thus to regain the advantages of having a single goal, this is not true for general constrained optimization.

4.1 Merit Functions for Constrained Minimization

Most (but not all, see Section 5.4.3) constrained optimization techniques overcome this dichotomy by introducing a merit function to try to balance the
two conflicting requirements of minimization and feasibility. Given parameters p, a composite function $\Phi(x, p)$ is a *merit function* if (some) minimizers
of $\Phi(x, p)$ with respect to x approach those of $f(x)$ subject to the constraints
as p approaches some set $\mathcal{P}$. Thus a merit function combines both optimality
requirements into a single "artificial" objective function. In principal, it then
only remains to use the best *unconstrained* minimization methods to solve
the constrained problem. If only life were that simple!

Consider the case of equality constrained minimization, that is finding x_*
to

$$\underset{x \in \mathbb{R}^n}{\text{minimize}} f(x) \text{ subject to } c(x) = 0. \tag{4.1}$$

A suitable merit function in this case is the *quadratic penalty function*

$$\Phi(x, \mu) = f(x) + \frac{1}{2\mu} \|c(x)\|_2^2, \tag{4.2}$$

where μ is a positive scalar parameter. It is easy to believe that if μ is small
and we try to minimize $\Phi(x, \mu)$ much of the effort will be concentrated on
making the second objective term $\frac{1}{2\mu} \|c(x)\|_2^2$ small, that is in forcing $c(x)$ to
be small. But as f has a slight presence in the merit function, any remaining
energy will be diverted to making $f(x)$ small amongst all of the values for
which $c(x)$ is. Formally, it is easy to show that, under modest conditions,
some minimizers of $\Phi(x, \mu)$ converge to solutions of (4.1) as μ approaches
the set $\{0\}$ from above. Unfortunately, it is possible that $\Phi(x, \mu)$ may have
other stationary points that are not solutions of (4.1)–indeed this must be
the case if $c(x) = 0$ are inconsistent. The quadratic penalty function is but
one of many merit functions for equality constrained minimization.

4.2 The Logarithmic Barrier Function for Inequality Constraints

For the inequality constrained problem

$$\underset{x \in \mathbb{R}^n}{\text{minimize}} f(x) \text{ subject to } c(x) \geq 0 \tag{4.3}$$

the best known merit function is the *logarithmic barrier function*

$$\Phi(x, \mu) = f(x) - \mu \sum_{i=1}^{m} \log c_i(x),$$

where μ is again a positive scalar *barrier parameter*. Each logarithmic term
$- \log c_i(x)$ becomes infinite as x approaches the boundary of the i-th inequality from the feasible side, and is undefined (effectively infinite) beyond there.

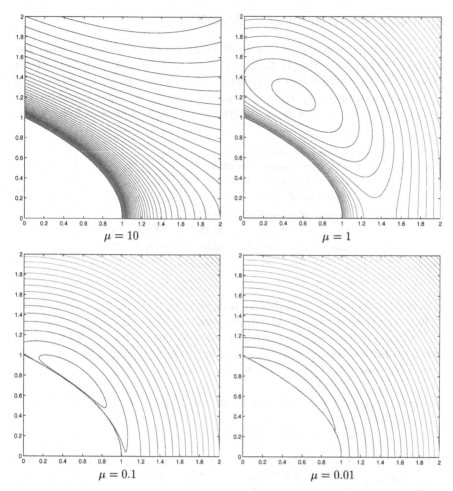

Fig. 4.1. The logarithmic barrier function for $\min x_1^2 + x_2^2$ subject to $x_1 + x_2^2 \geq 1$. The contours for $\mu = 0.01$ are visually indistinguishable from $f(x)$ for feasible points.

The size of the logarithmic term is mitigated when μ is small, and it is then possible to get close to the boundary of the feasible region before its effect is felt, any minimization effort being directed towards reducing the objective. Once again, it is easy to show that, under modest conditions, some minimizers of $\Phi(x, \mu)$ converge to solutions of (4.3) as μ approaches the set $\{0\}$ from above. And once again a possible defect is that $\Phi(x, \mu)$ may have other, useless stationary points. The contours of a typical example are shown in Figure 4.1.

4.3   A Basic Barrier-Function Algorithm

The logarithmic barrier function is different in one vital aspect from the quadratic penalty function in that it requires that there is a *strictly* interior point. If we apply the obvious sequential minimization algorithm to $\Phi(x,\mu)$, a strictly interior starting point is required, and all subsequent iterates will be strictly interior. The obvious "interior-point" algorithm is as follows.

Given $\mu_0 > 0$, set $k = 0$.
Until "convergence", iterate:
    Find x_k^s for which $c(x_k^s) > 0$.
    Starting from x_k^s, use an unconstrained
       minimization algorithm to find an
       "approximate" minimizer x_k of $\Phi(x,\mu_k)$.
    Compute $\mu_{k+1} > 0$ smaller than μ_k such
       that $\lim_{k\to\infty} \mu_{k+1} = 0$. and increase k by 1.

In practice it is common to choose $\mu_{k+1} = 0.1\mu_k$ or even $\mu_{k+1} = \mu_k^2$, while perhaps the obvious choice for a subsequent starting point is $x_{k+1}^s = x_k$.

Fortunately, as we have hinted, basic convergence for the algorithm is easily established. Recall that the *active set* $\mathcal{A}(x)$ at a point x is $\mathcal{A}(x) = \{i \mid c_i(x) = 0\}$. Then we have the following.

Theorem 4.1. *Suppose that f, $c \in C^2$, that $(y_k)_i \overset{\text{def}}{=} \mu_k/c_i(x_k)$ for $i = 1,\ldots,m$, that*

$$\|\nabla_x \Phi(x_k,\mu_k)\|_2 \leq \epsilon_k$$

where ϵ_k converges to zero as $k \to \infty$, and that x_k converges to x_* for which $\{a_i(x_*)\}_{i \in \mathcal{A}(x_*)}$ are linearly independent. Then x_* satisfies the first-order necessary optimality conditions for the problem

$$\underset{x \in \mathbb{R}^n}{\text{minimize}} \, f(x) \ \text{subject to} \ c(x) \geq 0$$

and $\{y_k\}$ converge to the associated Lagrange multipliers y_*.

Notice here how the algorithm delivers something unexpected, namely estimates of the Lagrange multipliers. Also see the role played by the linearly independence of the active constraint gradients, regrettably quite a strong constraint qualification.

4.4   Potential Difficulties

As we now know that it suffices to (approximately) minimize $\Phi(x,\mu)$, how should we proceed? As $\Phi(x,\mu)$ is a smooth function, we can immediately appeal to the methods we discussed in Sections 2 and 3. But we need to be careful. Very, very careful.

We could use a linesearch method. Of note here is the fact that the barrier function has logarithmic singularities, indeed is undefined for infeasible points. Thus it makes sense to design a specialized linesearch to cope with the singularity of the log. Alternatively, we could use a trust-region method. Here we need to be able to instantly reject candidate steps for which $c(x_k + s_k) \not> 0$. More importantly, while all (consistent) trust-region norms are equivalent, (ideally) we should "shape" the trust region for any barrier-function model to cope with the contours of the singularity. This implies that the trust-region shape may vary considerably from iteration to iteration, with its shape reflecting the eigenvalues arising from the singularity.

4.4.1 Potential Difficulty I: Ill-Conditioning of the Barrier Hessian

At the heart of both linesearch and trust-region methods is, of course, the Newton (second-order) model and related Newton direction. The computation of a Newton model/direction for the logarithmic barrier function is vital, and the resulting equations have a lot of (exploitable) structure. The gradient of the barrier function is

$$\nabla_x \Phi(x, \mu) = g(x) - \mu \sum_i a_i(x)/c_i(x) = g(x) - A^T(x)y(x) = g(x, y(x)),$$

where $y_i(x) \overset{\text{def}}{=} \mu/c_i(x)$ and $g(x, y)$ is the gradient of the Lagrangian function for (4.3). Likewise, the Hessian is

$$\nabla_{xx} \Phi(x, \mu) = H(x, y(x)) + \mu A^T(x) C^{-2}(x) A(x),$$

where $H(x, y(x)) = H(x) - \sum_{i=1}^{m} y_i(x) H_i(x)$ and $C(x) = \text{diag}(c_1(x), \ldots, c_m(x))$, the diagonal matrix whose entries are the $c_i(x)$. Thus the Newton correction s^P from x for the barrier function satisfies

$$(H(x, y(x)) + \mu A^T(x) C^{-2}(x) A(x)) s^P = -g(x, y(x)). \qquad (4.4)$$

Since $y(x) = \mu C^{-1}(x)e$, (4.4) is sometimes written as

$$\left(H(x, y(x)) + A^T(x) C^{-1}(x) Y(x) A(x) \right) s^P = -g(x, y(x)), \qquad (4.5)$$

or

$$\left(H(x, y(x)) + A^T(x) Y^2(x) A(x)/\mu \right) s^P = -g(x, y(x)), \qquad (4.6)$$

where $Y(x) = \text{diag}(y_1(x), \ldots, y_m(x))$.

This is where we need to be careful. For we have the following estimates of the eigenvalues of the barrier function as we approach a solution.

Theorem 4.2. *Suppose that the assumptions of Theorem 4.1 are satisfied, that $A_{\mathcal{A}}$ is the matrix whose rows are $\{a_i^T(x_*)\}_{i \in \mathcal{A}(x_*)}$, that $m_a = |\mathcal{A}(x_*)|$, and that x_* is non-degenerate, that is $(y_*)_i > 0$ for all $i \in \mathcal{A}(x_*)$. Then the Hessian matrix of the barrier function, $\nabla_{xx}\Phi(x_k, \mu_k)$, has m_a eigenvalues*

$$\lambda_i(A_{\mathcal{A}}^T Y_{\mathcal{A}}^2 A_{\mathcal{A}})/\mu_k + O(1) \text{ for } i = 1, \ldots, m_a$$

and the remaining $n - m_a$ eigenvalues

$$\lambda_i(N_{\mathcal{A}}^T H(x_*, y_*) N_{\mathcal{A}}) + O(\mu_k) \text{ for } i = 1, \ldots, n - m_a$$

as $k \to \infty$, where $\lambda_i(.)$ denotes the i-th eigenvalue of its matrix argument, $Y_{\mathcal{A}}$ is the diagonal matrix of active Lagrange multipliers at x_* and $N_{\mathcal{A}} = is$ an orthogonal basis for the null-space of $A_{\mathcal{A}}$.

This demonstrates that the condition number of $\nabla_{xx}\Phi(x_k, \mu_k)$ is $O(1/\mu_k)$ as μ_k shrinks to zero, and suggests that it may not be straightforward to find the minimizer numerically. Look at how the contours around x_* in Figure 4.1 bunch together as μ approaches zero.

4.4.2 Potential Difficulty II: Poor Starting Points

As if this potential defect isn't serious enough, there is a second significant difficulty with the naive method we described earlier. This is that $x_{k+1}^S = x_k$ appears to be a very poor starting point for a Newton step just after the (small) barrier parameter is reduced. To see this suppose, as will be the case at the end of the minimization for the k-th barrier subproblem, that

$$0 \approx \nabla_x \Phi(x_k, \mu_k) = g(x_k) - \mu_k A^T(x_k) C^{-1}(x_k) e$$
$$\approx g(x_k) - \mu_k A_{\mathcal{A}}^T(x_k) C_{\mathcal{A}}^{-1}(x_k) e,$$

the approximation being true because the neglected terms involve $y(x_k) = \mu_k / c_i(x_k)$ which converge to zero for inactive constraints. Then in the non-degenerate case, again roughly speaking, the Newton correction s^P for the new barrier parameter satisfies

$$\mu_{k+1} A_{\mathcal{A}}^T(x_k) C_{\mathcal{A}}^{-2}(x_k) A_{\mathcal{A}}(x_k) s^P \approx (\mu_{k+1} - \mu_k) A_{\mathcal{A}}^T(x_k) C_{\mathcal{A}}^{-1}(x_k) e \qquad (4.7)$$

since

$$\nabla_x \Phi(x_k, \mu_{k+1}) \approx g(x_k) - \mu_{k+1} A_{\mathcal{A}}^T(x_k) C_{\mathcal{A}}^{-1}(x_k) e$$
$$\approx (\mu_{k+1} - \mu_k) A_{\mathcal{A}}^T(x_k) C_{\mathcal{A}}^{-1}(x_k) e$$

and the $\mu_{k+1} A_{\mathcal{A}}^T(x_k) C_{\mathcal{A}}^{-2}(x_k) A_{\mathcal{A}}(x_k)$ term dominates $\nabla_{xx}\Phi(x_k, \mu_{k+1})$. If $A_{\mathcal{A}}(x_k)$ is full rank, then multiplying the approximation (4.7) from the left first by the generalized inverse, $(A_{\mathcal{A}} A_{\mathcal{A}}^T)^{-1} A_{\mathcal{A}}$ of $A_{\mathcal{A}}$ and then by $C_{\mathcal{A}}^2$ implies that

$$A_{\mathcal{A}}(x_k) s^P \approx \left(1 - \frac{\mu_k}{\mu_{k+1}}\right) c_{\mathcal{A}}(x_k)$$

from which a Taylor expansion of $c_A(x_k + s^P)$ reveals that

$$c_A(x_k + s^P) \approx c_A(x_k) + A_A(x_k)s^P \approx \left(2 - \frac{\mu_k}{\mu_{k+1}}\right)c_A(x_k) < 0$$

whenever $\mu_{k+1} < \frac{1}{2}\mu_k$. Hence a Newton step will asymptotically be infeasible for anything but the most modest decrease in μ, and thus the method is unlikely to converge fast.

We will return to both of these issues shortly, but first we need to examine barrier methods in a seemingly different light.

4.5 A Different Perspective: Perturbed Optimality Conditions

We now consider what, superficially, appears to be a completely different approach to inequality-constrained optimization. Recall from Theorem (1.8) that the first order optimality conditions for (4.3) are that there are Lagrange multipliers (or, as they are sometimes called, dual variables) y for which

$$\begin{aligned} g(x) - A^T(x)y &= 0 \quad &\text{(dual feasibility)} \\ C(x)y &= 0 \quad &\text{(complementary slackness) and} \end{aligned}$$
$$c(x) \geq 0 \text{ and } y \geq 0.$$

Now consider the "perturbed" problem

$$\begin{aligned} g(x) - A^T(x)y &= 0 \quad &\text{(dual feasibility)} \\ C(x)y &= \mu e \quad &\text{(\emph{perturbed} complementary slackness) and} \end{aligned}$$
$$c(x) > 0 \text{ and } y > 0,$$

where $\mu > 0$.

Primal-dual path-following methods aim to track solutions to the system

$$g(x) - A^T(x)y = 0 \text{ and } C(x)y - \mu e = 0 \tag{4.8}$$

as μ shrinks to zero, while maintaining $c(x) > 0$ and $y > 0$. This approach has been amazingly successful when applied to linear programming problems, and has been extended to many other classes of convex optimization problems. Since (4.8) is simply a nonlinear system, an obvious (locally convergent) way to solve the system is, as always, to use Newton's method. It is easy to show that the Newton correction (s^{PD}, w) to (x, y) satisfies

$$\begin{pmatrix} H(x,y) & -A^T(x) \\ YA(x) & C(x) \end{pmatrix} \begin{pmatrix} s^{PD} \\ w \end{pmatrix} = - \begin{pmatrix} g(x) - A^T(x)y \\ C(x)y - \mu e \end{pmatrix}. \tag{4.9}$$

Using the second equation to eliminate w gives that

$$\begin{aligned} \left(H(x,y) + A^T(x)C^{-1}(x)YA(x)\right) s^{PD} &= -\left(g(x) - \mu A^T(x)C^{-1}(x)e\right) \\ &= g(x, y(x)), \tag{4.10} \end{aligned}$$

where, as before, $y(x) = \mu C^{-1}(x)e$. But now compare this with the Newton barrier system (4.5). Amazingly, the only difference is that the (left-hand-side) coefficient matrix in (4.5) mentions the specific $y(x)$ while that for (4.10) uses a generic y. And it is this difference that turns out to be crucial. The freedom to choose y in $H(x,y) + A^T(x)C^{-1}(x)YA(x)$ for the primal-dual approach proves to be vital. Making the primal choice $y(x) = \mu C^{-1}(x)e$ can be poor, while using a more flexible approach in which y is chosen by other means, such as through the primal-dual correction $y + w$ is often highly successful.

We now return to the potential difficulties with the primal approach we identified in Sections 4.4.1 and 4.4.2.

4.5.1 Potential Difficulty II ... Revisited

We first show that, despite our reservations in Section 4.4.2, the value $x_{k+1}^s = x_k$ can be a good starting point. The problem with the primal correction s^P is that the primal method has to choose $y = y(x_k^s) = \mu_{k+1}C^{-1}(x_k)e$, and this is a factor μ_{k+1}/μ_k too small to be a good Lagrange multiplier estimate–recall that Theorem 4.1 shows that $\mu_k C^{-1}(x_k)e$ converges to y_*.

But now suppose instead that we use the primal-dual correction s^{PD} and choose the "proper" $y = \mu_k C^{-1}(x_k)e$ rather than $y(x_k^s)$–we know that this is a good choice insofar as this Newton step should decrease the dual infeasibility and complementary slackness since $(x_k, \mu_k C^{-1}(x_k)e)$ are already good estimates. In this case, arguing as before, in the non-degenerate case, the correction s^{PD} satisfies

$$\mu_k A_{\mathcal{A}}^T(x_k)C_{\mathcal{A}}^{-2}(x_k)A_{\mathcal{A}}(x_k)s^{PD} \approx (\mu_{k+1} - \mu_k)A_{\mathcal{A}}^T(x_k)C_{\mathcal{A}}^{-1}(x_k)e,$$

and thus if $A_{\mathcal{A}}(x_k)$ is full rank,

$$A_{\mathcal{A}}(x_k)s^{PD} \approx \left(\frac{\mu_{k+1}}{\mu_k} - 1\right)c_{\mathcal{A}}(x_k).$$

Then using a Taylor expansion of $c_{\mathcal{A}}(x_k + s^{PD})$ reveals that

$$c_{\mathcal{A}}(x_k + s^{PD}) \approx c_{\mathcal{A}}(x_k) + A_{\mathcal{A}}(x_k)s^{PD} \approx \frac{\mu_{k+1}}{\mu_k}c_{\mathcal{A}}(x_k) > 0,$$

and thus $x_k + s^{PD}$ is feasible–the result is easy to show for inactive constraints. Hence, simply by using a different model Hessian we can compute a useful Newton correction from $x_{k+1}^s = x_k$ that both improves the violation of the optimality conditions (and ultimately leads to fast convergence) and stays feasible.

4.5.2 Primal-Dual Barrier Methods

In order to globalize the primal-dual iteration, we simply need to build an appropriate model of the logarithmic barrier function within either a linesearch or trust-region framework for minimizing $\Phi(x, \mu_k)$. As we have already pointed out the disadvantages of only allowing the (primal) Hessian approximation $\nabla_{xx}\Phi(x_k, \mu_k)$, we instead prefer the more flexible search-direction model problem to (approximately)

$$\underset{s \in \mathbb{R}^n}{\text{minimize}} \; \langle s, g(x, y(x)) \rangle + \tfrac{1}{2} \left\langle s, \left(H(x,y) + A^T(x)C^{-1}(x)YA(x) \right) s \right\rangle, \quad (4.11)$$

possibly subject to a trust-region constraint. We have already noticed that the first-order term $g(x, y(x)) = \nabla_x \Phi(x, \mu)$ as $y(x) = \mu C^{-1}(x)e$, and thus the model gradient is that of the barrier function as required by our global convergence analyses of linesearch and trust-region methods. We have discounted always choosing $y = y(x)$ in (4.11), and have suggested that the choice $y = (\mu_{k-1}/\mu_k)y(x)$ when changing the barrier parameter results in good use of the starting point. Another possibility is to use $y = y^{\text{OLD}} + w^{\text{OLD}}$, where w^{OLD} is the primal-dual correction to the previous dual-variable estimates y^{OLD}. However, this needs to be used with care since there is no *a priori* assurance that $y^{\text{OLD}} + w^{\text{OLD}} > 0$, and indeed it is usual to prefer $y = \max(y^{\text{OLD}} + w^{\text{OLD}}, \epsilon(\mu_k)e)$ for some "small" $\epsilon(\mu_k) > 0$. The choice $\epsilon(\mu_k) = \mu_k^{1.5}$ leads to a realistic primal-dual method, although other precautions need sometimes to be taken.

4.5.3 Potential Difficulty I ... Revisited

We now return to the other perceived difficulty with barrier or primal-dual path-following methods, namely that the inherent ill-conditioning in the barrier Hessian makes it hard to generate accurate Newton steps when the barrier parameter is small. Let $\mathcal{I}$ be the set of inactive constraints at x_*, and denote the active and inactive components of c and y with suffices $\mathcal{A}$ and $\mathcal{I}$ respectively. Thus $c_{\mathcal{A}}(x_*) = 0$ and $c_{\mathcal{I}}(x_*) > 0$, while if the solution is non-degenerate, $y_{\mathcal{A}}(x_*) > 0$ and $y_{\mathcal{I}}(x_*) = 0$. As we have seen, the Newton correction s^{PD} satisfies (4.9), while the equivalent system (4.10) clearly has a condition number that approaches infinity as x and y reach their limits because $c_{\mathcal{A}}(x)$ approaches zero while $y_{\mathcal{A}}(x)$ approaches $y_{\mathcal{A}}(x_*) > 0$.

But now suppose that we separate (4.9) into

$$\begin{pmatrix} H(x,y) & -A_{\mathcal{A}}^T(x) & -A_{\mathcal{I}}^T(x) \\ Y_{\mathcal{A}}A_{\mathcal{A}}(x) & C_{\mathcal{A}}(x) & 0 \\ Y_{\mathcal{I}}A_{\mathcal{A}}(x) & 0 & C_{\mathcal{I}}(x) \end{pmatrix} \begin{pmatrix} s^{\text{PD}} \\ w_{\mathcal{A}} \\ w_{\mathcal{I}} \end{pmatrix} = - \begin{pmatrix} g(x) - A^T(x)y \\ C_{\mathcal{A}}(x)y_{\mathcal{A}} - \mu e \\ C_{\mathcal{I}}(x)y_{\mathcal{I}} - \mu e \end{pmatrix},$$

and then eliminate the variables $w_{\mathcal{I}}$, multiply the second equation by $Y_{\mathcal{A}}^{-1}$ and use $C_{\mathcal{I}}(x)y_{\mathcal{I}} = \mu e$, we obtain

$$\begin{pmatrix} H(x,y) + A_{\mathcal{I}}^T(x)C_{\mathcal{I}}(x)^{-1}Y_{\mathcal{I}}A_{\mathcal{I}}(x) & -A_{\mathcal{A}}^T(x) \\ A_{\mathcal{A}}(x) & C_{\mathcal{A}}(x)Y_{\mathcal{A}}^{-1} \end{pmatrix} \begin{pmatrix} s^{\text{PD}} \\ w_{\mathcal{A}} \end{pmatrix}$$
$$= - \begin{pmatrix} g(x) - A_{\mathcal{A}}^T(x)y_{\mathcal{A}} - \mu A_{\mathcal{I}}^T(x)C_{\mathcal{I}}^{-1}(x)e \\ c_{\mathcal{A}}(x) - \mu Y_{\mathcal{A}}^{-1}e \end{pmatrix}. \quad (4.12)$$

But then we see that the terms involving inverses, $C_{\mathcal{I}}^{-1}(x)$ and $Y_{\mathcal{A}}^{-1}$, remain bounded, and indeed in the limit the system becomes

$$\begin{pmatrix} H(x,y) & -A_{\mathcal{A}}^T(x) \\ A_{\mathcal{A}}(x) & 0 \end{pmatrix} \begin{pmatrix} s^{\mathrm{PD}} \\ w_{\mathcal{A}} \end{pmatrix} = - \begin{pmatrix} g(x) - A_{\mathcal{A}}^T(x)y_{\mathcal{A}} - \mu A_{\mathcal{I}}^T(x)C_{\mathcal{I}}^{-1}(x)e \\ 0 \end{pmatrix}$$

which is well behaved. Thus just because (4.10) is ill conditioned, this does not preclude us from finding s^{PD} from an equivalent, perfectly well-behaved system like (4.12).

4.6   A Practical Primal-Dual Method

Following on from the above, we now give the skeleton of a reasonable primal-dual method.

> Given $\mu_0 > 0$ and feasible (x_0^s, y_0^s), set $k = 0$.
> Until "convergence", iterate:
>   *Inner minimization*: starting from (x_k^s, y_k^s), use an
>   unconstrained minimization algorithm to find (x_k, y_k) for which
> $\|C(x_k)y_k - \mu_k e\| \le \mu_k$ and $\|g(x_k) - A^T(x_k)y_k\| \le \mu_k^{1.00005}$.
>   Set $\mu_{k+1} = \min(0.1\mu_k, \mu_k^{.9999})$.
>   Find (x_{k+1}^s, y_{k+1}^s) using a primal-dual Newton step from (x_k, y_k).
>   If (x_{k+1}^s, y_{k+1}^s) is infeasible, reset (x_{k+1}^s, y_{k+1}^s) to (x_k, y_k).
>   Increase k by 1.

The inner minimization will be performed by either a linesearch or trust-region method for minimizing $\Phi(x, \mu_k)$, the stopping rules $\|C(x_k)y_k - \mu_k e\| \le \mu_k$ and $\|g(x_k) - A^T(x_k)y_k\| \le \mu_k^{1.00005}$ certainly being attainable as the first-order optimality condition for minimizing $\Phi(x, \mu_k)$ is that $g(x) - A^T(x)y = 0$, where $C(x)y = \mu_k e$. The extra step, in which the starting point is computed by performing a primal-dual Newton step from (x_k, y_k), is simply included to generate a value that is already close to first order critical, and the stopping tolerances are specially chosen to encourage this. Indeed we have the following asymptotic convergence result.

Theorem 4.3. *Suppose that $f, c \in C^2$, that a subsequence $\{(x_k, y_k)\}$, $k \in \mathcal{K}$, of the practical primal-dual method converges to (x_*, y_*) satisfying second-order sufficiency conditions, that $A_{\mathcal{A}}(x_*)$ is full-rank, and that $(y_*)_{\mathcal{A}} > 0$. Then the starting point satisfies the inner-minimization termination test (i.e., $(x_k, y_k) = (x_k^s, y_k^s)$) for all k sufficiently large, and the whole sequence $\{(x_k, y_k)\}$ converges to (x_*, y_*) at a superlinear rate (with a Q-factor at least 1.9998).*

This is a highly acceptable result, the convergence being essentially quadratic (which would correspond to a Q-factor of two–any sequence $\{\sigma_k\}$ is said to converge to σ_* with *Q-factor* at least q if $|\sigma_{k+1} - \sigma_*| \leq \gamma|\sigma_k - \sigma_*|^q$ for some $\gamma > 0$).

Primal-dual interior-point methods have the potential for both excellent theoretical and practical behaviour. There are polynomial interior-point algorithms for linear, (convex) quadratic and semi-definite programming. While it is unlikely that this is true for more general (nonconvex) problems, the barrier function globalization is most effective in practice, and the asymptotic behaviour is normally just as for the convex case. From a global perspective, it is very important that iterates are kept away from constraint boundary until near to convergence, as otherwise very slow progress will be made–this is certainly born out in practice. Finally, while the methods we have discussed in this section have all required an interior starting point, it is possible to find one (if there is one!) by solving the "phase-one" problem to

$$\underset{(x,\gamma)}{\text{minimize}} \ \gamma \ \text{subject to} \ c(x) + \gamma e \geq 0;$$

any feasible point (x, γ) for this auxiliary problem for which $\gamma < 0$ is suitable, for then $c(x) > 0$.

It is quite common in practice to replace the inequality $c_i(x) \geq 0$ by the equation $c_i(x) - s_i = 0$, and simple bound $s_i \geq 0$ on the *slack* variable s_i. This has the algebraic advantage that the inequality constraints are then all simple bounds and thus that barrier terms only appear on the diagonal of the Hessian model, but arguably the disadvantages that the dimensionality of the problem has been artificially increased, and that we now need to use some means of coping with equality constraints. We consider this latter point next.

5   SQP Methods for Equality Constrained Optimization

In this final section, having already investigated very good methods for dealing with inequality constraints, we now turn our attention to the problem (4.1), in which there are only equality constraints on the variables. Of course in practice, there are frequently both equations and inequalities, and composite methods using the barrier/interior-point methods discussed in Section 4 and the SQP methods we shall consider here are often used. Alternatively, SQP methods themselves may easily be generalized to handle inequality constraints. For brevity we shall not consider such extensions further here.

5.1   Newton's Method for First-Order Optimality

Sequential Quadratic Programming (SQP) methods (sometimes called successive or recursive quadratic programming methods) are most naturally derived by considering the first-order necessary conditions for (4.1)–we will see where the names come from shortly. Recall at optimality we expect to have

$$g(x, y) \equiv g(x) - A^T(x)y = 0 \ \text{and} \ c(x) = 0. \tag{5.1}$$

This is a system of nonlinear equations in the variables x and the Lagrange multipliers y. Notice that the system is actually linear in y so that if x were known it would be straightforward to find y.

Suppose now that (x, y) is an approximation to a solution of (5.1). Then, as always, we might apply Newton's method to try to improve (x, y), and this leads us to construct a correction (s, w) for which

$$\begin{pmatrix} H(x,y) & -A^T(x) \\ A(x) & 0 \end{pmatrix} \begin{pmatrix} s \\ w \end{pmatrix} = - \begin{pmatrix} g(x,y) \\ c(x) \end{pmatrix}. \tag{5.2}$$

Newton's method would then apply the same procedure to the "improved" estimate $(x_+, y_+) = (x + s, y + w)$.

There are a number of alternative formulations of (5.2). Firstly (5.2) may be written as the symmetric system of equations

$$\begin{pmatrix} H(x,y) & A^T(x) \\ A(x) & 0 \end{pmatrix} \begin{pmatrix} s \\ -w \end{pmatrix} = - \begin{pmatrix} g(x,y) \\ c(x) \end{pmatrix};$$

notice here that the coefficient matrix is indefinite because of its zero 2,2 block. Secondly, on writing $y_+ = y + w$, the equation becomes

$$\begin{pmatrix} H(x,y) & -A^T(x) \\ A(x) & 0 \end{pmatrix} \begin{pmatrix} s \\ y_+ \end{pmatrix} = - \begin{pmatrix} g(x) \\ c(x) \end{pmatrix},$$

or finally, in symmetric form,

$$\begin{pmatrix} H(x,y) & A^T(x) \\ A(x) & 0 \end{pmatrix} \begin{pmatrix} s \\ -y_+ \end{pmatrix} = - \begin{pmatrix} g(x) \\ c(x) \end{pmatrix}.$$

In practice we might prefer to approximate $H(x, y)$ by some symmetric B, and instead solve

$$\begin{pmatrix} B & A^T(x) \\ A(x) & 0 \end{pmatrix} \begin{pmatrix} s \\ -y_+ \end{pmatrix} = - \begin{pmatrix} g(x) \\ c(x) \end{pmatrix} = \begin{pmatrix} B & -A^T(x) \\ A(x) & 0 \end{pmatrix} \begin{pmatrix} s \\ y_+ \end{pmatrix}. \tag{5.3}$$

One could imagine solving these related systems by finding an LU factorization of the coefficient matrix in the unsymmetric case, or a symmetric-indefinite (a generalization of Cholesky) factorization in the symmetric case. Alternatively, if B is invertible, s and y_+ might be found successively by solving

$$A(x)B^{-1}A(x)^T y = -c + A(x)B^{-1}g \text{ and then } Bs = A(x)^T y - g$$

using symmetric factorizations of B and $A(x)B^{-1}A(x)^T$. For very large problems, iterative methods might be preferred, and here GMRES(k) or QMR, for the unsymmetric case, or MINRES or conjugate-gradients (restricted to the null-space of $A(x)$), for the symmetric case, have all been suggested. Thus there are many ways to solve the system(s) of linear equations that arise from SQP methods, and there is currently much interest in exploiting the structure in such systems to derive very efficient methods.

But where does the name "sequential quadratic programming" come from?

5.2   The Sequential Quadratic Programming Iteration

A *quadratic program* is a problem involving the optimization of a quadratic function subject to a set of linear inequality and/or equality constraints. Consider the quadratic programming problem

$$\underset{s \in \mathbb{R}^n}{\text{minimize}} \ \langle s, g(x) \rangle + \tfrac{1}{2}\langle s, Bs \rangle \ \text{subject to} \ A(x)s = -c(x). \quad (5.4)$$

Why this problem? Well, Theorem 1.2 indicates that $c(x) + A(x)s$ is a first-order (Taylor) approximation to the constraint function $c(x + s)$, while a potential second-order model of the decrease $f(x + s) - f(x)$ is $\langle s, g(x) \rangle + \tfrac{1}{2}\langle s, Bs \rangle$. Thus one can argue that (5.4) gives a suitable (at least first-order) model of (4.1). An objection might be that really we should be aiming for true second-order approximations to all functions concerned, but this would lead to the significantly-harder minimization of a quadratic function subject to quadratic constraints–constraint curvature is a major obstacle.

The interesting feature of (5.4) is that it follows immediately from Theorem 1.6 that any first-order critical point of (5.4) is given by (5.3). Thus Newton-like methods for first-order optimality are equivalent to the solution of a sequence of related quadratic programs. Hence the name. Notice that if $B = H(x, y)$, solving (5.4) is actually Newton's method for (5.1), and this suggests that B should be an approximation to the Hessian of the Lagrangian function, not the objective function. Clearly the constraint curvature that we would have liked to have added to the linear approximations of the constraints has worked its way into the objective function!

To summarize, the basic SQP iteration is as follows.

Given (x_0, y_0), set $k = 0$.
Until "convergence" iterate:
    Compute a suitable symmetric B_k using (x_k, y_k).
    Find

$$s_k = \arg \underset{s \in \mathbb{R}^n}{\min} \langle s, g_k \rangle + \tfrac{1}{2}\langle s, B_k s \rangle \ \text{subject to} \ A_k s = -c_k \quad (5.5)$$

    along with associated Lagrange multiplier estimates y_{k+1}.
    Set $x_{k+1} = x_k + s_k$ and increase k by 1.

The SQP method is both simple and fast. If $B_k = H(x_k, y_k)$, the method is Newton's method for (5.1), and thus is quadratically convergent provided that (x_0, y_0) is sufficiently close to a first-order critical point (x_*, y_*) of (4.1) for which

$$\begin{pmatrix} H(x_*, y_*) & A^T(x_*) \\ A(x_*) & 0 \end{pmatrix}$$

is non-singular. Moreover, the method is superlinearly convergent when B_k is a "good" approximation to $H(x_k, y_k)$, and there is even no necessity that this be so for fast convergence. It should also be easy for the reader to believe that had we wanted to solve the problem (4.3) involving inequality constraints, the suitable SQP subproblem would be

$$\underset{s \in \mathbb{R}^n}{\text{minimize}} \ \langle s, g(x) \rangle + \tfrac{1}{2} \langle s, Bs \rangle \ \text{subject to} \ A(x)s \geq -c(x)$$

in which the nonlinear inequalities have been linearized.

But, as the reader will already have guessed, this basic iteration also has drawbacks, leading to a number of vital questions. For a start it is a Newton-like iteration, and thus may diverge from poor starting points. So how do we globalize this iteration? How should we pick B_k? What should we do if (5.4) is unbounded from below? And precisely when is it unbounded?

The problem (5.4) only has a solution if the constraints $A(x)s = -c(x)$ are consistent. This is certainly the case if $A(x)$ is full rank, but may not be so if $A(x)$ is rank deficient–we shall consider alternatives that deal with this deficiency later. Applying Theorem 1.7 to (5.4), we deduce that any stationary point (s, y_+) satisfying (5.3) solves (5.4) only if B is positive semi-definite on the manifold $\{s : A(x)s = 0\}$–if B is positive definite on the manifold (s, y_+) is the unique solution to the problem. If the m by n matrix $A(x)$ is full rank and the columns of $N(x)$ form a basis for the null-space of $A(x)$, it is easy to show that B being positive (semi-)definite on the manifold $\{s : A(x)s = 0\}$ is equivalent to $N(x)^T B N(x)$ being positive (semi-)definite which is in turn equivalent to the matrix

$$\begin{pmatrix} B & A^T(x) \\ A(x) & 0 \end{pmatrix}$$

(being non-singular and) having m negative eigenvalues. If B violates these assumptions, (5.4) is unbounded.

For the remainder of this section, we focus on methods to globalize the SQP iteration. And it should not surprise the reader that we shall do so by considering linesearch and trust-region schemes.

5.3   Linesearch SQP Methods

The obvious way to embed the SQP step s_k within a linesearch framework is to pick $x_{k+1} = x_k + \alpha_k s_k$, where the step $\alpha_k > 0$ is chosen so that

$$\Phi(x_k + \alpha_k s_k, p_k) \ \text{``<''} \ \Phi(x_k, p_k), \tag{5.6}$$

and where $\Phi(x, p)$ is a "suitable" merit function depending on parameters p_k. Of course it is then vital that s_k be a descent direction for $\Phi(x, p_k)$ at x_k, as otherwise there may be no α_k for which (5.6) is satisfied. As always with linesearch methods, this limits the choice of B_k, and it is usual to insist that B_k be positive definite–the reader may immediately object that this is imposing an unnatural requirement, since B_k is supposed to be approximating the (usually) indefinite matrix $H(x_k, y_k)$, and we can only sympathise with such a view!

What might a suitable merit function be? One possibility is to use the quadratic penalty function (4.2). In this case, we have the following result.

Theorem 5.1. *Suppose that B_k is positive definite, and that (s_k, y_{k+1}) are the SQP search direction and its associated Lagrange multiplier estimates for the problem*

$$\underset{x \in \mathbb{R}^n}{\text{minimize}}\, f(x) \quad \text{subject to}\quad c(x) = 0$$

at x_k. Then if x_k is not a first-order critical point, s_k is a descent direction for the quadratic penalty function $\Phi(x, \mu_k)$ at x_k whenever

$$\mu_k \leq \frac{\|c(x_k)\|_2}{\|y_{k+1}\|_2}.$$

We know that the parameter μ_k for the quadratic penalty function needs to approach zero for its minimizers to converge to those of (4.1), so Theorem 5.1 simply confirms this by suggesting how to adjust the parameter.

The quadratic penalty function has another role to play if the constraints are inconsistent. For consider the quadratic (Newton-like) model

$$\underset{s \in \mathbb{R}^n}{\text{minimize}}\, \langle s, g_k + A_k^T c_k / \mu_k \rangle + \tfrac{1}{2}\langle s, (B_k + 1/\mu_k A_k^T A_k)s \rangle$$

that might be used to compute a step s_k^Q from x_k. Stationary points of this model satisfy

$$(B_k + 1/\mu_k A_k^T A_k)s_k^Q = -(g_k + A_k^T c_k / \mu_k)$$

or, on defining $y_k^Q \overset{\text{def}}{=} -\mu_k^{-1}(c_k + A_k s_k^Q)$,

$$\begin{pmatrix} B_k & A_k^T \\ A_k & -\mu_k I \end{pmatrix} \begin{pmatrix} s_k^Q \\ -y_k^Q \end{pmatrix} = -\begin{pmatrix} g_k \\ c_k \end{pmatrix}. \tag{5.7}$$

But now compare this system with (5.3) that which defines the SQP step: the only difference is the vanishingly small 2,2 block $-\mu_k I$ in the coefficient matrix. While this indicates that Newton-like directions for the quadratic penalty function will become increasingly good approximations to SQP steps (and, incidentally, it can be shown that a Newton iteration for (4.2) with well chosen μ_k converges superlinearly under reasonable assumptions), the main point of the alternative (5.7) is that rank-deficiency in A_k is neutralised by the presence of 2,2 block term $-\mu_k I$. Nevertheless, the quadratic penalty function is rarely used, its place often being taken by non-differentiable exact penalty functions.

The *non-differentiable exact penalty function* is given by

$$\Phi(x, \rho) = f(x) + \rho \|c(x)\| \qquad (5.8)$$

for any norm $\| \cdot \|$ and scalar $\rho > 0$. Notice that the function is non-different-iable particularly when $c(x) = 0$, the very values we hope to attain! The following result helps explain why such a function is considered so valuable.

Theorem 5.2. *Suppose that $f, c \in C^2$, and that x_* is an isolated local min-imizer of $f(x)$ subject to $c(x) = 0$, with corresponding Lagrange multipli-ers y_*. Then x_* is also an isolated local minimizer of $\Phi(x, \rho)$ provided that $\rho > \|y_*\|_D$, where the dual norm $\|y\|_D = \sup_{x \neq 0} \frac{\langle y, x \rangle}{\|x\|}$.*

Notice that the fact that ρ merely needs to be larger than some critical value for $\Phi(x, \rho)$ to be usable to try to identify solutions to (4.1) is completely different to the quadratic penalty function, for which the parameter had to take on a limiting value.

More importantly, as we now see, $\Phi(x, \rho)$ may be used as a merit function for the SQP step.

Theorem 5.3. *Suppose that B_k is positive definite, and that (s_k, y_{k+1}) are the SQP search direction and its associated Lagrange multiplier estimates for the problem*

$$\underset{x \in \mathbb{R}^n}{\text{minimize}} \, f(x) \ \text{ subject to } \ c(x) = 0$$

at x_k. Then if x_k is not a first-order critical point, s_k is a descent di-rection for the non-differentiable penalty function $\Phi(x, \mu_k)$ at x_k whenever $\rho_k \geq \|y_{k+1}\|_D$.

Once again, this theorem indicates how ρ_k needs to be adjusted for use within a linesearch SQP framework.

Thus far, everything looks perfect. We have methods for globalizing the SQP iteration, an iteration that should ultimately converge very fast. But unfortunately, it is not as simple as that. For consider the example in Fig-ure 5.1. Here the current iterate lies close to (actually on) the constraint, the SQP step moves tangentially from it, and thus moves away as the constraint is nonlinear, but unfortunately, at the same time, the value of the objective function rises. Thus any merit function like (4.2) or (5.8) composed simply from positive combinations of the objective and (powers) of norms of con-straint violations will increase after such an SQP step, and thus necessarily $\alpha_k \neq 1$ in (5.6)–worse still, this behaviour can happen arbitrarily close to the minimizer. This has the unfortunate side effect that it may happen that the expected fast convergence achievable by Newton-like methods will be thwarted by the merit function. That is, there is a serious mismatch between the global and local convergence needs of the SQP method. The fact that the merit function may prevent acceptance of the full SQP step is known as the *Maratos effect*.

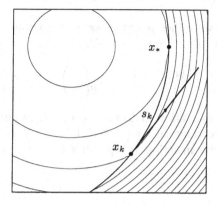

Fig. 5.1. ℓ_1 non-differentiable exact penalty function ($\rho = 1$): $f(x) = 2(x_1^2 + x_2^2 - 1) - x_1$ and $c(x) = x_1^2 + x_2^2 - 1$. Solution: $x_* = (1,0)$, $y_* = \frac{3}{2}$. The SQP direction using the optimal Hessian $H(x_*, y_*) = I$. Notice how the merit function increases at the point $x_k + s_k$.

The Maratos effect occurs because the curvature of the constraints is not adequately represented by linearization in the SQP model. In particular,

$$c(x_k + s_k) = O(\|s_k\|^2).$$

This suggests that we need to correct for this curvature. We may do this by computing a *second-order correction* from $x_k + s_k$, that is an extra step s_k^C for which

$$c(x_k + s_k + s_k^C) = o(\|s_k\|^2). \tag{5.9}$$

Since we do not want to destroy potential for fast convergence, we must also insist that the correction is small relative to the SQP step, and thus that

$$s_k^C = o(s_k). \tag{5.10}$$

There are a number of ways to compute a second-order correction. The first is simply to move back as quickly as possible towards the constraints. This suggests we compute a minimum (ℓ_2-)norm solution to $c(x_k + s_k) + A(x_k + s_k)s_k^C = 0$. It is easy to check that the required solution satisfies

$$\begin{pmatrix} I & A^T(x_k + s_k) \\ A(x_k + s_k) & 0 \end{pmatrix} \begin{pmatrix} s_k^C \\ -y_{k+1}^C \end{pmatrix} = -\begin{pmatrix} 0 \\ c(x_k + s_k) \end{pmatrix}.$$

Since this requires that we re-evaluate the constraints *and* their Jacobian at $x_k + s_k$, we might hope instead to find a minimum norm solution to $c(x_k + s_k) + A(x_k)s_k^C = 0$, and thus that

$$\begin{pmatrix} I & A^T(x_k) \\ A(x_k) & 0 \end{pmatrix} \begin{pmatrix} s_k^C \\ -y_{k+1}^C \end{pmatrix} = -\begin{pmatrix} 0 \\ c(x_k + s_k) \end{pmatrix}.$$

A third amongst many other possibilities is to compute another SQP step from $x_k + s_k$, that is to compute s_k^{C} so that

$$\begin{pmatrix} B_k^{\text{C}} & A^T(x_k + s_k) \\ A(x_k + s_k) & 0 \end{pmatrix} \begin{pmatrix} s_k^{\text{C}} \\ -y_{k+1}^{\text{C}} \end{pmatrix} = - \begin{pmatrix} g(x_k + s_k) \\ c(x_k + s_k) \end{pmatrix},$$

where B_k^{C} is an approximation to $H(x_k + s_k, y_k^{+})$. It can easily be shown that all of the above corrections satisfy (5.9)–(5.10). In Figure 5.2, we illustrate a second-order correction in action. It is possible to show that, under reasonable

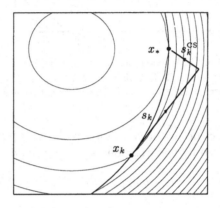

Fig. 5.2. ℓ_1 non-differentiable exact penalty function ($\rho = 1$): $f(x) = 2(x_1^2 + x_2^2 - 1) - x_1$ and $c(x) = x_1^2 + x_2^2 - 1$ solution: $x_* = (1,0)$, $y_* = \frac{3}{2}$. See that the second-order correction s_k^{CS} helps avoid the Maratos effect for the above problem with the ℓ_1-penalty function. Notice how s_k^{CS} more than compensates for the increase in the merit function at the point $x_k + s_k$, and how much closer $x_k + s_k + s_k^{\text{CS}}$ is to x_* than is x_k.

assumptions, any step $x_k + s_k + s_k^{\text{CS}}$ made up from the SQP step s_k and a second-order correction s_k^{CS} satisfying (5.9)–(5.10) will ultimately reduce (5.8). So now we can have both global and very fast asymptotic convergence at the expense of extra problem evaluations. Of course, we have stressed that a second SQP step gives a second-order correction, so another way of viewing this is to require that the merit function decreases at least every second iteration, and to tolerate non-monotonic behaviour in the interim.

5.4   Trust-Region SQP Methods

The main disadvantage of (at least naive) linesearch SQP methods is the un-natural requirement that B_k be positive definite. We saw the same restriction in the unconstrained case, although at least then there was some expectation that ultimately the true Hessian H_k would be positive (semi-) definite. In the unconstrained case, indefinite model Hessians were better handled in a trust-region framework, and the same is true in the constrained case.

The obvious trust-region generalization of the basic SQP step-generation subproblem (5.4) is to find

$$s_k = \arg\min_{s\in\mathbb{R}^n}\langle s, g_k\rangle + \tfrac{1}{2}\langle s, B_k s\rangle \text{ subject to } A_k s = -c_k \text{ and } \|s\| \le \Delta_k.$$
(5.11)

Since we do not require that B_k be positive definite, this allows us to use $B_k = H(x_k, y_k)$ if we so desire. However a few moments reflection should make it clear that such an approach has a serious flaw. Let Δ^{CRIT} be the least distance to the linearized constraints, i.e.

$$\Delta^{\mathrm{CRIT}} \stackrel{\text{def}}{=} \min \|s\| \text{ subject to } A_k s = -c_k.$$

The difficulty is that if $\Delta_k < \Delta^{\mathrm{CRIT}}$, then there is *no solution* to the trust-region subproblem (5.11). This implies that unless $c_k = 0$, the subproblem is meaningless for all sufficiently small trust-region radius (see Figure 5.3). Thus we need to consider alternatives. In this section, we shall review the

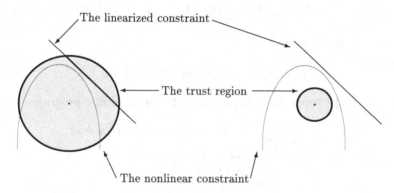

The linearized constraint

The trust region

The nonlinear constraint

Fig. 5.3. The intersection between the linearization of a nonlinear constraint and a spherical trust region. In the left figure, the trust-region radius is sufficiently large for the trust region and the linearized constraint to intersect. This is not so for the smaller trust region illustrated in the right figure.

$S\ell_p QP$ method of Fletcher, the composite step SQP methods due to Vardi, to Byrd and Omojokun, and to Celis, Dennis and Tapia, and the filter-SQP approach of Fletcher and Leyffer.

5.4.1   The $S\ell_p QP$ Method Our first trust-region approach is to try to minimize the ℓ_p-(exact) penalty function

$$\Phi(x, \rho) = f(x) + \rho\|c(x)\|_p$$
(5.12)

for sufficiently large $\rho > 0$ and some ℓ_p norm ($1 \le p \le \infty$). We saw in Section 5.3 that feasible minimizers of (5.12) may be solutions to (4.1) so long as $\rho > 0$ is large enough. Of course, as $\Phi(x, \rho)$ is non-differentiable, we cannot simply apply one of the unconstrained trust-region methods discussed in Section 3, but must instead build a specialized method.

Since we are discussing trust-region methods, a suitable model problem is the $\ell_\mathbf{p}QP$

$$\underset{s \in \mathbb{R}^n}{\text{minimize}}\ f_k + \langle s, g_k \rangle + \tfrac{1}{2}\langle s, B_k s \rangle + \rho \|c_k + A_k s\|_p \text{ subject to } \|s\| \le \Delta_k.$$

This has the major advantage that the model problem is always consistent, since now the only constraint is the trust-region bound. In addition, when ρ and Δ_k are large enough, it can be shown that the model minimizer *is* the SQP direction so long as $A_k s = -c_k$ is consistent. Moreover, when the norms are polyhedral (e.g., the ℓ_1 or ℓ_∞ norms), $\ell_\mathbf{p}QP$ is equivalent to a quadratic program.

To see this, consider for example the $\ell_1 QP$ model problem with an ℓ_∞ trust region

$$\underset{s \in \mathbb{R}^n}{\text{minimize}}\ \langle s, g_k \rangle + \tfrac{1}{2}\langle s, B_k s \rangle + \rho \|c_k + A_k s\|_1 \text{ subject to } \|s\|_\infty \le \Delta_k.$$

But we can always write

$$c_k + A_k s = u - v, \quad \text{where } (u, v) \ge 0.$$

Hence the $\ell_1 QP$ subproblem is equivalent to the quadratic program

$$\begin{aligned}
\underset{s \in \mathbb{R}^n,\ u,v \in \mathbb{R}^m}{\text{minimize}} \quad & \langle s, g_k \rangle + \tfrac{1}{2}\langle s, B_k s \rangle + \rho \langle e, u + v \rangle \\
\text{subject to} \quad & A_k s - u + v = -c_k \\
& u \ge 0, \ v \ge 0 \\
\text{and} \quad & -\Delta_k e \le s \le \Delta_k e.
\end{aligned}$$

Notice that the QP involves inequality constraints, but there are good methods (especially of the interior-point variety) for solving such problems. In particular, it is possible to exploit the structure of the u and v variables.

In order to develop a practical $S\ell_1 QP$ method, it should not surprise the reader that we need to ensure that every step we generate achieves as much reduction in the model $f_k + \langle s, g_k \rangle + \tfrac{1}{2}\langle s, B_k s \rangle + \rho \|c_k + A_k s\|_p$ as would have been achieved at a Cauchy point. One such Cauchy point requires the solution to $\ell_1 LP$ model

$$\underset{s \in \mathbb{R}^n}{\text{minimize}}\ \langle s, g_k \rangle + \rho \|c_k + A_k s\|_1 \text{ subject to } \|s\|_\infty \le \Delta_k,$$

which may be reformulated as a linear program. Fortunately approximate solutions to both $\ell_1 LP$ and $\ell_1 QP$ subproblems suffice. In practice it is also important to adjust ρ as the method progresses so as to ensure that ρ is larger than the (as yet unknown) $\|y_*\|_D$, and this may be achieved by using the available Lagrange multiplier estimates y_k. Such a scheme is globally convergent, but there is still a need for a second-order correction to prevent the Maratos effect and thus allow fast asymptotic convergence. If $c(x) = 0$ are inconsistent, the method converges to (locally) least value of the infeasibility $\|c(x)\|$ provided $\rho \to \infty$.

The alert reader will have noticed that in this section we have replaced the ℓ_2 trust-region of the unconstraint trust-region method by a box or ℓ_∞ trust-region. The reason for this apparent lack of consistency is that minimizing a quadratic subject to linear constraints *and* an additional quadratic trust-region is too hard. On the other hand, adding box-constraints does not increase the complexity of the resulting (quadratic programming) trust-region subproblem.

5.4.2   Composite-Step Methods

An alternative approach to avoid the difficulties caused by inconsistent QP subproblems is to separate the computation of the step into two stages. The aim of a *composite-step* method is to find

$$s_k = n_k + t_k,$$

where the *normal step* n_k moves towards feasibility of the linearized constraints (within the trust region), while the *tangential step* t_k reduces the model objective function (again within the trust-region) without sacrificing feasibility obtained from n_k. Of course since the normal step is solely concerned with feasibility, the model objective may get worse, and indeed it may not recover during the tangential step. The fact that the tangential step is required to maintain any gains in (linearized) feasibility achieved during the normal step implies that

$$A_k(n_k + t_k) = A_k n_k \text{ and hence that } A_k t_k = 0.$$

We illustrate possible normal and tangential steps in Figure 5.4.

5.4.2.1   Constraint Relaxation–Vardi's Method

Vardi's approach is an early composite-step method. The normal step is found by relaxing the requirement

$$A_k s = -c_k \text{ and } \|s\| \leq \Delta_k$$

to

$$A_k n = -\sigma_k c_k \text{ and } \|n\| \leq \Delta_k,$$

where $\sigma_k \in [0, 1]$ is small enough so that there is a feasible n_k. Clearly $s = 0$ is feasible if $\sigma_k = 0$, and the largest possible $\sigma_{\max}$ may be found by computing

$$\max_{\sigma \in (0,1]} \left[\min_{\|s\| \leq \Delta_k} \|A_k s + \sigma c_k\| = 0 \right].$$

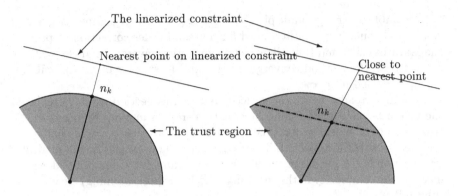

Fig. 5.4. Computing the normal step. The left-hand figure shows the largest possible normal step. The right-hand figure illustrates a shorter normal step n, and the freedom this then allows for the tangential step–any point on the dotted line is a potential tangential step.

In practice, some value between zero and $\sigma_{\max}$ is chosen, since this gives some "elbow-room" in which to compute the tangential step. The main defect with the approach is that there may be no normal step if the linearized constraints are inconsistent.

Once a normal step has been determined, the tangential step is computed as the

$$\text{(approximate) arg min}_{t\in\mathbb{R}^n} \ \langle t, g_k + B_k n_k\rangle + \tfrac{1}{2}\langle t, B_k t\rangle$$
$$\text{subject to} \quad A_k t = 0 \ \text{ and } \ \|n_k + t\| \le \Delta_k.$$

Although of historical interest, the method has been effectively superseded by the Byrd–Omojokun approach we describe next.

5.4.2.2 Constraint Reduction–the Byrd–Omojokun Method

The Byrd–Omojokun method aims to cope with the inconsistency issue that afflicts Vardi's approach. Rather than relaxing the constraints, the normal step is now computed as

$$\text{approximately minimize } \|A_k n + c_k\| \text{ subject to } \|n\| \le \Delta_k,$$

in order to achieve a reasonable improvement in linearized infeasibility that is consistent with the trust-region. The tangential step is then computed exactly as in Vardi's method.

An important aspect is that it is possible to use the conjugate gradient method to solve both subproblems. This provides Cauchy points in both cases and allows the method to be used to solve large problems. The method has been shown to be globally convergent (under reasonable assumptions) using an ℓ_2 merit function, and is the basis of the successful KNITRO software package.

5.4.2.3   Constraint Lumping–the Celis–Dennis–Tapia Method

A third method which might be considered to be of the composite-step variety is that due to Celis, Dennis and Tapia. In this approach, the requirement that $A_k s = -c_k$ is replaced by requiring that

$$\|A_k s + c_k\| \leq \sigma_k$$

for some $\sigma_k \in [0, \|c_k\|]$. The value of σ_k is chosen so that the normal step n_k satisfies

$$\|A_k n + c_k\| \leq \sigma_k \text{ and } \|n\| \leq \Delta_k.$$

Having found a suitable normal step, the tangential step is found as an

(approximate) $\arg \min_{t \in \mathbb{R}^n} \langle t, g_k + B_k n_k \rangle + \frac{1}{2}\langle t, B_k t \rangle$
subject to $\|A_k t + A_k n_k + c_k\| \leq \sigma_k$ and $\|t + n_k\| \leq \Delta_k$.

While finding a suitable σ_k is inconvenient, the real Achilles' heel of this approach is that the tangential step subproblem is (much) harder than those we have considered so far. If the ℓ_2-norm is used for the constraints, we need to find the minimizer of a quadratic objective within the intersection of two "spherical" regions. Unlike the case involving a single sphere (recall Section 3.5.1), it is not known if there is an efficient algorithm in the two-sphere case. Alternatively, if polyhedral (ℓ_1 or ℓ_∞) norms are used and B_k is indefinite, the subproblem becomes a nonconvex quadratic program for which there is unlikely to be an efficient general-purpose algorithm–in the special case where B_k is positive semi-definite and the ℓ_∞ norm is used, the subproblem is a convex QP. For this reason, the Celis–Dennis–Tapia approach is rarely used in practice.

5.4.3   Filter Methods

The last SQP method we shall consider is the most recent. The approach taken is quite radical in that, unlike all of the methods we have considered so far, it makes no use of a merit function to force global convergence. The main objection to merit functions is that they depend, to a large degree, on arbitrary or *a priori* unknown parameters. A secondary objection is that they tend to be overly conservative in accepting promising potential iterates. But if we wish to avoid merit functions, we need some other device to encourage convergence. The new idea is to use a "filter"

Let $\theta(x) = \|c(x)\|$ be some norm of the constraint violation at x. A *filter* is a set of pairs $\{(\theta_k, f_k)\}$ of violations and objective values such that no member dominates another, i.e., it does not happen that

$$\theta_i \text{``}<\text{''} \theta_j \text{ and } f_i \text{``}<\text{''} f_j$$

for any pair of filter points $i \neq j$–the "$<$" here informally means "very slightly smaller than". We illustrate a filter in Figure 5.5. A potential new entry to

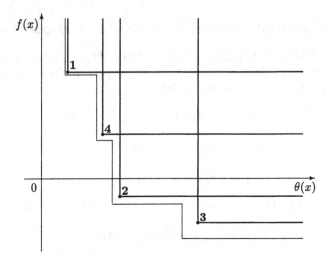

Fig. 5.5. A filter with four entries.

the "north-east" of any of the existing filter entries would not be permitted, and the forbidden region is the intersection of the solid horizontal and vertical lines emanating to the right and above each filter point. For theoretical reasons (akin to requiring sufficient decrease), we slightly enlarge the forbidden region by putting a small margin around each filter point, and this is illustrated in the figure by the dotted lines.

And now it is clear how to use a filter. Any potential SQP (or other) iterate $x_k + s_k$ will immediately be rejected if it lies in the forbidden filter region accumulated during the previous k iterations. This may be embedded in a trust-region framework, and a typical iteration might be as follows:

If possible find

$$s_k = \arg\min_{s \in \mathbb{R}^n} \langle s, g_k \rangle + \tfrac{1}{2}\langle s, B_k s \rangle$$

$$\text{subject to } A_k s = -c_k \text{ and } \|s\| \leq \Delta_k,$$

but otherwise, find s_k such that

$$\theta(x_k + s_k) \text{``<''} \theta_i \text{ for all } i \leq k.$$

If $x_k + s_k$ is "acceptable" for the filter, set $x_{k+1} = x_k + s_k$
and possibly add $(f((x_k + s_k), \theta(x_k + s_k))$ to the filter,
"prune" the filter, and increase Δ_k.
Otherwise reduce Δ_k and try again.

A few words of explanation are needed. The trust-region and linearized constraints will always be compatible if c_k is small enough so long as they are at $c(x) = 0$. Thus if the trust-region subproblem is incompatible, one remedy is simply to move closer to the constraints. This is known as a *restoration* step. By "pruning" the filter, we mean that a new point may completely dominate one or more existing filter points and, in this case, the dominated entry may be removed without altering the filter. For example, if a new entry were accepted to the "south-west" of point 4 in our figure, point 4 would be pruned.

While the basic filter idea is rather simple, in practice, it is significantly more complicated than this. In particular, there are theoretical reasons why some points that are acceptable to the filter should still be rejected if any decrease in the SQP model of the objective function is far from realized in practice.

6   Conclusion

We hope we have conveyed the impression that research into the design, convergence and implementation of algorithms for nonlinear optimization is an exciting and expanding area. We have only been able to outline the developments in the field, and have made no attempt to survey the vast literature that has built up over the last 50 years. Current algorithms for specialized problems like linear and quadratic programming and unconstrained optimization are well capable of solving problems involving millions of unknowns (and constraints), while those for generally constrained optimization routinely solve problems in the tens and, perhaps even, hundreds of thousands of unknowns and constraints. The next big goal is to be able to design algorithms that have some hope of finding global optima for large problems, the current state-of-the-art being for problems with tens or hundreds of unknowns. Clearly closing the gap between local and global optimization has some way to go!

Appendices

A   Seminal Books and Papers

The following books and papers are classics in the field. Although many of them cover topics outside the material we have described, they are all worth reading. This section constitutes a personal view of the most significant papers in the area. It is not meant to be a complete bibliography.

General Text Books

There are a large number of text books devoted to nonlinear (and even more for linear) programming. Those we find most useful and which emphasize practical methods are

J. Dennis and R. Schnabel, "Numerical Methods for Unconstrained Optimization and Nonlinear Equations", (republished by) SIAM (Classics in Applied Mathematics 16) (1996),

R. Fletcher, "Practical Methods of Optimization", 2nd edition Wiley (1987), (republished in paperback 2000),

P. Gill, W. Murray and M. Wright, "Practical Optimization", Academic Press (1981), and

J. Nocedal and S. Wright, "Numerical Optimization", Springer Verlag (1999).

The first of these concentrates on unconstrained optimization, while the remainder cover (continuous) optimization in general.

Early Quasi-Newton Methods

These methods were introduced by

W. Davidon, "Variable metric method for minimization", manuscript (1958), finally published *SIAM J. Optimization* **1** (1991) 1:17,

and championed by

R. Fletcher and M. J. D. Powell, "A rapidly convergent descent method for minimization", *Computer J.* (1963) 163:168.

Although the so-called DFP method has been superseded by the more reliable BFGS method, it paved the way for a number of classes of important updates.

More Modern Quasi-Newton Methods

Coincidentally, all of the papers

C. G. Broyden, "The convergence of a class of double-rank minimization algorithms", *J. Inst. Math. Applcs.*, **6** (1970) 76:90,

R. Fletcher, "A new approach to variable metric algorithms", *Computer J.* (1970) **13** (1970) 317:322,

D. Goldfarb, "A family of variable metric methods derived by variational means", *Math. Computation* **24** (1970) 23:26, and

D. F. Shanno, "Conditioning of quasi-Newton methods for function minimization", *Math. Computation* **24** (1970) 647:657

appeared in the same year. The aptly-named BFGS method has stood the test of time well, and is still regarded as possibly the best secant updating formula.

Quasi-Newton Methods for Large Problems

Limited memory methods are secant-updating methods that discard old information so as to reduce the amount of storage required when solving large problems. The methods first appeared in

> J. Nocedal, "Updating quasi-Newton matrices with limited storage", *Mathematics of Computation* **35** (1980) 773:782, and

> A. Buckley and A. Lenir, "QN-like variable storage conjugate gradients", *Math. Programming* **27** (1983) 155:175.

Secant updating formulae proved to be less useful for large-scale computation, but a successful generalization, applicable to what are known as partially separable functions, was pioneered by

> A. Griewank and Ph. Toint, "Partitioned variable metric updates for large structured optimization problems", *Numerische Mathematik* **39** (1982) 119:137, see also 429:448, as well as

> A. Griewank and Ph. Toint, "On the unconstrained optimization of partially separable functions", in *Nonlinear Optimization 1981* (Powell, M., ed.) Academic Press (1982)

Conjugate Gradient Methods for Large Problems

Generalizations of Conjugate Gradient methods for non-quadratic minimization were originally proposed by

> R. Fletcher and C. M. Reeves, "Function minimization by conjugate gradients", *Computer J.* (1964) 149:154, and

> E. Polak and G. Ribiére, "Note sur la convergence de méthodes de directions conjuguées", *Revue Française d'informatique et de recherche opérationelle* **16** (1969) 35:43.

An alternative is to attempt to solve the (linear) Newton system by a conjugate-gradient like method. Suitable methods for terminating such a procedure while still maintaining fast convergence were proposed by

> R. S. Dembo and T. Steihaug, "Truncated-Newton algorithms for large-scale unconstrained optimization", *Mathematical Programming* **26** (1983) 190:212.

Non-Monotone Methods

While it is usual to think of requiring that the objective function decreases at every iteration, this is not actually necessary for convergence so long as there is some overall downward trend. The first method along these lines was by

L. Grippo, F. Lampariello and S. Lucidi, "A nonmonotone line search technique for Newton's method", *SIAM J. Num. Anal.*, **23** (1986) 707:716.

Trust-Region Methods

The earliest methods that might be regarded as trust-region methods are those by

K. Levenberg, "A method for the solution of certain problems in least squares", *Quarterly J. Appl. Maths*, **2** (1944) 164:168, and

D. Marquardt, "An algorithm for least-squares estimation of nonlinear parameters" *SIAM J. Appl. Maths*, **11** (1963) 431:441

for the solution of nonlinear least-squares problems, although they are motivated from the perspective of modifying indefinite Hessians rather than restricting the step. Probably the first "modern" interpretation is by

S. Goldfeldt, R. Quandt and H. Trotter, "Maximization by quadratic hill-climbing", *Econometrica*, **34** (1966) 541:551.

Certainly, the earliest proofs of convergence are given by

M. Powell, "A New Algorithm for Unconstrained Optimization", in *Nonlinear Programming*, (Rosen, J., Mangasarian, O., and Ritter, K., eds.) Academic Press (1970),

while a good modern introduction is by

J. Moré, "Recent developments in algorithms and software for trust region methods", in *Mathematical Programming: The State of the Art*, (Bachem, A., Grötschel, M., and Korte, B., eds.) Springer Verlag (1983).

You might want to see our book

A. Conn, N. Gould and Ph. Toint, "Trust-region methods", SIAM (2000)

for a comprehensive history and review of the large variety of articles on trust-region methods.

Trust-Region Subproblems

Almost all you need to know about solving small-scale trust-region subproblems is contained in the paper

J. Moré and D. Sorensen, "Computing a trust region step", *SIAM J. Sci. Stat. Comp.* **4** (1983) 533:572.

Likewise

T. Steihaug, "The conjugate gradient method and trust regions in large scale optimization", *SIAM J. Num. Anal.* **20** (1983) 626:637

provides the basic truncated conjugate-gradient approach used so successfully for large-scale problems. More recently[1]

N. Gould, S. Lucidi, M. Roma and Ph. Toint, "Solving the trust-region subproblem using the Lanczos method", *SIAM J. Optimization* **9** (1999) 504:525

show how to improve on Steihaug's approach by moving around the trust-region boundary. A particularly nice new paper by

Y. Yuan, "On the truncated conjugate-gradient method", *Math. Programming*, **87** (2000) 561:573

proves that Steihaug's approximation gives at least 50% of the optimal function decrease when applied to convex problems.

The Symmetric Rank-One Quasi-Newton Approximation

Since trust-region methods allow nonconvex models, perhaps the simplest of all Hessian approximation methods, the Symmetric Rank-One update, is back in fashion. Although it is unclear who first suggested the method,

C. Broyden, "Quasi-Newton methods and their application to function minimization", *Math. Computation* **21** (1967) 577:593

is the earliest reference that we know of. Its revival in fortune is due[2] to

A. Conn, N. Gould and Ph. Toint, "Convergence of quasi-Newton matrices generated by the Symmetric Rank One update" *Math. Programming*, **50** (1991) 177:196 (see also *Math. Comp.* **50** (1988) 399:430), and

R. Byrd, H. Khalfan and R. Schnabel "Analysis of a symmetric rank-one trust region method" *SIAM J. Optimization* **6** (1996) 1025:1039,

and it has now taken its place alongside the BFGS method as the pre-eminent updating formula.

More Non-Monotone Methods

Non-monotone methods have also been proposed in the trust-region case. The basic reference here is the paper by

---

[1] We would hate to claim "seminal" status for one of our own papers!
[2] See previous footnote ...

Ph. Toint, "A non-monotone trust-region algorithm for nonlinear optimization subject to convex constraints", *Math. Programming*, **77** (1997) 69:94.

Barrier Function Methods

Although they appear to have originated in a pair of unpublished University of Oslo technical reports by K. Frisch in the mid 1950s, (logarithmic) barrier function were popularized by

A. Fiacco and G. McCormick, "The sequential unconstrained minimization technique for nonlinear programming: a primal-dual method", *Management Science* **10** (1964) 360:366; see also *ibid* (1964) 601:617.

A full early history is given in the book

A. Fiacco and G. McCormick, "Nonlinear programming: sequential unconstrained minimization techniques" (1968), republished as *Classics in Applied Mathematics 4*, SIAM (1990).

The worsening conditioning of the Hessian was first highlighted by

F. Lootsma, "Hessian matrices of penalty functions for solving constrained optimization problems", *Philips Research Reports*, **24** (1969) 322:331, and

W. Murray, "Analytical expressions for eigenvalues and eigenvectors of the Hessian matrices of barrier and penalty functions", *J. Optimization Theory and Applications*, **7** (1971) 189:196,

although recent work by

M. Wright, "Ill-conditioning and computational error in interior methods for nonlinear programming", *SIAM J. Optimization* **9** (1999) 84:111, and

S. Wright, "Effects of finite-precision arithmetic on interior-point methods for nonlinear programming", *SIAM J. Optimization* **12** (2001) 36:78

demonstrates that this "defect" is far from fatal.

Interior-Point Methods

The interior-point revolution was started by

N. Karmarkar, "A new polynomial-time algorithm for linear programming", *Combinatorica* **4** (1984) 373:395.

It did not take long for

P. Gill, W. Murray, M. Saunders, J. Tomlin and M. Wright, "On projected Newton barrier methods for linear programming and an equivalence to Karmarkar's projective method", *Math. Programming*, **36** (1986) 183:209

to realize that this radical "new" approach was actually something that nonlinear programmers had tried (but, most unfortunately, discarded) in the past.

SQP Methods

The first SQP method was proposed in the overlooked 1963 Harvard Master's thesis of R. Wilson. The generic linesearch SQP method is that of

B. Pschenichny, "Algorithms for general problems of mathematical programming", *Kibernetica*, **6** (1970) 120:125,

while there is a much larger variety of trust-region SQP methods, principally because of the constraint incompatibility issue.

Merit Functions for SQP

The first use of an exact penalty function to globalize the SQP method was by

S. Han, "A globally convergent method for nonlinear programming", *J. Optimization Theory and Applics*, **22** (1977) 297:309, and

M. Powell, "A fast algorithm for nonlinearly constrained optimization calculations", in *Numerical Analysis, Dundee 1977* (G. Watson, ed) Springer Verlag (1978) 144:157.

The fact that such a merit function may prevent full SQP steps was observed N. Maratos in his 1978 U. of London Ph. D. thesis, while methods for combating the Maratos effect were subsequently proposed by

R. Fletcher, "Second-order corrections for non-differentiable optimization", in *Numerical Analysis, Dundee 1981* (G. Watson, ed) Springer Verlag (1982) 85:114, and

R. Chamberlain, M. Powell, C. Lemaréchal, and H. Pedersen, "The watchdog technique for forcing convergence in algorithms for constrained optimization", *Math. Programming Studies*, **16** (1982) 1:17.

An SQP method that avoids the need for a merit function altogether by staying feasible is given by

J. Bonnans, E. Panier, A. Tits, and J. Zhou, "Avoiding the Maratos effect by means of a nonmonotone linesearch II. Inequality constrained problems–feasible iterates", *SIAM J. Num. Anal.*, **29** (1992) 1187:1202.

Hessian Approximations

There is a vast literature on suitable Hessian approximations for use in SQP methods. Rather than point at individual papers, a good place to start is

P. Boggs and J. Tolle, "Sequential quadratic programming", *Acta Numerica* **4** (1995) 1:51,

but see also our paper

N. Gould and Ph. Toint, "SQP methods for large-scale nonlinear programming", in *System modelling and optimization, methods, theory and applications* (M. Powell and S. Scholtes, eds.) Kluwer (2000) 149:178.

Trust-Region SQP Methods

Since the trust-region and the linearized constraints may be incompatible, almost all trust-region SQP methods modify the basic SQP method in some way. The $S\ell_1$QP method is due to

R. Fletcher, "A model algorithm for composite non-differentiable optimization problems", *Math. Programming Studies*, **17** (1982) 67:76.

Methods that relax the constraints include those proposed by

A. Vardi, "A trust region algorithm for equality constrained minimization: convergence properties and implementation", *SIAM J. Num. Anal.*, **22** (1985) 575:591, and

M. Celis, J. Dennis and R. Tapia, "A trust region strategy for nonlinear equality constrained optimization", in *Numerical Optimization 1984* (P. Boggs, R. Byrd and R. Schnabel, eds), SIAM (1985) 71:82,

as well as a method that appeared in the 1989 U. of Colorado at Boulder Ph. D. thesis of E. Omojokun, supervised by R. Byrd. The Filter-SQP approach may be found in[3]

R. Fletcher and S. Leyffer, "Nonlinear programming without a penalty function", *Math. Programming*, **91** (2002) 239:269.

Modern Methods for Nonlinear Programming

Many modern methods for nonlinearly constrained optimization tend to be SQP-interior-point hybrids. A good example is due to

R. Byrd, J. Gilbert and J. Nocedal, "A trust region method based on interior point techniques for nonlinear programming", *Math. Programming A* **89** (2000) 149:185,

and forms the basis for the excellent KNITRO package.

---

[3] Once again, see previous footnote ...

B Optimization Resources on the World-Wide-Web

B.1 Answering Questions on the Web

A good starting point for finding out more about optimization are the two
lists of Frequently Asked Questions (FAQs) on optimization. The Linear Programming FAQ,

> `www-unix.mcs.anl.gov/otc/Guide/faq/linear-programming-faq.html`

is dedicated to question on linear optimization problems as well as certain
aspects of mixed integer linear programming. The Nonlinear Programming
FAQ,

> `www-unix.mcs.anl.gov/otc/Guide/faq/nonlinear-programming-faq.html`

offers a concise introduction to nonlinear optimization. The NEOS guide,

> `www-fp.mcs.anl.gov/otc/Guide`

provides an overview of optimization and the solvers available. It contains
the optimization tree,

> `www-fp.mcs.anl.gov/otc/Guide/OptWeb`

a dichotomy of optimization problems. Both sites are maintained by the Optimization Technology Center

> `www.ece.nwu.edu/OTC`

a loose collaboration between Argonne National Laboratory and Northwestern University in the USA.

Hans Mittelmann of Arizona State University maintains a decision tree
for optimization software,

> `plato.la.asu.edu/guide.html`

and he also provides a useful set of benchmarks for optimization software,

> `plato.la.asu.edu/bench.html`

Harvey Greenberg's Mathematical Programming Glossary,

> `www.cudenver.edu/~hgreenbe/glossary/glossary.html`

contains brief definitions of commonly used expressions in optimization and
operations research. The usenet newsgroup

> `sci.op-research`

is dedicated to answering questions on optimization and operations research.
Brian Borchers edits a weekly digest of postings to it. You can receive the
digest by sending an email to

> `listserv@listserv.okstate.edu`

with the message

> `SUBSCRIBE ORCS-L Your Name`

B.2   Solving Optimization Problems on the Web

B.2.1   The NEOS Server Probably the most important and useful optimization site on the web is the NEOS server[4] at

`www-neos.mcs.anl.gov/neos`

which allows you to solve optimization problems over the internet. NEOS handles several thousand (!) submissions per week. The server provides a wide choice of state-of-the-art optimization software which can be used remotely without the need to install or maintain any software.

The problems should preferably be formulated in a modelling language such as AMPL[5] or GAMS[6] (see Section B.2.3). However, some solvers also accept problem descriptions in other formats such as C or fortran source code or the verbose linear programming MPS format.

There are a number of solvers implementing algorithms for nonlinearly constrained optimization problems. Most are hybrids, and thus capable of handling both equality and inequality constraints. There are at least three interior point solvers (see Section 4).

KNITRO (with a silent "K"), is a primal-dual interior-point method which uses trust regions.

LOQO is based on an infeasible primal-dual interior-point method. It uses a linesearch and a version of a filter to enforce global convergence.

MOSEK can only be used to solve *convex* large-scale smooth nonlinear optimization problems. It does not work for *nonconvex* problems.

There are at least three solvers implementing SQP algorithms (see Section 5).

DONLP2 implements a linesearch SQP algorithm with an exact non-differentiable ℓ_1-penalty function as a merit function. It uses dense linear algebra.

FILTER implements a trust-region SQP algorithm which is suitable for solving large nonlinearly constrained problems with small degrees of freedom. It uses a filter (see Section 5.4.3) to promote global convergence.

SNOPT implements a linesearch SQP algorithm which uses an augmented Lagrangian as a merit function. It maintains a positive definite limited memory approximation of the Hessian of the Lagrangian.

There is also a range of other solvers not covered in this article.

[4] J. Czyzyk, M. Mesnier and J. Moré. The NEOS server. *IEEE Journal on Computational Science and Engineering*, 5:68–75, 1998.

[5] R. Fourer, D. Gay and B. Kernighan. *AMPL: A modelling Language for Mathematical Programming*. Boyd & Fraser Publishing Company, Massachusetts, 1993.

[6] A. Brooke, D. Kendrick, A. Meeraus and R. Raman. *GAMS A user's guide*. GAMS Developments Corporation, 1217 Potomac Street, N.W., Washington DC 20007, USA, December 1998.

CONOPT is a feasible path method based on the generalized reduced gradient algorithm.

LANCELOT implements an augmented Lagrangian algorithm. It uses a trust-region to promote global convergence.

MINOS implements a sequential linearly constrained algorithm. Steplength control is heuristic (for want of a suitable merit function), but superlinear convergence is often achieved.

PATHNLP finds stationary points for the nonlinear problem by solving the Karush-Kuhn-Tucker conditions (see Theorems 1.6 and 1.8), written as a mixed complementarity problem, using the PATH solver.

Consult the NEOS guide (see Section B.1) for appropriate contacts.

A wide range of other optimization problems can also be solved such as semi-infinite optimization, mixed integer linear and nonlinear optimization, semidefinite optimization, complementarity problems, non-differentiable optimization, and unconstrained and stochastic optimization problems. The fact that the server maintains state-of-the-art optimization software makes is suitable for medium to large scale applications.

Users with their own copy of the modelling systems AMPL or GAMS can even invoke the NEOS solvers out of their local AMPL or GAMS session using KESTREL,

`www-neos.mcs.anl.gov/neos/kestrel.html`

This is very convenient as it makes it possible to post- or pre-process the models using a local copy of the modelling tool.

B.2.2   Other Online Solvers The system www-Nimbus, from

`nimbus.mit.jyu.fi`

is designed to solve (small) multi-objective optimization problems. It consists of a sequence of menus to input the multi-objective problem as well as some facilities for displaying the solution. It requires the user to interactively guide the optimization and requires some familiarity with multi-objective terminology. An online tutorial guides the user through the process. Certain topology optimization problems can be solved at

`www.topopt.dtu.dk`

The input is via a GUI and the solution is also display graphically. The system Baron,

`archimedes.scs.uiuc.edu/baron/availability.html`

allows the solution of small global optimization problems online.

B.2.3   Useful Sites for Modelling Problems Prior to Online Solution AMPL (A Mathematical Programming Language)

www.ampl.com

is a modelling language for optimization problems. The site lists extensions to the book, allows the solution of example models and contains a list of available solvers. Further AMPL models can be found at the following sites: NLP models by Bob Vanderbei:

www.sor.princeton.edu/~rvdb/ampl/nlmodels

MINLP and MPEC models by Sven Leyffer:

www.maths.dundee.ac.uk/~sleyffer/MacMINLP

www.maths.dundee.ac.uk/~sleyffer/MacMPEC

The COPS collection of Jorge Moré:

www-unix.mcs.anl.gov/~more/cops

These sites are especially useful to help with your own modelling exercises. GAMS (the General Algebraic Modelling System)

www.gams.com

is another modelling language. The site contains documentation on GAMS and some example models. More GAMS models can be found on the GAMS-world pages. These are sites, dedicated to important modelling areas, see

www.gamsworld.org

It also offers a translation service from one modelling language to another. Recently, optimization solvers have also been interfaced to matlab at

tomlab.biz/

B.2.4   Free Optimization Software

An extension of MPS to nonlinear optimization, SIF (standard input format), can be used to model optimization problems. The reference document can be found at

www.numerical.rl.ac.uk/lancelot/sif/sifhtml.html

A collection of optimization problems in SIF is available at CUTEr can be found via

www.cse.clrc.ac.uk/Activity/CUTEr

Two solvers, LANCELOT

www.cse.clrc.ac.uk/Activity/LANCELOT

and GALAHAD

www.cse.clrc.ac.uk/Activity/GALAHAD

are available freely for non-commercial users.

AMPL and some solvers are also available freely in limited size student versions, which allow the solution of problems of up to 300 variables and constraints, see

netlib.bell-labs.com/netlib/ampl/student/

B.3   Optimization Reports on the Web

Optimization online,

www.optimization-online.org

is an e-print site for papers on optimization. It is sponsored by the Mathematical Programming Society. It allows you to search for preprints on certain subjects. A monthly digest summarizes all monthly submissions.

The two main optimization journals, Mathematical Programming and SIAM Journal on Optimization maintain free sites with access to titles and abstracts, see

link.springer.de/link/service/journals/10107/

and

www.siam.org/journals/siopt/siopt.htm

C   Sketches of Proofs

Theorems 1.1 and 1.2 can be found in any good book on analysis. Theorem 1.1 follows directly by considering the remainders of truncated Taylor expansions of the univariate function $f(x + \alpha s)$ with $\alpha \in [0, 1]$, while Theorem 1.2 uses the Newton formula

$$F(x + s) = F(x) + \int_0^1 \nabla_x F(x + \alpha s)s\,d\alpha.$$

Proof (of Theorem 1.3). Suppose otherwise, that $g(x_*) \neq 0$. A Taylor expansion in the direction $-g(x_*)$ gives

$$f(x_* - \alpha g(x_*)) = f(x_*) - \alpha\|g(x_*)\|^2 + O(\alpha^2).$$

For sufficiently small α, $\frac{1}{2}\alpha\|g(x_*)\|^2 \geq O(\alpha^2)$, and thus

$$f(x_* - \alpha g(x_*)) \leq f(x_*) - \frac{1}{2}\alpha\|g(x_*)\|^2 < f(x_*).$$

This contradicts the hypothesis that x_* is a local minimizer. $\square$

180 Nicholas I. M. Gould and Sven Leyffer

Proof (of Theorem 1.4). Again, suppose otherwise that $\langle s, H(x_*)s \rangle < 0$. A Taylor expansion in the direction s gives

$$f(x_* + \alpha s) = f(x_*) + \tfrac{1}{2}\alpha^2 \langle s, H(x_*)s \rangle + O(\alpha^3),$$

since $g(x_*) = 0$. For sufficiently small α, $-\tfrac{1}{4}\alpha^2 \langle s, H(x_*)s \rangle \geq O(\alpha^3)$, and thus

$$f(x_* + \alpha s) \leq f(x_*) + \tfrac{1}{4}\alpha^2 \langle s, H(x_*)s \rangle < f(x_*).$$

Once again, this contradicts the hypothesis that x_* is a local minimizer. $\square$

Proof (of Theorem 1.5). By continuity $H(x)$ is positive definite for all x in a open ball $\mathcal{N}$ around x_*. The generalized mean value theorem then says that if $x_* + s \in \mathcal{N}$, there is a value z between the points x_* and $x_* + s$ for which

$$f(x_* + s) = f(x_*) + \langle s, g(x_*) \rangle + \tfrac{1}{2}\langle s, H(z)s \rangle = f(x_*) + \tfrac{1}{2}\langle s, H(z)s \rangle > f(x_*)$$

for all nonzero s, and thus x_* is an isolated local minimizer. $\square$

Proof (of Theorem 1.6). We consider feasible perturbations about x_*. Consider a vector valued C^2 (C^3 for Theorem 1.7) function $x(\alpha)$ of the scalar α for which $x(0) = x_*$ and $c(x(\alpha)) = 0$. (The constraint qualification is that all such feasible perturbations are of this form). We may then write

$$x(\alpha) = x_* + \alpha s + \tfrac{1}{2}\alpha^2 p + O(\alpha^3) \tag{C.1}$$

and we require that

$$\begin{aligned}
0 &= c_i(x(\alpha)) = c_i(x_* + \alpha s + \tfrac{1}{2}\alpha^2 p + O(\alpha^3)) \\
&= c_i(x_*) + \langle a_i(x_*), \alpha s + \tfrac{1}{2}\alpha^2 p \rangle + \tfrac{1}{2}\alpha^2 \langle s, H_i(x_*)s \rangle + O(\alpha^3) \\
&= \alpha \langle a_i(x_*), s \rangle + \tfrac{1}{2}\alpha^2 \left(\langle a_i(x_*), p \rangle + \langle s, H_i(x_*)s \rangle \right) + O(\alpha^3)
\end{aligned}$$

using Taylor's theorem. Matching similar asymptotic terms, this implies that for such a feasible perturbation

$$A(x_*)s = 0 \tag{C.2}$$

and

$$\langle a_i(x_*), p \rangle + \langle s, H_i(x_*)s \rangle = 0 \tag{C.3}$$

for all $i = 1, \ldots, m$. Now consider the objective function

$$\begin{aligned}
f(x(\alpha)) &= f(x_* + \alpha s + \tfrac{1}{2}\alpha^2 p + O(\alpha^3)) \\
&= f(x_*) + \langle g(x_*), \alpha s + \tfrac{1}{2}\alpha^2 p \rangle + \tfrac{1}{2}\alpha^2 \langle s, H(x_*)s \rangle + O(\alpha^3) \\
&= f(x_*) + \alpha \langle g(x_*), s \rangle + \tfrac{1}{2}\alpha^2 \left(\langle g(x_*), p \rangle + \langle s, H(x_*)s \rangle \right) + O(\alpha^3).
\end{aligned} \tag{C.4}$$

This function is unconstrained along $x(\alpha)$, so we may deduce, as in Theorem 1.3, that

$$\langle g(x_*), s \rangle = 0 \text{ for all } s \text{ such that } A(x_*)s = 0. \tag{C.5}$$

If we let S be a basis for the null-space of $A(x_*)$, we may write

$$g(x_*) = A^T(x_*)y_* + Sz_* \qquad (C.6)$$

for some y_* and z_*. Since, by definition, $A(x_*)S = 0$, and as it then follows from (C.5) that $g^T(x_*)S = 0$, we have that

$$0 = S^T g(x_*) = S^T A^T(x_*)y_* + S^T Sz_* = S^T Sz_*.$$

Hence $S^T Sz_* = 0$ and thus $z_* = 0$ since S is of full rank. Thus (C.6) gives

$$g(x_*) - A^T(x_*)y_* = 0. \quad \square \qquad (C.7)$$

Proof (of Theorem 1.7). We have shown that

$$f(x(\alpha)) = f(x_*) + \tfrac{1}{2}\alpha^2 \left(\langle p, g(x_*) \rangle + \langle s, H(x_*)s \rangle \right) + O(\alpha^3) \qquad (C.8)$$

for all s satisfying $A(x_*)s = 0$, and that (C.7) holds. Hence, necessarily,

$$\langle p, g(x_*) \rangle + \langle s, H(x_*)s \rangle \geq 0 \qquad (C.9)$$

for all s and p satisfying (C.2) and (C.3). But (C.7) and (C.3) combine to give

$$\langle p, g(x_*) \rangle = \sum_{i=1}^{m} (y_*)_i \langle p, a_i(x_*) \rangle = - \sum_{i=1}^{m} (y_*)_i \langle s, H_i(x_*)s \rangle$$

and thus (C.9) is equivalent to

$$\left\langle s, \left(H(x_*) - \sum_{i=1}^{m} (y_*)_i H_i(x_*) \right) s \right\rangle \equiv \langle s, H(x_*, y_*)s \rangle \geq 0$$

for all s satisfying (C.2). $\quad \square$

Proof (of Theorem 1.8). As in the proof of Theorem 1.6, we consider feasible perturbations about x_*. Since any constraint that is inactive at x_* (i.e., $c_i(x_*) > 0$) will remain inactive for small perturbations, we need only consider perturbations that are constrained by the constraints active at x_*, (i.e., $c_i(x_*) = 0$). Let $\mathcal{A}$ denote the indices of the active constraints. We then consider a vector valued C^2 (C^3 for Theorem 1.9) function $x(\alpha)$ of the scalar α for which $x(0) = x_*$ and $c_i(x(\alpha)) \geq 0$ for $i \in \mathcal{A}$. In this case, assuming that $x(\alpha)$ may be expressed as (C.1), we require that

$$\begin{aligned}
0 \leq c_i(x(\alpha)) &= c(x_* + \alpha s + \tfrac{1}{2}\alpha^2 p + O(\alpha^3)) \\
&= c_i(x_*) + \langle a_i(x_*), \alpha s + \tfrac{1}{2}\alpha^2 p \rangle + \tfrac{1}{2}\alpha^2 \langle s, H_i(x_*)s \rangle + O(\alpha^3) \\
&= \alpha \langle a_i(x_*), s \rangle + \tfrac{1}{2}\alpha^2 \left(\langle a_i(x_*), p \rangle + \langle s, H_i(x_*)s \rangle \right) + O(\alpha^3)
\end{aligned}$$

for all $i \in \mathcal{A}$. Thus

$$\langle s, a_i(x_*) \rangle \geq 0 \qquad (C.10)$$

and

$$\langle p, a_i(x_*)\rangle + \langle s, H_i(x_*)s\rangle \geq 0 \quad \text{when} \quad \langle s, a_i(x_*)\rangle = 0 \qquad \text{(C.11)}$$

for all $i \in \mathcal{A}$. The expansion of $f(x(\alpha))$ (C.4) then implies that x_* can only be a local minimizer if

$$\mathcal{S} = \{s \mid \langle s, g(x_*)\rangle < 0 \text{ and } \langle s, a_i(x_*)\rangle \geq 0 \text{ for } i \in \mathcal{A}\} = \emptyset.$$

But then the result follows directly from Farkas' Lemma–a proof of this famous result is given, for example, as Lemma 9.2.4 in R. Fletcher "Practical Methods of Optimization", Wiley (1987, 2nd edition).

Farkas' Lemma. Given any vectors g and a_i, $i \in \mathcal{A}$, the set

$$\mathcal{S} = \{s \mid \langle s, g\rangle < 0 \text{ and } \langle s, a_i\rangle \geq 0 \text{ for } i \in \mathcal{A}\}$$

is empty if and only if

$$g = \sum_{i\in\mathcal{A}} y_i a_i$$

for some $y_i \geq 0$, $i \in \mathcal{A}$

Proof (of Theorem 1.9). The expansion (C.4) for the change in the objective function will be dominated by the first-order term $\alpha\langle s, g(x_*)\rangle$ for feasible perturbations unless $\langle s, g(x_*)\rangle = 0$, in which case the expansion (C.8) is relevant. Thus we must have that (C.9) holds for all feasible s for which $\langle s, g(x_*)\rangle = 0$. The latter requirement gives that

$$0 = \langle s, g(x_*)\rangle = \sum_{i\in\mathcal{A}} y_i \langle s, a_i(x_*)\rangle,$$

and hence that either $y_i = 0$ or $\langle s, a_i(x_*)\rangle = 0$ (or both).

We now focus on the *subset* of all feasible arcs that ensure $c_i(x(\alpha)) = 0$ if $y_i > 0$ and $c_i(x(\alpha)) \geq 0$ if $y_i = 0$ for $i \in \mathcal{A}$. For those constraints for which $c_i(x(\alpha)) = 0$, we have that (C.2) and (C.3) hold, and thus for such perturbations $s \in \mathcal{N}_+$. In this case

$$\langle p, g(x_*)\rangle = \sum_{i\in\mathcal{A}} y_i \langle p, a_i(x_*)\rangle = \sum_{\substack{i\in\mathcal{A}\\y_i>0}} y_i \langle p, a_i(x_*)\rangle$$

$$= -\sum_{\substack{i\in\mathcal{A}\\y_i>0}} y_i \langle s, H_i(x_*)s\rangle = -\sum_{i\in\mathcal{A}} y_i \langle s, H_i(x_*)s\rangle$$

This combines with (C.9) to give that

$$\langle s, H(x_*, y_*)s \rangle$$
$$\equiv \left\langle s, \left(H(x_*) - \sum_{i=1}^{m}(y_*)_i H_i(x_*) \right) s \right\rangle = \langle p, g(x_*) \rangle + \langle s, H(x_*)s \rangle \geq 0.$$

for all $s \in \mathcal{N}_+$, which is the required result. □

Proof (of Theorem 1.10). Consider any feasible arc $x(\alpha)$. We have seen that (C.10) and (C.11) hold, and that first-order feasible perturbations are characterized by $\mathcal{N}_+$. It then follows from (C.11) that

$$\langle p, g(x_*) \rangle = \sum_{i \in \mathcal{A}} y_i \langle p, a_i(x_*) \rangle = \sum_{\substack{i \in \mathcal{A} \\ \langle s, a_i(x_*) \rangle = 0}} y_i \langle p, a_i(x_*) \rangle$$
$$\geq - \sum_{\substack{i \in \mathcal{A} \\ \langle s, a_i(x_*) \rangle = 0}} y_i \langle s, H_i(x_*)s \rangle = - \sum_{i \in \mathcal{A}} y_i \langle s, H_i(x_*)s \rangle,$$

and hence by assumption that

$$\langle p, g(x_*) \rangle + \langle s, H(x_*)s \rangle$$
$$\geq \left\langle s, \left(H(x_*) - \sum_{i=1}^{m}(y_*)_i H_i(x_*) \right) s \right\rangle \equiv \langle s, H(x_*, y_*)s \rangle > 0$$

for all $s \in \mathcal{N}_+$. But this then combines with (C.4) and (C.10) to show that $f(x(\alpha)) > f(x_*)$ for all sufficiently small α. □

Proof (of Theorem 2.1). From Taylor's theorem (Theorem 1.1), and using the bound

$$\alpha \leq \frac{2(\beta - 1)\langle p, g(x) \rangle}{\gamma(x)\|p\|_2^2},$$

we have that

$$f(x + \alpha p) \leq f(x) + \alpha\langle p, g(x) \rangle + \tfrac{1}{2}\gamma(x)\alpha^2\|p\|^2$$
$$\leq f(x) + \alpha\langle p, g(x) \rangle + \alpha(\beta - 1)\langle p, g(x) \rangle$$
$$= f(x) + \alpha\beta\langle p, g(x) \rangle. \quad □$$

Proof (of Corollary 2.2). Theorem 2.1 shows that the linesearch will terminate as soon as $\alpha^{(l)} \leq \alpha_{\max}$. There are two cases to consider. Firstly, it may be that α_{init} satisfies the Armijo condition, in which case $\alpha_k = \alpha_{\text{init}}$. If not, there must be a last linesearch iteration, say the lth, for which $\alpha^{(l)} > \alpha_{\max}$ (if the linesearch has not already terminated). Then $\alpha_k \geq \alpha^{(l+1)} = \tau\alpha^{(l)} > \tau\alpha_{\max}$. Combining these two cases gives the required result. □

Proof (of Theorem 2.3). We shall suppose that $g_k \neq 0$ for all k and that

$$\lim_{k\to\infty} f_k > -\infty.$$

From the Armijo condition, we have that

$$f_{k+1} - f_k \leq \alpha_k \beta \langle p_k, g_k \rangle$$

for all k, and hence summing over the first j iterations

$$f_{j+1} - f_0 \leq \sum_{k=0}^{j} \alpha_k \beta \langle p_k, g_k \rangle.$$

Since the left-hand side of this inequality is, by assumption, bounded below, so is the sum on right-hand-side. As this sum is composed of negative terms, we deduce that

$$\lim_{k\to\infty} \alpha_k \langle p_k, g_k \rangle = 0.$$

Now define the two sets

$$\mathcal{K}_1 = \left\{ k \mid \alpha_{\text{init}} > \frac{2\tau(\beta-1)\langle p_k, g_k \rangle}{\gamma \|p_k\|_2^2} \right\}$$

and

$$\mathcal{K}_2 = \left\{ k \mid \alpha_{\text{init}} \leq \frac{2\tau(\beta-1)\langle p_k, g_k \rangle}{\gamma \|p_k\|_2^2} \right\},$$

where γ is the assumed uniform Lipschitz constant. For $k \in \mathcal{K}_1$,

$$\alpha_k \geq \frac{2\tau(\beta-1)\langle p_k, g_k \rangle}{\gamma \|p_k\|_2^2}$$

in which case

$$\alpha_k \langle p_k, g_k \rangle \leq \frac{2\tau(\beta-1)}{\gamma} \left(\frac{\langle p_k, g_k \rangle}{\|p_k\|} \right)^2 < 0.$$

Thus

$$\lim_{k\in\mathcal{K}_1\to\infty} \frac{|\langle p_k, g_k \rangle|}{\|p_k\|_2} = 0. \tag{C.12}$$

For $k \in \mathcal{K}_2$,

$$\alpha_k \geq \alpha_{\text{init}}$$

in which case

$$\lim_{k\in\mathcal{K}_2\to\infty} |\langle p_k, g_k \rangle| = 0. \tag{C.13}$$

Combining (C.12) and (C.13) gives the required result. □

Proof (of Theorem 2.4). Follows immediately from Theorem 2.3, since for $p_k = -g_k$,

$$\min\left(|\langle p_k, g_k\rangle|, |\langle p_k, g_k\rangle|/\|p_k\|_2\right) = \|g_k\|_2 \min\left(1, \|g_k\|_2\right)$$

and thus

$$\lim_{k\to\infty} \min\left(|\langle p_k, g_k\rangle|, |\langle p_k, g_k\rangle|/\|p_k\|_2\right) = 0$$

implies that $\lim_{k\to\infty} g_k = 0$. $\square$

Proof (of Theorem 2.5). Let $\lambda^{\min}(B_k)$ and $\lambda^{\max}(B_k)$ be the smallest and largest eigenvalues of B_k. By assumption, there are bounds $\lambda^{\min} > 0$ and $\lambda^{\max}$ such that

$$\lambda^{\min} \leq \lambda^{\min}(B_k) \leq \frac{\langle s, B_k s\rangle}{\|s\|^2} \leq \lambda^{\max}(B_k) \leq \lambda^{\max}$$

for any nonzero vector s. Thus

$$|\langle p_k, g_k\rangle| = |\langle g_k, B_k^{-1} g_k\rangle| \geq \lambda_{\min}(B_k^{-1})\|g_k\|_2^2 = \frac{1}{\lambda_{\max}(B_k)}\|g_k\|_2^2 \geq \lambda_{\max}^{-1}\|g_k\|_2^2.$$

In addition

$$\|p_k\|_2^2 = \langle g_k, B_k^{-2} g_k\rangle \leq \lambda_{\max}(B_k^{-2})\|g_k\|_2^2 = \frac{1}{\lambda_{\min}(B_k^2)}\|g_k\|_2^2 \leq \lambda_{\min}^{-2}\|g_k\|_2^2,$$

and hence

$$\|p_k\|_2 \leq \lambda_{\max}^{-1}\|g_k\|_2,$$

which leads to

$$\frac{|\langle p_k, g_k\rangle|}{\|p_k\|_2} \geq \frac{\lambda_{\min}}{\lambda_{\max}}\|g_k\|_2.$$

Thus

$$\min\left(|\langle p_k, g_k\rangle|, |\langle p_k, g_k\rangle|/\|p_k\|_2\right) \geq \lambda_{\max}^{-1}\|g_k\|_2 \min\left(\|g_k\|_2, \lambda_{\min}\right).$$

and hence

$$\lim_{k\to\infty} \min\left(|\langle p_k, g_k\rangle|, |\langle p_k, g_k\rangle|/\|p_k\|_2\right) = 0$$

implies, as before, that $\lim_{k\to\infty} g_k = 0$. $\square$

Proof (of Theorem 2.6). Consider the sequence of iterates x_k, $k \in \mathcal{K}$, whose limit is x_*. By continuity, H_k is positive definite for all such k sufficiently large. In particular, we have that there is a $k_0 \geq 0$ such that

$$\langle p_k, H_k p_k\rangle \geq \tfrac{1}{2}\lambda_{\min}(H_*)\|p_k\|_2^2$$

for all $k \in \mathcal{K} \geq k_0$, where $\lambda_{\min}(H_*)$ is the smallest eigenvalue of $H(x_*)$. We may then deduce that

$$|\langle p_k, g_k\rangle| = -\langle p_k, g_k\rangle = \langle p_k, H_k p_k\rangle \geq \tfrac{1}{2}\lambda_{\min}(H_*)\|p_k\|_2^2. \tag{C.14}$$

for all such k, and also that

$$\lim_{k \in \mathcal{K} \to \infty} p_k = 0$$

since Theorem 2.5 implies that at least one of the left-hand sides of (C.14) and

$$\frac{|\langle p_k, g_k \rangle|}{\|p_k\|_2} = -\frac{\langle p_k, g_k \rangle}{\|p_k\|_2} \geq \tfrac{1}{2}\lambda_{\min}(H_*)\|p_k\|_2$$

converges to zero for all such k.

From Taylor's theorem, there is a z_k between x_k and $x_k + p_k$ such that

$$f(x_k + p_k) = f_k + \langle p_k, g_k \rangle + \tfrac{1}{2}\langle p_k, H(z_k)p_k \rangle.$$

Thus, the Lipschitz continuity of H gives that

$$\begin{aligned}
& f(x_k + p_k) - f_k - \tfrac{1}{2}\langle p_k, g_k \rangle \\
&= \tfrac{1}{2}(\langle p_k, g_k \rangle + \langle p_k, H(z_k)p_k \rangle) \\
&= \tfrac{1}{2}(\langle p_k, g_k \rangle + \langle p_k, H_k p_k \rangle) + \tfrac{1}{2}\langle p_k, (H(z_k) - H_k)p_k \rangle \leq \tfrac{1}{2}\gamma\|z_k - x_k\|_2\|p_k\|_2^2 \\
&\leq \tfrac{1}{2}\gamma\|p_k\|_2^3
\end{aligned} \tag{C.15}$$

since $H_k p_k + g_k = 0$. Now pick k sufficiently large so that

$$\gamma\|p_k\|_2 \leq \lambda_{\min}(H_*)(1 - 2\beta).$$

In this case, (C.14) and (C.15) give that

$$\begin{aligned}
& f(x_k + p_k) - f_k \\
&\leq \tfrac{1}{2}\langle p_k, g_k \rangle + \tfrac{1}{2}\lambda_{\min}(H_*)(1 - 2\beta)\|p_k\|_2^2 \\
&\leq \tfrac{1}{2}(1 - (1 - 2\beta))\langle p_k, g_k \rangle = \beta\langle p_k, g_k \rangle,
\end{aligned}$$

and thus that a unit stepsize satisfies the Armijo condition, which proves (i).

To obtain the remaining results, note that $\|H_k^{-1}\|_2 \leq 2/\lambda_{\min}(H_*)$ for all sufficiently large $k \in \mathcal{K}$. The iteration gives

$$\begin{aligned}
& x_{k+1} - x_* \\
&= x_k - x_* - H_k^{-1}g_k = x_k - x_* - H_k^{-1}(g_k - g(x_*)) \\
&= H_k^{-1}(g(x_*) - g_k - H_k(x_* - x_k)).
\end{aligned}$$

But Theorem 1.2 gives that

$$\|g(x_*) - g_k - H_k(x_* - x_k)\|_2 \leq \gamma\|x_* - x_k\|_2^2.$$

Hence

$$\|x_{k+1} - x_*\|_2 \leq \gamma\|H_k^{-1}\|_2\|x_* - x_k\|_2^2$$

which is (iii) when $\kappa = 2\gamma/\lambda_{\min}(H_*)$. Result (ii) follows since once an iterate becomes sufficiently close to x_*, (iii) implies that the next is even closer. $\square$

Conjugate Gradient Methods (Section 2)

All of the results given here are easy to verify, and may be found in any of the books of suggested background reading material. The result that any $p_k = p^i$ is a descent direction follows immediately since the fact that p^i minimizes $q(p)$ in $\mathcal{D}^i$ implies that

$$p^i = p^{i-1} - \frac{\langle g_k, d^{i-1} \rangle}{\langle d^{i-1}, B_k d^{i-1} \rangle d^{i-1}}.$$

Thus

$$\langle g_k, p^i \rangle = \langle g_k, p^{i-1} \rangle - \frac{(\langle g_k, d^{i-1} \rangle)^2}{\langle d^{i-1}, B_k d^{i-1} \rangle},$$

from which it follows that $\langle g_k, p^i \rangle < \langle g_k, p^{i-1} \rangle$. The result then follows by induction, since

$$\langle g_k, p^1 \rangle = -\frac{\|g_k\|_2^2}{\langle g_k, B_k g_k \rangle} < 0.$$

Proof (of Theorem 3.1). Firstly note that, for all $\alpha \geq 0$,

$$m_k(-\alpha g_k) = f_k - \alpha \|g_k\|_2^2 + \tfrac{1}{2}\alpha^2 \langle g_k, B_k g_k \rangle. \tag{C.16}$$

If g_k is zero, the result is immediate. So suppose otherwise. In this case, there are three possibilities:

(i) the curvature $\langle g_k, B_k g_k \rangle$ is not strictly positive; in this case $m_k(-\alpha g_k)$ is unbounded from below as α increases, and hence the Cauchy point occurs on the trust-region boundary.

(ii) the curvature $\langle g_k, B_k g_k \rangle > 0$ and the minimizer of $m_k(-\alpha g_k)$ occurs at or beyond the trust-region boundary; once again, the the Cauchy point occurs on the trust-region boundary.

(iii) the curvature $\langle g_k, B_k g_k \rangle > 0$ and the minimizer of $m_k(-\alpha g_k)$, and hence the Cauchy point, occurs before the trust-region is reached.

We consider each case in turn;

Case (i). In this case, since $\langle g_k, B_k g_k \rangle \leq 0$, (C.16) gives

$$m_k(-\alpha g_k) = f_k - \alpha \|g_k\|_2^2 + \tfrac{1}{2}\alpha^2 \langle g_k, B_k g_k \rangle \leq f_k - \alpha \|g_k\|_2^2 \tag{C.17}$$

for all $\alpha \geq 0$. Since the Cauchy point lies on the boundary of the trust region

$$\alpha_k^C = \frac{\Delta_k}{\|g_k\|}. \tag{C.18}$$

Substituting this value into (C.17) gives

$$f_k - m_k(s_k^C) \geq \|g_k\|_2^2 \frac{\Delta_k}{\|g_k\|} \geq \kappa_s \|g_k\|_2 \Delta_k \geq \tfrac{1}{2}\kappa_s \|g_k\|_2 \Delta_k \tag{C.19}$$

since $\|g_k\|_2 \geq \kappa_s\|g_k\|$.

Case (ii). In this case, let α_k^* be the unique minimizer of (C.16); elementary calculus reveals that

$$\alpha_k^* = \frac{\|g_k\|_2^2}{\langle g_k, B_k g_k\rangle}. \tag{C.20}$$

Since this minimizer lies on or beyond the trust-region boundary (C.18) and (C.20) together imply that

$$\alpha_k^{\mathrm{c}}\langle g_k, B_k g_k\rangle \leq \|g_k\|_2^2.$$

Substituting this last inequality in (C.16), and using (C.18) and $\|g_k\|_2 \geq \kappa_s\|g_k\|$, it follows that

$$f_k - m_k(s_k^{\mathrm{c}})$$
$$= \alpha_k^{\mathrm{c}}\|g_k\|_2^2 - \tfrac{1}{2}[\alpha_k^{\mathrm{c}}]^2\langle g_k, B_k g_k\rangle \geq \tfrac{1}{2}\alpha_k^{\mathrm{c}}\|g_k\|_2^2 = \tfrac{1}{2}\|g_k\|_2^2\frac{\Delta_k}{\|g_k\|} \geq \tfrac{1}{2}\kappa_s\|g_k\|_2\Delta_k.$$

Case (iii). In this case, $\alpha_k^{\mathrm{c}} = \alpha_k^*$, and (C.16) becomes

$$f_k - m_k(s_k^{\mathrm{c}}) = \frac{\|g_k\|_2^4}{\langle g_k, B_k g_k\rangle} - \tfrac{1}{2}\frac{\|g_k\|_2^4}{\langle g_k, B_k g_k\rangle} = \tfrac{1}{2}\frac{\|g_k\|_2^4}{\langle g_k, B_k g_k\rangle} \geq \tfrac{1}{2}\frac{\|g_k\|_2^2}{1+\|B_k\|},$$

where

$$|\langle g_k, B_k g_k\rangle| \leq \|g_k\|_2^2\|B_k\|_2 \leq \|g_k\|_2^2(1+\|B_k\|_2)$$

because of the Cauchy-Schwarz inequality.

The result follows since it is true in each of the above three possible cases. Note that the "1+" is only needed to cover case where $B_k = 0$, and that in this case, the "min" in the theorem might actually be replaced by $\kappa_s\Delta_k$. □

Proof (of Corollary 3.2). Immediate from Theorem 3.1 and the requirement that $m_k(s_k) \leq m_k(s_k^{\mathrm{c}})$. □

Proof (of Lemma 3.3). The mean value theorem gives that

$$f(x_k + s_k) = f(x_k) + \langle s_k, \nabla_x f(x_k)\rangle + \tfrac{1}{2}\langle s_k, \nabla_{xx} f(\xi_k)s_k\rangle$$

for some ξ_k in the segment $[x_k, x_k + s_k]$. Thus

$$|f(x_k + s_k) - m_k(s_k)|$$
$$= \tfrac{1}{2}|\langle s_k, H(\xi_k)s_k\rangle - \langle s_k, B_k s_k\rangle| \leq \tfrac{1}{2}|\langle s_k, H(\zeta_k)s_k\rangle| + \tfrac{1}{2}|\langle s_k, B_k s_k\rangle|$$
$$\leq \tfrac{1}{2}(\kappa_h + \kappa_b)\|s_k\|_2^2 \leq \tfrac{1}{2}\kappa_l^2(\kappa_h + \kappa_b)\|s_k\|^2 \leq \kappa_d\Delta_k^2$$

using the triangle and Cauchy-Schwarz inequalities. □

Proof (of Lemma 3.4). By definition,

$$1 + \|B_k\|_2 \le \kappa_h + \kappa_b,$$

and hence for any radius satisfying the given (first) bound,

$$\kappa_s \Delta_k \le \frac{\|g_k\|_2}{\kappa_h + \kappa_b} \le \frac{\|g_k\|_2}{1 + \|B_k\|_2}.$$

As a consequence, Corollary 3.2 gives that

$$f_k - m_k(s_k) \ge \tfrac{1}{2}\|g_k\|_2 \min\left[\frac{\|g_k\|_2}{1 + \|B_k\|_2}, \kappa_s \Delta_k\right] = \tfrac{1}{2}\kappa_s\|g_k\|_2\Delta_k. \qquad \text{(C.21)}$$

But then Lemma 3.3 and the assumed (second) bound on the radius gives that

$$|\rho_k - 1| = \left|\frac{f(x_k + s_k) - m_k(s_k)}{f_k - m_k(s_k)}\right| \le 2\frac{\kappa_d \Delta_k^2}{\kappa_s\|g_k\|_2\Delta_k} = \frac{2\kappa_d}{\kappa_s}\frac{\Delta_k}{\|g_k\|_2} \le 1 - \eta_v. \qquad \text{(C.22)}$$

Therefore, $\rho_k \ge \eta_v$ and the iteration is very successful. $\square$

Proof (of Lemma 3.5). Suppose otherwise that Δ_k can become arbitrarily small. In particular, assume that iteration k is the first such that

$$\Delta_{k+1} \le \kappa_\epsilon. \qquad \text{(C.23)}$$

Then since the radius for the previous iteration must have been larger, the iteration was unsuccessful, and thus $\gamma_d \Delta_k \le \Delta_{k+1}$. Hence

$$\Delta_k \le \epsilon \min\left(\frac{1}{\kappa_s(\kappa_h + \kappa_b)}, \frac{\kappa_s(1 - \eta_v)}{2\kappa_d}\right)$$
$$\le \|g_k\| \min\left(\frac{1}{\kappa_s(\kappa_h + \kappa_b)}, \frac{\kappa_s(1 - \eta_v)}{2\kappa_d}\right).$$

But this contradicts the assertion of Lemma 3.4 that the k-th iteration must be very successful. $\square$

Proof (of Lemma 3.6). The mechanism of the algorithm ensures that $x_* = x_{k_0+1} = x_{k_0+j}$ for all $j > 0$, where k_0 is the index of the last successful iterate. Moreover, since all iterations are unsuccessful for sufficiently large k, the sequence $\{\Delta_k\}$ converges to zero. If $\|g_{k_0+1}\| > 0$, Lemma 3.4 then implies that there must be a successful iteration of index larger than k_0, which is impossible. Hence $\|g_{k_0+1}\| = 0$. $\square$

Proof (of Theorem 3.7). Lemma 3.6 shows that the result is true when there are only a finite number of successful iterations. So it remains to consider the case where there are an infinite number of successful iterations. Let $\mathcal{S}$ be the index set of successful iterations. Now suppose that

$$\|g_k\| \geq \epsilon \qquad (C.24)$$

for some $\epsilon > 0$ and all k, and consider a successful iteration of index k. The fact that k is successful, Corollary 3.2, Lemma 3.5, and the assumption (C.24) give that

$$f_k - f_{k+1} \geq \eta_s[f_k - m_k(s_k)] \geq \delta_\epsilon \overset{\text{def}}{=} \tfrac{1}{2}\eta_s\epsilon \min\left[\frac{\epsilon}{1+\kappa_b}, \kappa_s\kappa_\epsilon\right]. \qquad (C.25)$$

Summing now over all successful iterations from 0 to k, it follows that

$$f_0 - f_{k+1} = \sum_{\substack{j=0 \\ j\in\mathcal{S}}}^{k}[f_j - f_{j+1}] \geq \sigma_k\delta_\epsilon,$$

where σ_k is the number of successful iterations up to iteration k. But since there are infinitely many such iterations, it must be that

$$\lim_{k\to\infty}\sigma_k = +\infty.$$

Thus (C.24) can only be true if f_{k+1} is unbounded from below, and conversely, if f_{k+1} is bounded from below, (C.24) must be false, and there is a subsequence of the $\|g_k\|$ converging to zero. $\square$

Proof (of Corollary 3.8). Suppose otherwise that f_k is bounded from below, and that there is a subsequence of successful iterates, indexed by $\{t_i\} \subseteq \mathcal{S}$, such that

$$\|g_{t_i}\| \geq 2\epsilon > 0 \qquad (C.26)$$

for some $\epsilon > 0$ and for all i. Theorem 3.7 ensures the existence, for each t_i, of a first successful iteration $\ell_i > t_i$ such that $\|g_{\ell_i}\| < \epsilon$. That is to say that there is another subsequence of $\mathcal{S}$ indexed by $\{\ell_i\}$ such that

$$\|g_k\| \geq \epsilon \text{ for } t_i \leq k < \ell_i \text{ and } \|g_{\ell_i}\| < \epsilon. \qquad (C.27)$$

We now restrict our attention to the subsequence of successful iterations whose indices are in the set

$$\mathcal{K} \overset{\text{def}}{=} \{k \in \mathcal{S} \mid t_i \leq k < \ell_i\},$$

where t_i and ℓ_i belong to the two subsequences defined above.

The subsequences $\{t_i\}$, $\{\ell_i\}$ and $\mathcal{K}$ are all illustrated in Figure 3.1, where, for simplicity, it is assumed that all iterations are successful. In this figure, we have marked position j in each of the subsequences represented in abscissa when j belongs to that subsequence. Note in this example that $\ell_0 = \ell_1 =$

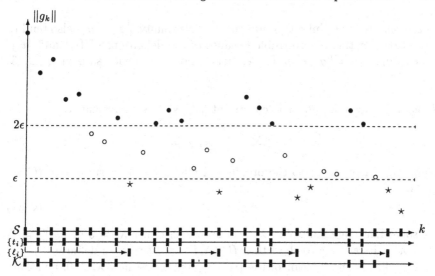

Fig. 3.1. The subsequences of the proof of Corollary 3.8
$\ell_2 = \ell_3 = \ell_4 = \ell_5 = 8$, which we indicated by arrows from $t_0 = 0$, $t_1 = 1$,
$t_2 = 2$, $t_3 = 3$, $t_4 = 4$ and $t_5 = 7$ to $k = 9$, and so on.

As in the previous proof, it immediately follows that

$$f_k - f_{k+1} \geq \eta_s[f_k - m_k(s_k)] \geq \tfrac{1}{2}\eta_s\epsilon \min\left[\frac{\epsilon}{1 + \kappa_b}, \kappa_s\Delta_k\right] \qquad \text{(C.28)}$$

holds for all $k \in \mathcal{K}$ because of (C.27). Hence, since $\{f_k\}$ is, by assumption, bounded from below, the left-hand side of (C.28) must tend to zero when k tends to infinity, and thus that

$$\lim_{\substack{k \to \infty \\ k \in \mathcal{K}}} \Delta_k = 0.$$

As a consequence, the second term dominates in the minimum of (C.28) and it follows that, for $k \in \mathcal{K}$ sufficiently large,

$$\Delta_k \leq \frac{2}{\epsilon\eta_s\kappa_s}[f_k - f_{k+1}].$$

We then deduce from this bound that, for i sufficiently large,

$$\|x_{t_i} - x_{\ell_i}\| \leq \sum_{\substack{j=t_i \\ j\in\mathcal{K}}}^{\ell_i-1} \|x_j - x_{j+1}\| \leq \sum_{\substack{j=t_i \\ j\in\mathcal{K}}}^{\ell_i-1} \Delta_j \leq \frac{2}{\epsilon\eta_s\kappa_s}[f_{t_i} - f_{\ell_i}]. \qquad \text{(C.29)}$$

But, because $\{f_k\}$ is monotonic and, by assumption, bounded from below, the right-hand side of (C.29) must converge to zero. Thus $\|x_{t_i} - x_{\ell_i}\|$ tends to zero as i tends to infinity, and hence, by continuity, $\|g_{t_i} - g_{\ell_i}\|$ also tend to zero. However this is impossible because of the definitions of $\{t_i\}$ and $\{\ell_i\}$, which imply that $\|g_{t_i} - g_{\ell_i}\| \geq \epsilon$. Hence, no subsequence satisfying (C.26) can exist. □

Proof (of Theorem 3.9). The constraint $\|s\|_2 \leq \Delta$ is equivalent to

$$\tfrac{1}{2}\Delta^2 - \tfrac{1}{2}\langle s, s\rangle \geq 0. \tag{C.30}$$

Applying Theorem 1.7 to the problem of minimizing $q(s)$ subject to (C.30) gives

$$g + Bs_* = -\lambda_* s_* \tag{C.31}$$

for some Lagrange multiplier $\lambda_* \geq 0$ for which either $\lambda_* = 0$ or $\|s_*\|_2 = \Delta$ (or both). It remains to show that $B + \lambda_* I$ is positive semi-definite.

If s_* lies in the interior of the trust-region, necessarily $\lambda_* = 0$, and Theorem 1.8 implies that $B + \lambda_* I = B$ must be positive semi-definite. Likewise if $\|s_*\|_2 = \Delta$ and $\lambda_* = 0$, it follows from Theorem 1.8 that necessarily $\langle v, Bv\rangle \geq 0$ for all $v \in \mathcal{N}_+ = \{v | \langle s_*, v\rangle \geq 0\}$. If $v \notin \mathcal{N}_+$, then $-v \in \mathcal{N}_+$, and thus $\langle v, Bv\rangle \geq 0$ for all v. Thus the only outstanding case is where $\|s_*\|_2 = \Delta$ and $\lambda_* > 0$. In this case, Theorem 1.8 shows that $\langle v, (B + \lambda_* I)v\rangle \geq 0$ for all $v \in \mathcal{N}_+ = \{v | \langle s_*, v\rangle = 0\}$, so it remains to consider $\langle v, Bv\rangle$ when $\langle s_*, v\rangle \neq 0$.

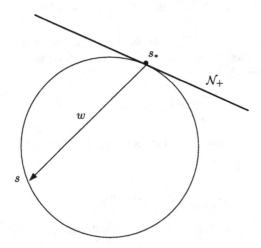

Fig. 3.2. Construction of "missing" directions of positive curvature.

Let s be any point on the boundary of the trust-region, and let $w = s - s_*$, as in Figure 3.2. Then

$$-\langle w, s_*\rangle = \langle s_* - s, s_*\rangle = \tfrac{1}{2}\langle s_* - s, s_* - s\rangle = \tfrac{1}{2}\langle w, w\rangle \tag{C.32}$$

since $\|s\|_2 = \Delta = \|s_*\|_2$. Combining this with (C.31) gives

$$q(s) - q(s_*)$$
$$= \langle w, g + Bs_* \rangle + \tfrac{1}{2}\langle w, Bw \rangle = -\lambda_*\langle w, s_* \rangle + \tfrac{1}{2}\langle w, Bw \rangle$$
$$= \tfrac{1}{2}\langle w, (B + \lambda_* I)w \rangle, \tag{C.33}$$

and thus necessarily $\langle w, (B + \lambda_* I)w \rangle \geq 0$ since s_* is a global minimizer. It is easy to show that

$$s = s_* - 2\frac{\langle s_*, v \rangle}{\langle v, v \rangle} v$$

lies on the trust-region boundary, and thus for this s, w is parallel to v from which it follows that $\langle v, (B + \lambda_* I)v \rangle \geq 0$.

When $B + \lambda_* I$ is positive definite, $s_* = -(B + \lambda_* I)^{-1}g$. If this point is on the trust-region boundary, while s is any value in the trust-region, (C.32) and (C.33) become $-\langle w, s_* \rangle \geq \tfrac{1}{2}\langle w \cdot w \rangle$ and $q(s) \geq q(s_*) + \tfrac{1}{2}\langle w, (B + \lambda_* I)w \rangle$ respectively. Hence, $q(s) > q(s_*)$ for any $s \neq s_*$. If s_* is interior, $\lambda_* = 0$, B is positive definite, and thus s_* is the unique unconstrained minimizer of $q(s)$.

Newton's Method for the Secular Equation (Section 3)

Recall that the Newton correction at λ is $-\phi(\lambda)/\phi'(\lambda)$. Since

$$\phi(\lambda) = \frac{1}{\|s(\lambda)\|_2} - \frac{1}{\Delta} = \frac{1}{(\langle s(\lambda), s(\lambda) \rangle)^{\frac{1}{2}}} - \frac{1}{\Delta},$$

it follows, on differentiating, that

$$\phi'(\lambda) = -\frac{\langle s(\lambda), \nabla_\lambda s(\lambda) \rangle}{(\langle s(\lambda), s(\lambda) \rangle)^{\frac{3}{2}}} = -\frac{\langle s(\lambda), \nabla_\lambda s(\lambda) \rangle}{\|s(\lambda)\|_2^3}.$$

In addition, on differentiating the defining equation

$$(B + \lambda I)s(\lambda) = -g,$$

it must be that

$$(B + \lambda I)\nabla_\lambda s(\lambda) + s(\lambda) = 0.$$

Notice that, rather than the value of $\nabla_\lambda s(\lambda)$, merely the numerator

$$\langle s(\lambda), \nabla_\lambda s(\lambda) \rangle = -\langle s(\lambda), (B + \lambda I)(\lambda)^{-1} s(\lambda) \rangle$$

is required in the expression for $\phi'(\lambda)$. Given the factorization $B + \lambda I = L(\lambda)L^T(\lambda)$, the simple relationship

$$\langle s(\lambda), (B + \lambda I)^{-1} s(\lambda) \rangle$$
$$= \langle s(\lambda), L^{-T}(\lambda)L^{-1}(\lambda)s(\lambda) \rangle = \langle L^{-1}(\lambda)s(\lambda), L^{-1}(\lambda)s(\lambda) \rangle = \|w(\lambda)\|_2^2$$

where $L(\lambda)w(\lambda) = s(\lambda)$ then justifies the Newton step. $\square$

Proof (of Theorem 3.10). We first show that

$$\langle d^i, d^j \rangle = \frac{\|g^i\|_2^2}{\|g^j\|_2^2} \|d^j\|_2^2 > 0 \tag{C.34}$$

for all $0 \leq j \leq i \leq k$. For any i, (C.34) is trivially true for $j = i$. Suppose it is also true for all $i \leq l$. Then, the update for d^{l+1} gives

$$d^{l+1} = -g^{l+1} + \frac{\|g^{l+1}\|_2^2}{\|g^l\|_2^2} d^l.$$

Forming the inner product with d^j, and using the fact that $\langle d^j, g^{l+1} \rangle = 0$ for all $j = 0, \ldots, l$, and (C.34) when $j = l$, reveals

$$\langle d^{l+1}, d^j \rangle$$
$$= -\langle g^{l+1}, d^j \rangle + \frac{\|g^{l+1}\|_2^2}{\|g^l\|_2^2} \langle d^l, d^j \rangle = \frac{\|g^{l+1}\|_2^2}{\|g^l\|_2^2} \frac{\|g^l\|_2^2}{\|g^j\|_2^2} \|d^j\|_2^2 = \frac{\|g^{l+1}\|_2^2}{\|g^j\|_2^2} \|d^j\|_2^2 > 0.$$

Thus (C.34) is true for $i \leq l + 1$, and hence for all $0 \leq j \leq i \leq k$.

We now have from the algorithm that

$$s^i = s^0 + \sum_{j=0}^{i-1} \alpha^j d^j = \sum_{j=0}^{i-1} \alpha^j d^j$$

as, by assumption, $s^0 = 0$. Hence

$$\langle s^i, d^i \rangle = \left\langle \sum_{j=0}^{i-1} \alpha^j d^j, d^i \right\rangle = \sum_{j=0}^{i-1} \alpha^j \langle d^j, d^i \rangle > 0 \tag{C.35}$$

as each $\alpha^j > 0$, which follows from the definition of α^j, since $\langle d^j, H d^j \rangle > 0$, and from relationship (C.34). Hence

$$\|s^{i+1}\|_2^2 = \langle s^{i+1}, s^{i+1} \rangle = \langle s^i + \alpha^i d^i, s^i + \alpha^i d^i \rangle$$
$$= \langle s^i, s^i \rangle + 2\alpha^i \langle s^i, d^i \rangle + \alpha^{i\,2} \langle d^i, d^i \rangle > \langle s^i, s^i \rangle = \|s^i\|_2^2$$

follows directly from (C.35) and $\alpha^i > 0$ which is the required result. □

Proof (of Theorem 3.11). The proof is elementary but rather complicated. See

Y. Yuan, "On the truncated conjugate-gradient method", *Mathematical Programming*, **87** (2000) 561:573

for full details. □

Proof (of Theorem 4.1). Let $\mathcal{A} = \mathcal{A}(x_*)$, and $\mathcal{I} = \{1, \ldots, m\} \setminus \mathcal{A}$ be the indices of constraints that are active and inactive at x_*. Furthermore let subscripts $\mathcal{A}$ and $\mathcal{I}$ denote the rows of matrices/vectors whose indices are indexed by these sets. Denote the left generalized inverse of $A_{\mathcal{A}}^T(x)$ by

$$A_{\mathcal{A}}^+(x) = \left(A_{\mathcal{A}}(x)A_{\mathcal{A}}^T(x)\right)^{-1} A_{\mathcal{A}}(x)$$

at any point for which $A_{\mathcal{A}}(x)$ is full rank. Since, by assumption, $A_{\mathcal{A}}(x_*)$ is full rank, these generalized inverses exists, and are bounded and continuous in some open neighbourhood of x_*. .

Now let

$$(y_k)_i = \frac{\mu_k}{c_i(x_k)}$$

for $i = 1, \ldots, m$, as well as

$$(y_*)_{\mathcal{A}} = A_{\mathcal{A}}^+(x_*)g(x_*)$$

and $(y_*)_{\mathcal{I}} = 0$. If $\mathcal{I} \neq \emptyset$, then

$$\|(y_k)_{\mathcal{I}}\|_2 \leq 2\mu_k \sqrt{|\mathcal{I}|}/ \min_{i \in \mathcal{I}} |c_i(x_*)| \tag{C.36}$$

for all sufficiently large k. It then follows from the inner-iteration termination test that

$$\|g(x_k) - A_{\mathcal{A}}^T(x_k)(y_k)_{\mathcal{A}}\|_2 \leq \|g(x_k) - A^T(x_k)y_k\|_2 + \|A_{\mathcal{I}}^T(x_k)(y_k)_{\mathcal{I}}\|_2$$
$$\leq \bar{\epsilon}_k \overset{\text{def}}{=} \epsilon_k + \mu_k \frac{2\sqrt{|\mathcal{I}|}\|A_{\mathcal{I}}\|_2}{\min_{i \in \mathcal{I}} |c_i(x_*)|}. \tag{C.37}$$

Hence

$$\|A_{\mathcal{A}}^+(x_k)g(x_k) - (y_k)_{\mathcal{A}}\|_2$$
$$= \|A_{\mathcal{A}}^+(x_k)(g(x_k) - A_{\mathcal{A}}^T(x_k)(y_k)_{\mathcal{A}})\|_2 \leq 2\|A_{\mathcal{A}}^+(x_*)\|_2\bar{\epsilon}_k.$$

Then

$$\|(y_k)_{\mathcal{A}} - (y_*)_{\mathcal{A}}\|_2$$
$$\leq \|A_{\mathcal{A}}^+(x_*)g(x_*) - A_{\mathcal{A}}^+(x_k)g(x_k)\|_2 + \|A_{\mathcal{A}}^+(x_k)g(x_k) - (y_k)_{\mathcal{A}}\|_2$$

which, in combination with (C.36) and convergence of x_k, implies that $\{y_k\}$ converges to y_*. In addition, continuity of the gradients and (C.37) implies that

$$g(x_*) - A^T(x_*)y_* = 0$$

while the fact that $c(x_k) > 0$ for all k, the definition of y_k and y_* (and the implication that $c_i(x_k)(y_k)_i = \mu_k$) shows that $c(x_*) \geq 0$, $y_* \geq 0$ and $c_i(x_*)(y_*)_i = 0$. Hence (x_*, y_*) satisfies the first-order optimality conditions. $\square$

Proof (of Theorem 4.2). A formal proof is given by

> W. Murray, "Analytical expressions for eigenvalues and eigenvectors of the Hessian matrices of barrier and penalty functions", *J. Optimization Theory and Applics*, **7** (1971) 189:196.

By way of a sketch, let $Q(x)$ and $N(x)$ be orthonormal bases for the range- and null-spaces of $A_{\mathcal{A}(x_*)}(x)$, and let $A_{\mathcal{I}}(x)$ be the matrix whose rows are $\{a_i^T(x)\}_{i \notin \mathcal{A}(x_*)}$. As we have shown, the required Hessian may be expressed (in decreasing terms of asymptotic dominance) as

$$\nabla_{xx}\Phi(x,\mu) = A_{\mathcal{A}}^T(x)Y_{\mathcal{A}}^2(x)A_{\mathcal{A}}(x)/\mu + H(x,y(x)) + \mu A_{\mathcal{I}}^T(x)C_{\mathcal{I}}^{-2}(x)A_{\mathcal{I}}(x).$$

Since the eigenvalues of $\nabla_{xx}\Phi(x,\mu)$ are not affected by orthonormal transformations, on pre- and post-multiplying $\nabla_{xx}\Phi(x,\mu)$ by $(Q(x) \ N(x))$ and its transpose, we see that the required eigenvalues are those of

$$\begin{pmatrix} Q(x)^T A_{\mathcal{A}}^T(x)Y_{\mathcal{A}}^2(x)A_{\mathcal{A}}(x)Q(x)/\mu + Q(x)^T HQ(x) & Q(x)^T HN(x) \\ N(x)^T HQ(x) & N(x)^T HN(x) \end{pmatrix} + O(\mu),$$

$$(C.38)$$

where we have used $A(x)N(x) = 0$ and note that $H = H(x,y(x))$. The dominant eigenvalues are those arising from the 1,1 block of (C.38), and these are those of $A_{\mathcal{A}}^T(x)Y_{\mathcal{A}}^2(x)A_{\mathcal{A}}(x)/\mu$ with an $O(1)$ error. Since the remaining eigenvalues must occur for eigenvectors orthogonal to those giving the 1,1 block, they will asymptotically be those of the 2,2 block, and thus those of $N(x)^T H(x,y(x))N(x)$ with an $O(\mu)$ term. □

Proof (of Theorem 4.3). The proof of this result is elementary, but rather long and involved. See

> N. Gould, D. Orban, A. Sartenaer and Ph. L. Toint, "Superlinear convergence of primal-dual interior point algorithms for nonlinear programming", *SIAM J. Optimization*, **11**(4) (2001) 974:1002

for full details. □

Proof (of Theorem 5.1). The SQP search direction s_k and its associated Lagrange multiplier estimates y_{k+1} satisfy

$$B_k s_k - A_k^T y_{k+1} = -g_k \qquad (C.39)$$

and

$$A_k s_k = -c_k. \qquad (C.40)$$

Pre-multiplying (C.39) by s_k and using (C.40) gives that

$$\langle s_k, g_k \rangle = -\langle s_k, B_k s_k \rangle + \langle s_k, A_k^T y_{k+1} \rangle = -\langle s_k, B_k s_k \rangle - \langle c_k, y_{k+1} \rangle. \quad (C.41)$$

Likewise (C.40) gives

$$\frac{1}{\mu_k}\langle s_k, A_k^T c_k \rangle = -\frac{\|c_k\|_2^2}{\mu_k}. \tag{C.42}$$

Combining (C.41) and (C.42), and using the positive definiteness of B_k, the Cauchy-Schwarz inequality and the fact that $s_k \neq 0$ if x_k is not critical, yields

$$\langle s_k, \nabla_x \Phi(x_k) \rangle = \left\langle s_k, g_k + \frac{1}{\mu_k} A_k^T c_k \right\rangle = -\langle s_k, B_k s_k \rangle - \langle c_k, y_{k+1} \rangle - \frac{\|c_k\|_2^2}{\mu_k}$$
$$< -\|c_k\|_2 \left(\frac{\|c_k\|_2}{\mu_k} - \|y_{k+1}\|_2 \right) \leq 0$$

because of the required bound on μ_k. □

Proof (of Theorem 5.2). The proof is slightly complicated as it uses the calculus of non-differentiable functions. See Theorem 14.3.1 in R. Fletcher, "Practical Methods of Optimization", Wiley (1987, 2nd edition), where the converse result, that if x_* is an isolated local minimizer of $\Phi(x, \rho)$ for which $c(x_*) = 0$ then x_* solves the given nonlinear program so long as ρ is sufficiently large, is also given. Moreover, Fletcher showns (Theorem 14.3.2) that x_* cannot be a local minimizer of $\Phi(x, \rho)$ when $\rho < \|y_*\|_D$. □

Proof (of Theorem 5.3). For small steps α, Taylor's theorem applied separately to f and c, along with (C.40), gives that

$$\Phi(x_k + \alpha s_k, \rho_k) - \Phi(x_k, \rho_k)$$
$$= \alpha \langle s_k, g_k \rangle + \rho_k (\|c_k + \alpha A_k s_k\| - \|c_k\|) + O(\alpha^2)$$
$$= \alpha \langle s_k, g_k \rangle + \rho_k (\|(1-\alpha)c_k\| - \|c_k\|) + O(\alpha^2)$$
$$= \alpha (\langle s_k, g_k \rangle - \rho_k \|c_k\|) + O(\alpha^2).$$

Combining this with (C.41), and once again using the positive definiteness of B_k, the Hölder inequality (that is that $\langle u, v \rangle \leq \|u\| \|v\|_D$ for any u, v) and the fact that $s_k \neq 0$ if x_k is not critical, yields

$$\Phi(x_k + \alpha s_k, \rho_k) - \Phi(x_k, \rho_k)$$
$$= -\alpha (\langle s_k, B_k s_k \rangle + \langle c_k, y_{k+1} \rangle + \rho_k \|c_k\|) + O(\alpha^2)$$
$$< -\alpha (-\|c_k\| \|y_{k+1}\|_D + \rho_k \|c_k\|) + O(\alpha^2)$$
$$= -\alpha \|c_k\| (\rho_k - \|y_{k+1}\|_D) + O(\alpha^2)$$
$$< 0$$

because of the required bound on ρ_k, for sufficiently small α. Hence sufficiently small steps along s_k from non-critical x_k reduce $\Phi(x, \rho_k)$. □

GniCodes – Matlab Programs for Geometric Numerical Integration

Ernst Hairer[1] and Martin Hairer[2]

[1] Section de mathématiques, Univ. Genève, CH-1211 Genève 24, Switzerland
[2] Mathematics Institute, Univ. Warwick, Coventry CV4 7AL, England

Abstract. Geometric numerical integration is synonymous with structure-preserving integration of ordinary differential equations. These notes, prepared for the Durham summer school 2002, are complementary to the monograph of Hairer, Lubich and Wanner [12]. They give an introduction to the subject, and they discuss and explain the use of Matlab programs for experimenting with structure-preserving algorithms.

We start by presenting some typical classes of problems having properties that are important and should be conserved by the discretization (Section 1). The flow of Hamiltonian differential equations is symplectic and possesses conserved quantities. Conservative systems have a time-reversible flow. Differential equations with first integrals and problems on manifolds are also considered. We then introduce in Section 2 simple symplectic and symmetric integrators, (partitioned) Runge-Kutta methods, composition and splitting methods, linear multistep methods, and algorithms for Hamiltonian problems on manifolds. We briefly discuss their symplecticity and symmetry. The improved performance of such geometric integrators is best understood with the help of a backward error analysis (Section 3). We explain some implications for the long-time integration of Hamiltonian systems and of completely integrable problems.

Section 4 is devoted to a presentation and explanation of Matlab codes for implicit Runge-Kutta, composition, and multistep methods. The final Section 5 gives a comparison of the different methods and illustrates the use of these programs at some typical interesting situations: the computation of Poincaré sections, and the simulation of the motion of two bodies on a sphere. The Matlab codes as well as their Fortran 77 counterparts can be downloaded at

http://www.unige.ch/math/folks/hairer

under the item "software".

1   Problems to be Solved

For the numerical solution of ordinary differential equations there exist well-developed theories, and excellent general purpose codes are available and widely used. If the flow of the differential equation has a particular structure, then its preservation by the discretization scheme can considerably improve its performance and its qualitative behaviour. This article focuses on structure-preserving algorithms for some important classes of problems – Hamiltonian systems and reversible differential equations.

1.1   Hamiltonian Systems

For a smooth function $H(p,q)$ defined on an open set $D \subset \mathbf{R}^d \times \mathbf{R}^d$ we consider the differential equation

$$\dot{p}_i = -\frac{\partial H}{\partial q_i}(p,q), \quad \dot{q}_i = \frac{\partial H}{\partial p_i}(p,q), \quad i = 1,\ldots,d. \tag{1.1}$$

The dimension d of the vectors p and q is called the 'degree of freedom' of the system. We also use the more compact notation

$$\dot{p} = -\nabla_q H(p,q), \quad \dot{q} = \nabla_p H(p,q),$$

or

$$\dot{y} = J^{-1}\nabla H(y), \quad y = \begin{pmatrix} p \\ q \end{pmatrix}, \quad J = \begin{pmatrix} 0 & I \\ -I & 0 \end{pmatrix}, \tag{1.2}$$

where $\nabla_p H$, $\nabla_q H$ and ∇H denote the column vectors of partial derivatives with respect to the components of p,q and y respectively. The matrix J is the structure matrix of Hamiltonian systems in canonical form.

Throughout this article we denote by $\varphi_t(y)$ the exact flow of the system (1.2), i.e. $\varphi_t(y_0) = y(t)$ is the solution at time t of the problem (1.2) with initial value $y(0) = y_0$.

Example 1.1 (Classical Mechanical Systems). Consider a mechanical system that can be described with (minimal) coordinates $q \in \mathbf{R}^d$. Denote its kinetic energy by $T(q,\dot{q}) = \frac{1}{2}\dot{q}^T M(q)\dot{q}$ (with a symmetric positive definite matrix $M(q)$) and its potential energy by $U(q)$. The motion of the system is then given as the solution of the Euler-Lagrange equations

$$\frac{d}{dt}\Big(M(q)\dot{q}\Big) = \frac{\partial}{\partial q}\Big(T(q,\dot{q}) - U(q)\Big) \tag{1.3}$$

corresponding to the variational problem $\int (T(q,\dot{q}) - U(q))\,dt \to \min$. Introducing the new variables $p := M(q)\dot{q}$ (momenta or Poisson variables) the differential equation (1.3) is equivalent to the Hamiltonian system (1.1) with

$$H(p,q) = \frac{1}{2}p^T M(q)^{-1}p + U(q). \tag{1.4}$$

This is an immediate consequence from computing the partial derivatives of this function $H(p,q)$. A simple example, often used for illustrations, is the *mathematical pendulum* for which the Hamiltonian is

$$H(p,q) = \frac{1}{2}p^2 - \cos q. \tag{1.5}$$

Due to their special structure, Hamiltonian systems have several interesting properties:

- the Hamiltonian $H(p, q)$ is constant along solutions of (1.1); for classical mechanical systems this means that the total energy (sum of kinetic and potential energies) is a conserved quantity.
- for systems with one degree of freedom the flow φ_t is area-preserving; for the general case it is volume-preserving. This means that

$$\mu(\varphi_t(A)) = \mu(A) \quad \text{for } t \geq 0 \qquad (1.6)$$

for any compact set $A \subset \mathbf{R}^d \times \mathbf{R}^d$ (μ denotes the Lebesgue measure).
- the flow φ_t is a symplectic transformation, i.e.

$$\varphi_t'(y)^T J \varphi_t'(y) = J \quad \text{for } t \geq 0, \qquad (1.7)$$

where the prime in $\varphi_t'(y)$ denotes the derivation with respect to y.

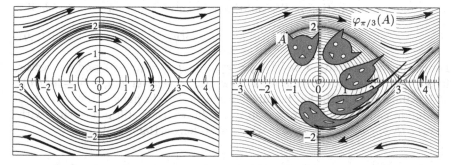

Fig. 1.1. Level curves $H(p, q) =$ constant for the pendulum problem (left picture), and area-preservation of its exact flow (right picture).

The first property is immediately verified by differentiating $\frac{d}{dt} H(p(t), q(t)) = \ldots = 0$. The solutions of the mathematical pendulum are therefore on the level curves of the Hamiltonian (1.5); see Figure 1.1. The second property is a consequence of the third, because (1.7) and the continuous dependence of $\varphi_t'(y)$ on t imply $\det \varphi_t'(y) = 1$. This together with the transformation formula for multiple integrals proves (1.6). The right picture of Figure 1.1 illustrates the area-preservation of the exact flow for the pendulum equation.

The symplecticity condition (1.7) has a nice geometric interpretation. It is equivalent to the property that

$$\omega(\varphi_t(A)) = \omega(A) \quad \text{for } t \geq 0$$

holds for any two-dimensional sub-manifold A of $\mathbf{R}^d \times \mathbf{R}^d$, where $\omega(A)$ denotes the sum of the oriented areas of the projections of A onto the (p_i, q_i)-plane. The important feature is that this property is characteristic for Hamiltonian systems (cf. [12, Chapter VI]), which means that whenever the flow of a differential equation $\dot{y} = f(y)$ is symplectic for all t and all y, then $f(y)$ is locally of the form $f(y) = J^{-1} \nabla H(y)$. This characteristic property of Hamiltonian systems motivates the search for discretizations that are symplectic.

1.2   Reversible Differential Equations

Consider first a mechanical system for which the equations of motion are given by the second order differential equation (1.3). Since $T(q, \dot{q})$ is quadratic in $\dot{q}$, they are equivalent to the system

$$\dot{q} = v, \quad \dot{v} = g(q, v), \tag{1.8}$$

satisfying $g(q, -v) = g(q, v)$. This implies the time-reversibility of the system that is whenever $\big(q(t), v(t)\big)$ is a solution of (1.8), then $\big(q(-t), -v(-t)\big)$ is a solution. For example, in the study of planetary motion, the same differential equation allows us to investigate the future and the past, one only has to change the sign of the velocity vector v.

More generally, we consider a differential equation $\dot{y} = f(y)$ and a linear invertible transformation ρ. We call the differential equation ρ-*reversible* if

$$(\rho \circ f)(y) = -(f \circ \rho)(y). \tag{1.9}$$

For the previous situation we have $y = (q, v)$, $\rho(q, v) = (q, -v)$, and the vector field $f(y) = \big(v, g(q, v)\big)$ indeed satisfies (1.9) whenever $g(q, -v) = g(q, v)$. This is illustrated in the left picture of Figure 1.2 at the hand of the perturbed pendulum equation $\dot{q} = v$, $\dot{v} = -\sin q - v^2/5$, which is still ρ-reversible with respect to $\rho(q, v) = (q, -v)$, but which is no longer Hamiltonian.

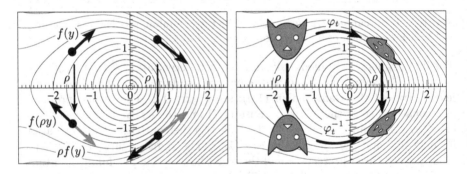

Fig. 1.2. The ρ-reversibility of the vector field $f(q, v) = (v, -\sin q - v^2/5)^T$ (left picture), and the ρ-reversibility of the corresponding flow.

The flow of a ρ-reversible differential equation has a remarkable property:

- it is ρ-reversible, i.e. it satisfies (see the right picture of Figure 1.2)

$$(\rho \circ \varphi_t)(y) = (\varphi_t^{-1} \circ \rho)(y) \quad \text{for all } t \text{ and all } y. \tag{1.10}$$

The proof of this statement is straightforward. One checks by differentiation that $(\rho \circ \varphi_t)(y)$ and $(\varphi_t^{-1} \circ \rho)(y) = (\varphi_{-t} \circ \rho)(y)$ are both solutions of the same differential equation $\dot{z} = -f(z)$, and are identical for $t = 0$. Formula (1.10) thus follows from the uniqueness of the solution of an initial value problem. Analogous to the situation of Hamiltonian problems, this property is characteristic for ρ-reversible differential equations. This means that whenever the flow of a differential equation $\dot{y} = f(y)$ satisfies (1.10), then (1.9) holds. It is thus natural to look for numerical methods that share this property.

Example 1.2 (Kepler Problem). The relative motion of two bodies which attract each other is described by the differential equation

$$\dot{q}_1 = v_1, \quad \dot{q}_2 = v_2, \quad \dot{v}_1 = -\frac{q_1}{(q_1^2 + q_2^2)^{3/2}}, \quad \dot{v}_2 = -\frac{q_2}{(q_1^2 + q_2^2)^{3/2}}. \quad (1.11)$$

Since it can be considered as a classical mechanical system, it is ρ-reversible for $\rho(q_1, q_2, v_1, v_2) = (q_1, q_2, -v_1, -v_2)$. However, there are more symmetries in this problem, and it is also seen to be ρ-reversible for $\rho(q_1, q_2, v_1, v_2) = (q_1, -q_2, -v_1, v_2)$.

Example 1.3 (Second Order Differential Equations). Many problems of practical applications lead to $\ddot{q} = g(q)$, or equivalently,

$$\dot{q} = v, \quad \dot{v} = g(q). \quad (1.12)$$

For example, all classical mechanical systems for which $M(q) = M$ is a constant matrix are of this form. The differential equation (1.12) is ρ-reversible for $\rho(q, v) = (q, -v)$ independent of the form of $g(q)$. It is Hamiltonian only if $g(q) = -\nabla_q U(q)$ for some potential function $U(q)$.

1.3   Hamiltonian and Reversible Systems on Manifolds

It is often difficult to find suitable minimal coordinates for describing the motion of mechanical systems. Moreover, they are in general only defined locally and frequent changes of coordinates may be necessary. To avoid this difficulty we consider coordinates $q \in \mathbf{R}^d$ that are subject to constraints $g(q) = 0$. Expressing the Euler-Lagrange equations and their Hamiltonian formulation in terms of these coordinates, we are led to a system of the form

$$\begin{aligned} \dot{p} &= -\nabla_q H(p, q) - \nabla_q g(q)\lambda \\ \dot{q} &= \nabla_p H(p, q), \quad 0 = g(q), \end{aligned} \quad (1.13)$$

where the additional term with the Lagrange multiplier λ forces the solution to satisfy $g(q) = 0$. Here, p and q are vectors in $\mathbf{R}^d$, $g(q) = (g_1(q), \ldots, g_m(q))^T$ is the vector of constraints, and $\nabla_q g = (\nabla_q g_1, \ldots, \nabla_q g_m)$ is the transposed Jacobian matrix of $g(q)$.

Differentiating the constraint $0 = g(q(t))$ with respect to time yields

$$0 = \nabla_q g(q)^T \nabla_p H(p, q) \quad (1.14)$$

(the so-called hidden constraint) which is an invariant of the flow of (1.13).
A second differentiation gives the relation

$$
\begin{aligned}
0 = \frac{\partial}{\partial q}\Big(\nabla_q g(q)^T \nabla_p H(p,q)\Big)\nabla_p H(p,q) \\
- \nabla_q g(q)^T \nabla_p^2 H(p,q)\Big(\nabla_q H(p,q) + \nabla_q g(q)\lambda\Big),
\end{aligned}
\tag{1.15}
$$

which allows us to express λ in terms of (p,q), if the matrix

$$
\nabla_q g(q)^T \nabla_p^2 H(p,q)\nabla_q g(q) \qquad \text{is invertible}
\tag{1.16}
$$

($\nabla_p^2 H$ denotes the Hessian matrix of H). Inserting the so-obtained function
$\lambda(p,q)$ into (1.13) gives a differential equation for (p,q) on the manifold

$$
\mathcal{M} = \{(p,q) \mid g(q) = 0, \ \nabla_q g(q)^T \nabla_p H(p,q) = 0\}.
\tag{1.17}
$$

This interpretation allows us to deduce the existence and uniqueness of the
solution from the standard theory for ordinary differential equations, provided
that the initial values satisfy $(p_0, q_0) \in \mathcal{M}$.

Important properties of the system (1.13) that should be conserved by a
discretization are the following:

- for $(p_0, q_0) \in \mathcal{M}$ the solution stays on the manifold $\mathcal{M}$ for all t; hence,
  the flow is a mapping $\varphi_t : \mathcal{M} \to \mathcal{M}$.
- the flow φ_t is a symplectic transformation on $\mathcal{M}$ which means that

$$
(\varphi_t'(y)\xi)^T J \varphi_t'(y)\eta = \xi^T J \eta \quad \text{for } \xi, \eta \in T_y\mathcal{M};
\tag{1.18}
$$

  here, the product $\varphi_t'(y)\xi$ is the directional derivative along the vector ξ
  belonging to the tangent space $T_y\mathcal{M}$ of $\mathcal{M}$ at y.
- for Hamiltonians satisfying $H(-p,q) = H(p,q)$ the flow φ_t is ρ-reversible
  for $\rho(p,q) = (-p,q)$ in the sense that (1.10) holds for $y = (p,q) \in \mathcal{M}$.

Example 1.4 (Two-Body Problem on the Sphere). We are interested in the
motion of two bodies which attract each other, but which are restricted to
stay on a sphere. Using Cartesian coordinates $q_1, q_2 \in \mathbf{R}^3$ for the positions of
the two bodies and $p_1, p_2 \in \mathbf{R}^3$ for their velocities, the Hamiltonian becomes
(after a suitable normalization)

$$
H(p_1, p_2, q_1, q_2) = \frac{1}{2}\big(p_1^T p_1 + p_2^T p_2\big) + U(q_1, q_2),
\tag{1.19}
$$

and the constraint equations $g(q_1, q_2) = 0$ with $g : \mathbf{R}^6 \to \mathbf{R}^2$ are given by

$$
q_1^T q_1 - 1 = 0, \quad q_2^T q_2 - 1 = 0.
\tag{1.20}
$$

According to Kozlov and Harin [18], we choose $U(q_1, q_2) = -\cos\vartheta / \sin\vartheta$ as the potential, where ϑ is the distance between the two bodies along a geodesics. We have $\cos\vartheta = q_1^T q_2$, so that the equations of motion become

$$\dot{q}_1 = p_1, \quad \dot{p}_1 = f(q_1^T q_2)q_2 - \lambda_1 q_1,$$
$$\dot{q}_2 = p_2, \quad \dot{p}_2 = f(q_1^T q_2)q_1 - \lambda_2 q_2, \tag{1.21}$$

together with (1.20), where

$$f(c) = \frac{1}{(1 - c^2)^{3/2}}.$$

The initial values have to lie on the manifold

$$\mathcal{M} = \{(p_1, p_2, q_1, q_2) \; ; \; q_1^T q_1 = 1, \; q_2^T q_2 = 1, \; q_1^T p_1 = 0, \; q_2^T p_2 = 0\},$$

and the solution stays on $\mathcal{M}$ for all t.

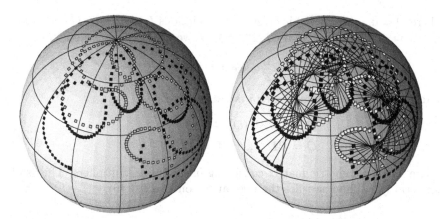

Fig. 1.3. A solution of the two-body problem on the sphere; initial values are indicated by larger symbols; the geodesic connection between the two bodies is plotted at every second time step in the right picture.

A particular solution is plotted in Figure 1.3. We have chosen

$$q_i = \big(\cos\phi_i \sin\theta_i, \sin\phi_i \sin\theta_i, \cos\theta_i\big)^T$$

with $(\phi_1, \theta_1) = (0.8, 0.6)$ and $(\phi_2, \theta_2) = (0.5, 1.5)$ as initial values for the positions, and

$$p_i = \big(-\dot{\phi}_i \sin\phi_i \sin\theta_i + \dot{\theta}_i \cos\phi_i \cos\theta_i, \; \dot{\phi}_i \cos\phi_i \sin\theta_i + \dot{\theta}_i \sin\phi_i \cos\theta_i, \; -\dot{\theta}_i \sin\theta_i\big)$$

with $(\dot{\phi}_1, \dot{\theta}_1) = (1.1, -0.2)$ and $(\dot{\phi}_2, \dot{\theta}_2) = (-0.8, 0.0)$ as initial values for the velocities. The two bodies are indicated by small squares in different greyscales. The right picture of Figure 1.3 also shows the geodesic connection between the two bodies.

Example 1.5 (Rigid Body Simulation). The motion of a rigid body with a
fixed point chosen at the origin can be described by an orthogonal matrix
$Q(t)$. Denoting by I_1, I_2, I_3 the moments of inertia of the body, its kinetic
energy is

$$T = \frac{1}{2}(I_1\Omega_1^2 + I_2\Omega_2^2 + I_3\Omega_3^2),$$

where the angular velocity $\Omega = (\Omega_1, \Omega_2, \Omega_3)^T$ of the body is defined by

$$\widehat{\Omega} = \begin{pmatrix} 0 & -\Omega_3 & \Omega_2 \\ \Omega_3 & 0 & -\Omega_1 \\ -\Omega_2 & \Omega_1 & 0 \end{pmatrix} = Q^T\dot{Q},$$

(see [2, Chapter 6]). In terms of Q, the kinetic energy on the manifold
$\{Q \,|\, Q^TQ = I\}$ becomes

$$T = \frac{1}{2}\,\mathrm{trace}(\widehat{\Omega}D\widehat{\Omega}^T) = \frac{1}{2}\,\mathrm{trace}(Q^T\dot{Q}D\dot{Q}^TQ) = \frac{1}{2}\,\mathrm{trace}(\dot{Q}D\dot{Q}^T),$$

where $D = \mathrm{diag}(d_1, d_2, d_3)$ is given by the relations $I_1 = d_2 + d_3$, $I_2 = d_3 + d_1$,
and $I_3 = d_1 + d_2$. With $P = \partial T/\partial\dot{Q} = \dot{Q}D$, we are thus concerned with

$$H(P, Q) = \frac{1}{2}\,\mathrm{trace}(PD^{-1}P^T) + U(Q),$$

and the constrained Hamiltonian system becomes

$$\begin{aligned} \dot{P} &= -\nabla_Q U(Q) - Q\Lambda, \\ \dot{Q} &= PD^{-1}, \qquad 0 = Q^TQ - I, \end{aligned} \tag{1.22}$$

where Λ is a symmetric matrix consisting of Lagrange multipliers. This is of
the form (1.13) and satisfies the regularity condition (1.16).

2   Symplectic and Symmetric Integrators

A numerical integrator is a family $\Phi_h(y)$ of maps on the phase space that
approximates the exact flow $\varphi_h(y)$ of the differential equation. It is the aim
of 'geometric integration' to construct and to study methods for which the
numerical solution, given by $y_{n+1} = \Phi_h(y_n)$, preserves the structure of the
problem. We are mainly interested in methods for which Φ_h is symplectic or
ρ-reversible, when it is applied to a Hamiltonian or ρ-reversible differential
equation, respectively.

2.1   Simple Symplectic Methods

The simplest numerical methods for general differential equations $\dot{y} = f(y)$
are the *explicit Euler method*

$$y_{n+1} = y_n + hf(y_n) \tag{2.1}$$

and the *implicit Euler method*

$$y_{n+1} = y_n + hf(y_{n+1}). \tag{2.2}$$

Here, h is the step size, and y_n is an approximation to the solution $y(t)$ at time $t = nh$. For Hamiltonian systems (1.1) we consider the method

$$p_{n+1} = p_n - h\nabla_q H(p_{n+1}, q_n), \quad q_{n+1} = q_n + h\nabla_p H(p_{n+1}, q_n), \tag{2.3}$$

which treats the p-variable by the implicit Euler method and the q-variable by the explicit Euler method. Similarly, we also consider

$$p_{n+1} = p_n - h\nabla_q H(p_n, q_{n+1}), \quad q_{n+1} = q_n + h\nabla_p H(p_n, q_{n+1}). \tag{2.4}$$

Both methods are called *symplectic Euler method*.

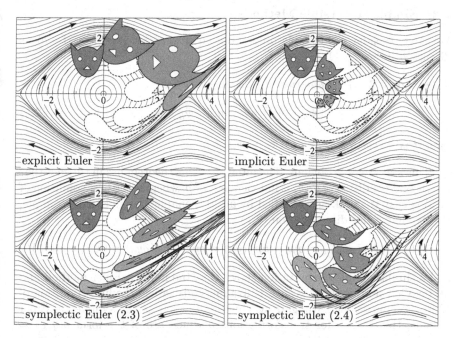

Fig. 2.1. Numerical flow with step size $h = \pi/3$ for the four 'Euler methods' of Section 2.1. The exact flow is included as a white shadow.

Example 2.1. We apply all four methods to the pendulum problem which is Hamiltonian with $H(p, q)$ given by (1.5), and we consider initial values in the set A of Figure 1.1. The numerical solution obtained with the large step size $h = \pi/3$ is illustrated in Figure 2.1. Neither the explicit nor the implicit Euler methods are area-preserving (i.e. symplectic). We shall see in the following theorem that both 'symplectic Euler methods' are area-preserving (hence the name symplectic). However, due to the large step size the numerical solution differs significantly from the exact solution which is included as a white shadow in the pictures (compare with Figure 1.1).

Theorem 2.1. *For the numerical schemes (2.3) and (2.4) the mapping*

$$\Phi_h \; : \; \begin{pmatrix} p_n \\ q_n \end{pmatrix} \mapsto \begin{pmatrix} p_{n+1} \\ q_{n+1} \end{pmatrix}$$

is a symplectic transformation.

The *proof* of this theorem is straightforward (de Vogelaere [42] and [12, p. 176]). One computes the Jacobian of Φ_h by implicit differentiation, and one checks the identity (1.7).

2.2   Simple Reversible Methods

We next consider ρ-reversible differential equations (i.e. $\rho \circ f = -f \circ \rho$) as discussed in Section 1.2.

Theorem 2.2. *If a numerical method Φ_h satisfies*

$$\rho \circ \Phi_h = \Phi_{-h} \circ \rho \quad \text{and} \quad \Phi_h = \Phi_{-h}^{-1}, \tag{2.5}$$

then it is ρ-reversible, i.e. $\rho \circ \Phi_h = \Phi_h^{-1} \circ \rho$.

This statement is obvious. The interest of this theorem lies in the fact that the second condition of (2.5) is independent of ρ, whereas the first condition of (2.5) is easy to check and satisfied by all 'reasonable' methods. For example, the explicit Euler discretization (2.1) yields

$$(\rho \circ \Phi_h)(y_n) = \rho y_{n+1} = \rho y_n + h\rho f(y_n) = \rho y_n - hf(\rho y_n) = (\Phi_{-h} \circ \rho)(y_n),$$

and a similar simple computation shows that the implicit Euler method and all (explicit and implicit) Runge-Kutta methods satisfy the first condition of (2.5). For partitioned Runge-Kutta methods, such as the symplectic Euler scheme, this is true for transformations ρ which are of the form $\rho(q, v) = (\rho_1(q), \rho_2(v))$.

If $\Phi_h(y)$ represents a numerical method of order at least one, i.e. $\Phi_h(y) = y + hf(y) + \mathcal{O}(h^2)$, then also $\Phi_{-h}^{-1}(y) = y + hf(y) + \mathcal{O}(h^2)$ and

$$\Phi_h^* := \Phi_{-h}^{-1} \tag{2.6}$$

is a numerical method of order at least one. It is called the *adjoint method* of Φ_h. Whenever an integrator satisfies

$$\Phi_h^* = \Phi_h, \tag{2.7}$$

it is called a *symmetric method*. The second condition in (2.5) of Theorem 2.2 is thus equivalent to the symmetry of the method Φ_h.

Exchanging $h \leftrightarrow -h$ and $y_n \leftrightarrow y_{n+1}$ in (2.1) shows that the adjoint of the explicit Euler method is the implicit Euler method and vice versa. Similarly, the adjoint of the symplectic Euler method (2.3) is the method (2.4). None of these methods is symmetric.

Using the notion of the adjoint method it is easy to construct symmetric methods: let Ψ_h be an arbitrary method of order at least one, then the compositions

$$\Psi_{h/2} \circ \Psi_{h/2}^* \quad \text{and} \quad \Psi_{h/2}^* \circ \Psi_{h/2} \tag{2.8}$$

are symmetric methods of order at least two. The symmetry follows from the properties $(\Phi_h \circ \Psi_h)^* = \Psi_h^* \circ \Phi_h^*$ and $(\Phi_h^*)^* = \Phi_h$, and order at least two is a consequence of the fact that symmetric method always have an even order.

For example, if we let Ψ_h be the explicit Euler method, then the methods of (2.8) are

$$y_{n+1} = y_n + h\, f\left(\frac{y_n + y_{n+1}}{2}\right), \tag{2.9}$$

the *implicit midpoint rule*, and

$$y_{n+1} = y_n + \frac{h}{2}\Big(f(y_n) + f(y_{n+1})\Big), \tag{2.10}$$

the *trapezoidal rule*, respectively.

2.3 Störmer/Verlet Scheme

We next consider Hamiltonian systems (1.1) and the symplectic Euler method (2.3) in the role of Ψ_h. The compositions (2.8) then yield

$$q_{n+1/2} = q_n + \frac{h}{2}\nabla_p H(p_n, q_{n+1/2})$$
$$p_{n+1} = p_n - \frac{h}{2}\Big(\nabla_q H(p_n, q_{n+1/2}) + \nabla_q H(p_{n+1}, q_{n+1/2})\Big) \tag{2.11}$$
$$q_{n+1} = q_{n+1/2} + \frac{h}{2}\nabla_p H(p_{n+1}, q_{n+1/2})$$

and

$$p_{n+1/2} = p_n - \frac{h}{2}\nabla_q H(p_{n+1/2}, q_n)$$
$$q_{n+1} = q_n + \frac{h}{2}\Big(\nabla_p H(p_{n+1/2}, q_n) + \nabla_p H(p_{n+1/2}, q_{n+1})\Big) \tag{2.12}$$
$$p_{n+1} = p_{n+1/2} - \frac{h}{2}\nabla_q H(p_{n+1/2}, q_{n+1})$$

respectively. For the important special case $H(p, q) = \frac{1}{2}p^2 + U(q)$, method (2.12) reduces to (after elimination of the p-variable)

$$q_{n+1} - 2q_n + q_{n-1} = -h^2 \nabla_q U(q_n). \tag{2.13}$$

This discretization of $\ddot{q} = -\nabla_q U(q)$ is attributed to Newton (cf. [13]), Delambre (cf. [25]), Encke, Störmer [36], and Verlet [41]. The methods (2.11) and (2.12) are nowadays often called *Störmer/Verlet*.

We collect the most important properties of the Störmer/Verlet scheme:

- the method is of order two;
- it is a symplectic method;
- it is a symmetric method;
- for separable Hamiltonians $T(p) + U(q)$ the method is explicit;
- the method exactly conserves quadratic first integrals $p^T C q$, e.g. the angular momentum in N-body problems.

The first four statements are immediate consequences of the above discussions. A proof of the last property is given in [12, p. 98].

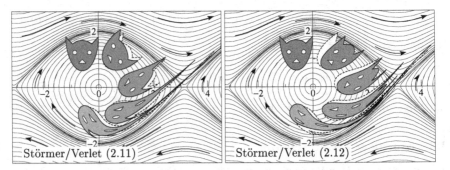

Fig. 2.2. Numerical flow with step size $h = \pi/3$ for the two versions of the Störmer/Verlet method. The exact flow is included as a white shadow.

In Figure 2.2 we repeat the experiment of Example 2.1, but this time with the two versions of the Störmer/Verlet method. We clearly observe the higher accuracy (compared to the first order methods) and the area-preservation.

The Störmer/Verlet scheme is an excellent geometric integrator and it is widely used, in particular in molecular dynamics where a correct qualitative simulation is of utmost importance. For long-time computations in astronomy, where a very high accuracy is demanded, the order two of the Störmer/Verlet scheme is too low.

2.4 Splitting Methods

A different approach for constructing simple geometric integrators is based on the idea of splitting the vector field as

$$\dot{y} = f^{[1]}(y) + f^{[2]}(y). \tag{2.14}$$

If by chance the exact flows $\varphi_t^{[1]}$ and $\varphi_t^{[2]}$ of the systems $\dot{y} = f^{[1]}(y)$ and $\dot{y} = f^{[2]}(y)$ can be calculated exactly, we can consider

$$\Phi_h = \varphi_h^{[1]} \circ \varphi_h^{[2]} \tag{2.15}$$

as simple numerical integrator. It follows from Taylor expansion that this method is of order one. Even more important is the symmetric (second order) composition

$$\Phi_h = \varphi_{h/2}^{[1]} \circ \varphi_h^{[2]} \circ \varphi_{h/2}^{[1]} \tag{2.16}$$

which is usually called *Strang splitting*. These splitting methods have the following obvious properties:

- if both, $f^{[1]}(y)$ and $f^{[2]}(y)$, are Hamiltonian vector fields, then the compositions (2.15) and (2.16) are *symplectic* integrators;
- if both, $f^{[1]}(y)$ and $f^{[2]}(y)$, are ρ-reversible, then the symmetric method (2.16) is ρ-*reversible*.

For some situations the splitting (2.14) is obvious. For example, if a Hamiltonian system has $H(p,q) = T(p) + U(q)$ as the Hamiltonian, then the flows corresponding to $H^{[1]}(p,q) = T(p)$ and $H^{[2]}(p,q) = U(q)$ are given explicitly by

$$\varphi_t^{[1]}(p,q) = \big(p, q + t\nabla_p T(p)\big), \quad \varphi_t^{[2]}(p,q) = \big(p - t\nabla_q U(q), q\big).$$

The resulting splitting methods (2.15) and (2.16) are then equivalent to the symplectic Euler method (2.3) and to the Störmer/Verlet scheme (2.12), respectively. However, in general it is an art to find a suitable splitting (cf. [25]).

2.5 High Order Geometric Integrators

We start this section with a numerical experiment that motivates the search for high order symplectic and symmetric numerical integrators. We consider the Kepler problem which is Hamiltonian with

$$H(p_1, p_2, q_1, q_2) = \frac{1}{2}\left(p_1^2 + p_2^2\right) - \frac{1}{\sqrt{q_1^2 + q_2^2}}, \tag{2.17}$$

and we take as initial values

$$q_1(0) = 1 - e, \quad q_2(0) = 0, \quad p_1(0) = 0, \quad p_2(0) = \sqrt{(1+e)(1-e)^{-1}},$$

such that the solution is an ellipse with eccentricity $e = 0.6$. Figure 2.3 shows the work precision diagrams (global error at the endpoint after 200 revolutions against the required number of function evaluations and the computer time, respectively) for the second order Störmer/Verlet scheme as well as for various methods of order eight. It clearly demonstrates that for high accuracy requirements (say 10 digits) the low order method cannot compete with the high order ones. It would need about 1000 times more cpu time. The irregularities at the right bottom corner of the pictures are due to round-off.

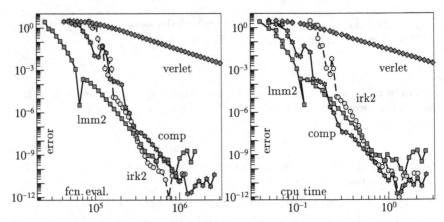

Fig. 2.3. Work precision diagrams for the Störmer/Verlet scheme and for three methods of order eight; implicit Runge-Kutta method (irk2), composition method (comp), and linear multistep method (lmm2).

Implicit Runge-Kutta Methods In the end of the 19th and the beginning of the 20th century Runge [30] and Kutta [19] introduced generalizations of the explicit Euler method with the aim of getting more accurate numerical approximations. These *explicit* methods can neither be symplectic nor symmetric as follows from the characterizations given below. Much more important for our purpose are *implicit* Runge-Kutta methods, introduced mainly in the work of Butcher [5]. For Hamiltonian systems or for general partitioned differential equations

$$\dot{q} = f(q, v), \quad \dot{v} = g(q, v) \tag{2.18}$$

we consider so-called *partitioned Runge-Kutta methods*, which treat the components of q and those of v by possibly different implicit Runge-Kutta methods. They are defined by

$$k_i = f\Big(q_n + h\sum_{j=1}^{s} a_{i,j}k_j, \, v_n + h\sum_{j=1}^{s} \widehat{a}_{i,j}\ell_j\Big),$$

$$\ell_i = g\Big(q_n + h\sum_{j=1}^{s} a_{i,j}k_j, \, v_n + h\sum_{j=1}^{s} \widehat{a}_{i,j}\ell_j\Big), \tag{2.19}$$

$$q_{n+1} = q_n + h\sum_{i=1}^{s} b_i k_i, \quad v_{n+1} = v_n + h\sum_{i=1}^{s} \widehat{b}_i \ell_i.$$

The equations for k_i, ℓ_i ($i = 1, \dots, s$) are nonlinear and have to be solved by fixed-point iteration, provided that the step size h is sufficiently small.

It turns out that the method (2.19) is *symplectic* for general Hamiltonian systems, if the following relations are satisfied:

$$b_i\widehat{a}_{i,j} + \widehat{b}_j a_{j,i} = b_i\widehat{b}_j \quad \text{for } i, j = 1, \dots, s,$$
$$b_i = \widehat{b}_i \quad \text{for } i = 1, \dots, s. \tag{2.20}$$

It is *symmetric*, if

$$a_{s+1-i,s+1-j} + a_{i,j} = b_j \qquad \text{for all } i, j,$$
$$\widehat{a}_{s+1-i,s+1-j} + \widehat{a}_{i,j} = \widehat{b}_j \qquad \text{for all } i, j. \tag{2.21}$$

If the method does not contain superfluous stages and if the stages are suitably ordered, the conditions (2.20) and (2.21) are also necessary for symplecticity and symmetry, respectively. These characterizations were originally obtained by Lasagni [22], Sanz-Serna [32] and Suris [37] for symplecticity, and by Stetter [35] and Wanner [43] for symmetry. They are discussed in detail in Chapters V and VI of [12].

For the important special case $\ddot{q} = g(q)$, i.e. $f(q, v) = v$ in (2.18) and $g(q, v)$ independent of v, the variables k_i can be eliminated explicitly and the method (2.19) reduces to

$$\ell_i = g\Big(q_n + hc_iv_n + h^2 \sum_{j=1}^{s} \widetilde{a}_{i,j}\ell_j\Big), \quad i = 1, \ldots s,$$

$$q_{n+1} = q_n + hv_n + h^2 \sum_{i=1}^{s} \widetilde{b}_i\ell_i, \qquad v_{n+1} = v_n + h\sum_{i=1}^{s} \widehat{b}_i\ell_i \tag{2.22}$$

where $c_i = \sum_{j=1}^{s} a_{i,j}$, and $\widetilde{b}_i$, $\widetilde{a}_{i,j}$ are the coefficients of $\widetilde{b}^T = b^T\widehat{A}$ and $\widetilde{A} = A\widehat{A}$.

Example 2.2 (Method used as 'irk2' in Figure 2.3). The most important symplectic implicit Runge-Kutta methods are the so-called *Gauss methods*. They are built on the Gaussian quadrature $(b_i, c_i)_{i=1}^{s}$, which is interpolatory and for which $c_1, \ldots, c_s$ are the zeros of the s-th shifted Legendre polynomial

$$\frac{d^s}{dx^s}\Big(x^s(x-1)^s\Big).$$

The coefficients $a_{i,j}$ are computed from the linear system

$$\sum_{j=1}^{s} a_{i,j}c_j^{k-1} = \frac{c_i^k}{k} \qquad \text{for } i, k = 1, \ldots, s.$$

We let $\widehat{b}_i = b_i$ and $\widehat{a}_{i,j} = a_{i,j}$ in (2.19), so that all components of the differential equation are treated by the same method.

The method was originally obtained and introduced in this way by Butcher [6], and it has many nice properties. It is of order $2s$ (which is maximal among all s-stage Runge-Kutta methods), it is symplectic and symmetric, so that it is extremely well suited in the context of geometric integration. The only disadvantage is that even for simple situations such as $\ddot{q} = g(q)$, it gives an implicit discretization. In the experiment of Figure 2.3 we use this method with $s = 4$. The '2' in 'irk2' indicates that the code is only for second order differential equations $\ddot{q} = g(q)$, and that it is implemented as (2.22).

Partitioned Multistep Methods Another extension of the Euler methods are linear multistep methods, originally introduced by Adams in 1855 and published in Bashforth [3]. Neither explicit nor implicit classical multistep methods have been successful in geometric integration. Lambert and Watson [21] considered special classes for second order differential equations $\ddot{q} = g(q)$, which have been revived by Quinlan and Tremaine [27] for the long-time integration of planetary orbits. For partitioned differential equations (2.18), which are more general than $\dot{q} = v$, $\dot{v} = g(q)$, these methods can be interpreted as partitioned linear multistep methods, defined by

$$
\begin{aligned}
\sum_{j=0}^{k} \alpha_j q_{n+j} &= h \sum_{j=0}^{k} \beta_j f(q_{n+j}, v_{n+j}), \\
\sum_{j=0}^{\widehat{k}} \widehat{\alpha}_j v_{n+j} &= h \sum_{j=0}^{\widehat{k}} \widehat{\beta}_j g(q_{n+j}, v_{n+j}).
\end{aligned}
\tag{2.23}
$$

It is not obvious how to discuss symplecticity and symmetry of multistep methods, because we are concerned with an algorithm $(y_n, \ldots, y_{n+k-1}) \mapsto y_{n+k}$ and not with a one-step method $y_{n+1} = \Phi_h(y_n)$ which is a transformation on the phase space. However, Kirchgraber [17] showed that to every consistent strictly stable multistep method one can associate a so-called *underlying one-step method* Φ_h which has the same long-time dynamics. More precisely, it satisfies the following properties:

- for every y_0, the sequence defined by $y_{n+1} = \Phi_h(y_n)$ is a solution of the multistep method;
- for an arbitrary starting approximation $y_0, \ldots, y_{k-1}$, the numerical approximation of the multistep method tends exponentially fast to a particular solution obtained by the underlying one-step method.

The existence of an underlying one-step method (as a formal series in powers of h) satisfying the first of these properties, can be shown for general consistent methods (2.23); see [12, Chapter XIV]. The second property cannot be fulfilled by methods that are not strictly stable. Assuming that for arbitrary starting approximations the multistep solution remains close to that obtained by the underlying one-step method, it is natural to call a method (2.23) *symplectic* and *symmetric*, if the underlying one-step method is symplectic and symmetric, respectively.

Unfortunately, it turns out that partitioned multistep methods cannot be symplectic (Tang [39]). However, they can be symmetric. In terms of the coefficients of the method (2.23), the symmetry of the underlying one-step method is equivalent to (assuming irreducibility of the methods)

$$
\begin{aligned}
\alpha_j &= -\alpha_{k-j}, & \beta_j &= \beta_{k-j} & \text{for } j = 0, \ldots, k, \\
\widehat{\alpha}_j &= -\widehat{\alpha}_{\widehat{k}-j}, & \widehat{\beta}_j &= \widehat{\beta}_{\widehat{k}-j} & \text{for } j = 0, \ldots, \widehat{k}.
\end{aligned}
\tag{2.24}
$$

For stable symmetric multistep methods the zeros of the generating polynomials $\rho(\zeta) = \sum_{j=0}^{k} \alpha_j \zeta^j$ and $\widehat{\rho}(\zeta) = \sum_{j=0}^{\widehat{k}} \widehat{\alpha}_j \zeta^j$ have to lie on the unit circle. Such methods cannot be strictly stable, and for this reason symmetric multistep methods have been disregarded for a long time.

Also for this class of methods we are mainly interested in the numerical solution of second order differential equations $\ddot{q} = g(q)$. Elimination of the v-variables in (2.23) yields the formula

$$\sum_{j=0}^{K} A_j q_{n+j} = h^2 \sum_{j=0}^{K} B_j g(q_{n+j}), \tag{2.25}$$

where the generating polynomials $R(\zeta) = \sum_{j=0}^{K} A_j \zeta^j$ and $S(\zeta) = \sum_{j=0}^{K} B_j \zeta^j$ are obtained from those of (2.23) by

$$R(\zeta) = \rho(\zeta) \cdot \widehat{\rho}(\zeta), \quad S(\zeta) = \sigma(\zeta) \cdot \widehat{\sigma}(\zeta).$$

Here, $\rho(\zeta), \widehat{\rho}(\zeta), \sigma(\zeta), \widehat{\sigma}(\zeta)$ are the generating polynomials of $\alpha_j, \widehat{\alpha}_j, \beta_j, \widehat{\beta}_j$, respectively. We recall that method (2.25) is of order p, if

$$R(e^h) - h^2 S(e^h) = \mathcal{O}(h^{p+2}) \quad \text{for} \quad h \to 0. \tag{2.26}$$

Formula (2.25) does not involve derivative approximations v_n. If they are needed, they can be obtained by finite differences from the position approximations q_n.

Example 2.3 (Method used as 'Imm2' in Figure 2.3). We put $K = 8$ and we let

$$R(\zeta) = (\zeta - 1)(\zeta^7 - 1) = (\zeta - 1)^2 (\zeta^6 + \zeta^5 + \zeta^4 + \zeta^3 + \zeta^2 + \zeta + 1),$$

so that all zeros lie on the unit circle and, apart from $\zeta = 1$, all zeros are simple. To get a method of order $p = 8$, the polynomial $S(\zeta)$ has to satisfy

$$S(\zeta) = R(\zeta)/\log^2 \zeta + \mathcal{O}((\zeta - 1)^p)$$

(cf. condition (2.26)). Expanding the right-hand expression into a Taylor series at $\zeta = 1$ and truncating to get a polynomial of degree 7, we obtain the generating polynomial

$$S(\zeta) = \frac{13207}{8640}(\zeta^7 + \zeta) - \frac{8934}{8640}(\zeta^6 + \zeta^2) + \frac{42873}{8640}(\zeta^5 + \zeta^3) - \frac{33812}{8640}\zeta^4.$$

The resulting method (2.25) is of order 8 for problems $\ddot{q} = g(q)$, symmetric, and explicit (because $B_K = 0$). An approximation to the derivative is obtained by symmetric differences as

$$\dot{y}_n = \frac{1}{840h}\Big(672\,(y_{n+1} - y_{n-1}) - 168\,(y_{n+2} - y_{n-2})$$
$$+ 32\,(y_{n+3} - y_{n-3}) - 3\,(y_{n+4} - y_{n-4})\Big).$$

Composition Methods We consider the composition of a given basic one-step method $\Phi_h(y)$ with different step sizes:

$$\Psi_h = \Phi_{\gamma_s h} \circ \ldots \circ \Phi_{\gamma_2 h} \circ \Phi_{\gamma_1 h}. \tag{2.27}$$

The aim is to increase the order (and hence the accuracy) while preserving desirable properties (symplecticity, symmetry) of the basic method. This idea has mainly been developed in the papers of Suzuki [38], Yoshida [44], and McLachlan [24]. For a recent comprehensive survey see [25] and Chapters II, III, and V of [12].

The reason for the success of composition methods within geometric integrators are the following properties:

- if Φ_h is symplectic, then the composition method Ψ_h is *symplectic*;
- if Φ_h is symmetric and if the step size parameters γ_i satisfy $\gamma_i = \gamma_{s+1-i}$, then the composition Ψ_h is *symmetric*.

The main problem consists of finding parameters γ_i such that the composition Ψ_h is of a given order. Suzuki [38] and Yoshida [44] propose general simple procedures that allow one to construct composition methods of arbitrarily high order. However, for orders higher than four they are not very efficient. One is therefore obliged to investigate and solve the set of order conditions for the γ_i which guarantee that the method Ψ_h of (2.27) has a certain order.

Example 2.4 (Method used as 'comp' in Figure 2.3). From the many published examples of composition methods, let us present the coefficients of a method of order 8 with $s = 17$ steps:

$$
\begin{aligned}
\gamma_1 = \gamma_{17} &= 0.13020248308889008087881763 \\
\gamma_2 = \gamma_{16} &= 0.56116298177510838456196441 \\
\gamma_3 = \gamma_{15} &= -0.38947496264484728640807860 \\
\gamma_4 = \gamma_{14} &= 0.15884190655515560089621075 \\
\gamma_5 = \gamma_{13} &= -0.39590389413323757733623154 \\
\gamma_6 = \gamma_{12} &= 0.18453964097831570709183254 \\
\gamma_7 = \gamma_{11} &= 0.25837438768632204729397911 \\
\gamma_8 = \gamma_{10} &= 0.29501172360931029887096624 \\
\gamma_9 &= -0.60550853383003451169892108
\end{aligned}
$$

This set of coefficients is due to Kahan and Li [15]. The little picture to the right illustrates the 17 steps necessary for obtaining order 8. The zig-zag behaviour is typical for composition methods. It is impossible to get high order without negative step sizes.

For the computations of Figure 2.3 we use the Störmer/Verlet scheme (2.12) as a basic integrator. The resulting composition method is symplectic and (due to $\gamma_i = \gamma_{18-i}$) symmetric.

2.6   Rattle for Constrained Hamiltonian Systems

Let us explain how the Störmer/Verlet method (2.12) can be generalized to solve constrained Hamiltonian systems of the form (1.13). Without taking much care of velocity approximations Ryckaert, Ciccotti and Berendsen [31] show how constraints $g(q) = 0$ can be included in the formulation (2.13). Anderson [1] reformulates their method and includes a velocity approximation that satisfies the hidden constraint (1.14). The resulting algorithm, still for separable Hamiltonians, is called 'Rattle'. Later, Jay [16] and Reich [28] observed that the Rattle algorithm can be extended to general Hamiltonians.

Recall that the exact flow of a constrained Hamiltonian system lies on the manifold $\mathcal{M}$, defined in (1.17). Assume therefore that an approximation $(p_n, q_n) \in \mathcal{M}$ is given. One step of the algorithm is defined as

$$
\begin{aligned}
p_{n+1/2} &= p_n - \frac{h}{2} \left(\nabla_q H(p_{n+1/2}, q_n) + \nabla_q g(q_n) \lambda_n \right) \\
q_{n+1} &= q_n + \frac{h}{2} \left(\nabla_p H(p_{n+1/2}, q_n) + \nabla_p H(p_{n+1/2}, q_{n+1}) \right) \\
0 &= g(q_{n+1}) \\
p_{n+1} &= p_{n+1/2} - \frac{h}{2} \left(\nabla_q H(p_{n+1/2}, q_{n+1}) + \nabla_q g(q_{n+1}) \mu_n \right) \\
0 &= \nabla_q g(q_{n+1})^T \nabla_p H(p_{n+1}, q_{n+1}).
\end{aligned}
\tag{2.28}
$$

For fixed λ_n, the first two equations uniquely define $p_{n+1/2}$ and q_{n+1}, if h is sufficiently small. The parameter λ_n has to be chosen to satisfy $g(q_{n+1}) = 0$. This is possible if the matrix (1.16) is invertible. In the last two equations, μ_n has to be chosen to satisfy the constraint for p_{n+1}.

Similar to the Störmer/Verlet method for unconstrained Hamiltonian systems, this algorithm has many nice properties that are useful within geometric integration:

- the numerical solution stays on the manifold $\mathcal{M}$; i.e. the method (2.28) defines a numerical flow $\Phi_h : \mathcal{M} \to \mathcal{M}$;
- the numerical flow $\Phi_h : \mathcal{M} \to \mathcal{M}$ is a symplectic transformation on $\mathcal{M}$;
- the method is symmetric;
- the method is convergent of order two.

The symplecticity of the numerical flow was first shown by Leimkuhler and Skeel [23]. The other properties are easy consequences of the definition of the method. This integrator is an ideal candidate as a basic method for compositions of the form (2.27). For elaborate proofs and for extensions to higher orders we refer to Section VII.1 of [12].

3   Theoretical Foundation of Geometric Integrators

Intuitively, it is quite obvious that a symplectic method should be preferred for the integration of Hamiltonian systems. Similarly, symmetric (more precisely, ρ-reversible) integrators should be preferred for ρ-reversible differential equations. This is motivated by the fact that the symplecticity of the flow is characteristic for Hamiltonian systems, and the ρ-reversibility of the flow is characteristic for ρ-reversible differential equations.

In this section we give some more precise statements on the long-time behaviour of geometric integrators. In particular, we discuss the idea of *backward error analysis* which is the key for a deeper understanding of most numerical phenomena. This idea was common to many numerical analysts already before a systematic study started with the work of Feng [8], Sanz-Serna [33], Yoshida [45], Hairer [9] and many others.

3.1   Backward Error Analysis

Consider an ordinary differential equation

$$\dot{y} = f(y) \tag{3.1}$$

and a numerical method $y_{n+1} = \Phi_h(y_n)$. The idea of backward error analysis consists of searching and studying a *modified differential equation*

$$\dot{y} = f(y) + h f_2(y) + h^2 f_3(y) + \dots, \tag{3.2}$$

such that the exact time-h flow $\widetilde{\varphi}_h(y)$ of (3.2) is equal to the numerical flow $\Phi_h(y)$. Already simple examples (e.g. trapezoidal rule applied to a quadrature problem $\dot{y} = f(t)$) show that the series in (3.2) cannot be expected to converge in general. The precise statement is the following:

Theorem 3.1. *Consider the differential equation (3.1) with an infinitely differentiable vector field $f(y)$. Assume that the numerical flow admits a Taylor series expansion of the form*

$$\Phi_h(y) = y + h f(y) + h^2 d_2(y) + h^3 d_3(y) + \dots . \tag{3.3}$$

Then, there exist unique vector fields $f_j(y)$ such that for any $N \geq 1$

$$\Phi_h(y) = \widetilde{\varphi}_{h,N}(y) + \mathcal{O}(h^{N+1}),$$

where $\widetilde{\varphi}_{t,N}$ is the exact flow of the truncated modified equation

$$\dot{y} = f(y) + h f_2(y) + \dots + h^{N-1} f_N(y)$$

(notice that the flow $\widetilde{\varphi}_{t,N}$ also depends on h, because h is a parameter in the modified differential equation).

Let us outline a constructive *proof*. Without taking care of convergence we expand the exact flow of (3.2) into a Taylor series

$$\tilde{\varphi}_h(y) = y + h\,\tilde{y}'(0) + \frac{h^2}{2!}\,\tilde{y}''(0) + \frac{h^3}{3!}\,\tilde{y}'''(0) + \dots$$

$$= y + h\big(f(y) + hf_2(y) + h^2 f_3(y) + \dots\big) \qquad (3.4)$$

$$+ \frac{h^2}{2!}\big(f'(y) + hf_2'(y) + \dots\big)\big(f(y) + hf_2(y) + \dots\big) + \dots$$

(where the prime denotes derivation with respect to time) and compare like powers of h in the expressions (3.4) and (3.3). This yields recurrence relations for the functions $f_j(y)$, namely,

$$f_2(y) = d_2(y) - \frac{1}{2!}f'f(y) \qquad (3.5)$$

$$f_3(y) = d_3(y) - \frac{1}{3!}\Big(f''(f,f)(y) + f'f'f(y)\Big) - \frac{1}{2!}\Big(f'f_2(y) + f_2'f(y)\Big).$$

Example 3.1. We consider the pendulum equation $\dot{q} = p,\ \dot{p} = -\sin q$ and apply the explicit Euler discretization (2.1). We have $d_j(y) = 0$ for all $j \geq 2$, so that (3.5) yields for the modified equation

$$\begin{pmatrix} \dot{q} \\ \dot{p} \end{pmatrix} = \begin{pmatrix} p \\ -\sin q \end{pmatrix} + \frac{h}{2}\begin{pmatrix} \sin q \\ p\cos q \end{pmatrix} + \frac{h^2}{12}\begin{pmatrix} -4p\cos q \\ (p^2 + 4\cos q)\sin q \end{pmatrix} + \dots . \qquad (3.6)$$

For the implicit Euler method (2.2) we get (3.6) with h replaced by $-h$. A similar computation yields for the symplectic Euler method (2.3) the modified differential equation

$$\begin{pmatrix} \dot{q} \\ \dot{p} \end{pmatrix} = \begin{pmatrix} p \\ -\sin q \end{pmatrix} + \frac{h}{2}\begin{pmatrix} -\sin q \\ p\cos q \end{pmatrix} + \frac{h^2}{12}\begin{pmatrix} 2p\cos q \\ (p^2 - 2\cos q)\sin q \end{pmatrix} + \dots , \qquad (3.7)$$

whereas the same equation with h replaced by $-h$ is obtained for the method (2.4). The four pictures of Figure 3.1 show the exact flow of the modified differential equations (truncated after the $\mathcal{O}(h^2)$ term) corresponding to these four Euler methods together with the numerical solution for the initial value $(p_0, q_0) = (-1.2, 0.7)$. We observe a surprisingly good agreement. This figure should be compared to the exact flow of the unperturbed system (cf. Figure 1.1).

The $\mathcal{O}(h)$ perturbation in (3.6) provokes the origin to become a source for the explicit Euler method, and a sink for the implicit Euler method. For the two symplectic discretizations we observe that the solutions of the modified equation are periodic, and that the numerical approximation lies near a closed curve. It has thus the correct qualitative behaviour. This is explained by the fact that the differential equation (3.7) is Hamiltonian with

$$\tilde{H}(p,q) = \frac{1}{2}p^2 - \cos q - \frac{h}{2}p\sin q + \frac{h^2}{12}(p^2 - \cos q)\cos q + \dots ,$$

so that the exact solutions stay on the level curves of $\tilde{H}(p,q)$.

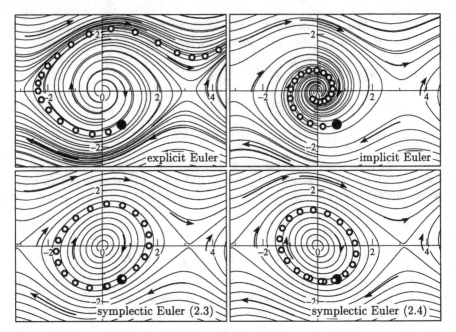

Fig. 3.1. Numerical solution with step size $h = 0.4$ for the four 'Euler methods' of Section 2.1 compared to the exact flow of their modified differential equations truncated after the $\mathcal{O}(h^2)$ term.

3.2   Properties of the Modified Equation

The previous example demonstrates that the numerical solution is extremely close to the exact solution of the modified differential equation. Therefore instead of studying the properties of the numerical solution, it is justified investigating the corresponding properties of the modified differential equation, and this is often much simpler. Let us collect some properties valid for general differential equations:

- if the method is of order r, i.e. $\Phi_h(y) - \varphi_h(y) = \mathcal{O}(h^{r+1})$, then we have $f_j(y) = 0$ for $j = 2, \ldots, r$;

- if $h^{r+1}\delta_{r+1}(y)$ is the leading term of the local truncation error, i.e. $\Phi_h(y) - \varphi_h(y) = h^{r+1}\delta_{p+1}(y) + \mathcal{O}(h^{r+2})$, then we have $f_{r+1}(y) = \delta_{r+1}(y)$;

- if $\Phi_h(y)$ has the modified equation (3.2), then the adjoint method has $f_j^*(y) = (-1)^{j+1} f_j(y)$ as coefficient functions of the modified equation;

- for symmetric methods the modified equation is an expansion in even powers of h; i.e. $f_{2k}(y) = 0$ for all k.

We now turn our attention to Hamiltonian systems and to ρ-reversible differential equations.

Theorem 3.2 (Local Modified Hamiltonian). *Consider a Hamiltonian system (1.1) with smooth Hamiltonian $H : D \to \mathbf{R}$ $(D \subset \mathbf{R}^{2d})$ and apply a symplectic numerical method $\Phi_h(y)$. Then, the vector fields $f_k(y)$ of the modified differential equation are locally Hamiltonian, i.e. locally we have $f_k(y) = J^{-1}\nabla H_k(y)$.*

The *proof* is by induction on k. Its ideas can be traced back to Moser [26], and it can be found in Benettin & Giorgilli [4], Tang [40], Reich [29], and in Chapter IX of [12]. We briefly outline the idea of proof since it is applicable to many other situations.

We assume (by induction) that the truncated modified equation

$$\dot{y} = f(y) + h f_2(y) + \ldots + h^{k-1} f_k(y) \tag{3.8}$$

is Hamiltonian. Its flow $\tilde{\varphi}_{t,k}(y)$ satisfies

$$\Phi_h(y) = \tilde{\varphi}_{h,k}(y) + h^{k+1} f_{k+1}(y) + \mathcal{O}(h^{k+2}).$$

Since Φ_h and $\tilde{\varphi}_{h,k}$ are symplectic transformations,

$$J = \Phi_h'(y)^T J \Phi_h'(y) = J + h^{k+1}\Big(f_{k+1}'(y)^T J + J f_{k+1}'(y)\Big) + \mathcal{O}(h^{k+2})$$

holds. Consequently, the matrix $J f_{k+1}'(y)$ is symmetric and the existence of $H_{k+1}(y)$ satisfying $f_{k+1}(y) = J^{-1}\nabla H_{k+1}(y)$ follows from the integrability lemma. □

If H and Φ_h are both defined and smooth on the whole of $\mathbf{R}^{2d}$ or on a simply connected domain D, the functions H_k of the modified Hamiltonian are globally defined. However, as shown by the following example, the functions H_k are in general not globally defined, and the above theorem cannot be used for the study of the long-time behaviour of numerical solutions.

Example 3.2. For the harmonic oscillator $\dot{p} = -q$, $\dot{q} = p$, consider the discretization

$$p_{n+1} = p_n - hq_n - h^2\gamma p_{n+1}, \quad q_{n+1} = q_n + hp_{n+1} - h^2\gamma q_n \tag{3.9}$$

where $\gamma = 0.25/(p_{n+1}^2+q_n^2)$. It is a $\mathcal{O}(h^2)$ perturbation of the symplectic Euler method and therefore it is a method of order 1. Its symplecticity follows from the fact that it can be written as

$$p_{n+1} = p_n - h\nabla_q S(p_{n+1},q_n), \quad q_{n+1} = q_n + h\nabla_p S(p_{n+1},q_n)$$

with $S(p,q) = \frac{1}{2}(p^2+q^2) - \frac{h}{4}\arg(q+ip)$. Its numerical approximation, plotted in the right picture of Figure 3.2, is disappointing and does not show the correct qualitative behaviour. This is due to the fact that $S(p,q)$, and hence $H_2(p,q)$, are not globally defined.

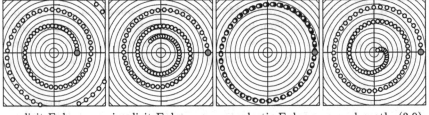

explicit Euler   implicit Euler   symplectic Euler   sympl. meth. (3.9)

Fig. 3.2. Numerical solution of different first order methods applied to the harmonic oscillator with step size $h = 0.15$.

Theorem 3.3 (Global Modified Hamiltonian). *Consider a Hamiltonian system (1.1) with smooth Hamiltonian* $H : D \to \mathbf{R}$ *($D \subset \mathbf{R}^{2d}$) and apply the symplectic method*

$$p_{n+1} = p_n - h\,\nabla_q S(p_{n+1}, q_n), \quad q_{n+1} = q_n + h\,\nabla_p S(p_{n+1}, q_n)$$

with generating function

$$S(p,q) = S_1(p,q) + h\,S_2(p,q) + h^2 S_3(p,q) + \dots,$$

where all $S_k(p,q)$ *are globally defined on* D. *Then, the vector fields* $f_k(y)$ *of the modified differential equation are globally Hamiltonian, i.e. we have* $f_k(y) = J^{-1}\nabla H_k(y)$ *with smooth* $H_k : D \to \mathbf{R}$.

The *proof* of this theorem is based on the Hamilton-Jacobi differential equation (cf. Section IX.3.2 of [12]). Let us mention that all previous methods (symplectic Euler, Störmer/Verlet, symplectic partitioned Runge-Kutta methods, composition methods) satisfy the assumption of Theorem 3.3.

Theorem 3.4 (ρ-Reversible Modified Vector Field). *Consider a ρ-reversible differential equation (cf. Section 1.2) and apply a ρ-reversible numerical method* $\Phi_h(y)$. *Then, the vector fields* $f_k(y)$ *of the modified differential equation are ρ-reversible, i.e. they satisfy (1.9).*

The *proof* uses the same ideas as that of Theorem 3.2.

3.3 Long-Time Behaviour of Geometric Integrators

Using backward error analysis and in particular the results of Theorems 3.3 and 3.4, we shall show that symplectic integrators (for Hamiltonian systems) and ρ-reversible integrators (for ρ-reversible differential equations) have an improved long-time behaviour. We study the conservation of the Hamiltonian and of general first integrals, and the error growth for integrable systems.

Conservation of the Hamiltonian We know that the Hamiltonian $H(p, q)$ is constant along exact solutions of the Hamiltonian system (energy conservation for mechanical systems). Since the local error of an r-th order integrator is of size $\mathcal{O}(h^{r+1})$, we have $H(p_{n+1}, q_{n+1}) - H(p_n, q_n) = \mathcal{O}(h^{r+1})$. Summing up these errors, we obtain $H(p_n, q_n) - H(p_0, q_0) = \mathcal{O}(nh^{r+1}) = \mathcal{O}(th^r)$ for $t = nh$, because no cancellation of errors can be expected for general integrators. However, for symplectic integrators we have the much more favourable estimate

$$H(p_n, q_n) - H(p_0, q_0) = \mathcal{O}(h^r) \quad \text{for} \quad nh \leq T \tag{3.10}$$

with an extremely large T (in practice it can be considered as infinity), provided that the numerical solution stays in a compact set. This can be explained with the help of Theorem 3.3 as follows: the modified differential equation is Hamiltonian with

$$\widetilde{H}(p, q) = H(p, q) + h^r H_{r+1}(p, q) + h^{r+1} H_{r+2}(p, q) + \ldots . \tag{3.11}$$

The exact flow of the modified equation, and hence also the numerical solution, keep the modified Hamiltonian $\widetilde{H}(p, q)$ exactly constant. If the numerical solution stays in a compact set, the functions $H_j(p, q)$ are bounded along the numerical solution so that (3.10) holds. This argument is not yet rigorous, because the series (3.11) usually does not converge. If one truncates the series suitably, one can rigorously prove (3.10) on exponentially long time intervals, i.e. for $T = \mathcal{O}(e^{\gamma/h})$ with some positive γ (cf. [4], [11], and Section IX.7 of [12]).

It is also natural to study whether other first integrals can be well conserved by numerical integrators. Recall that $I(y)$ is a *first integral* of $\dot{y} = f(y)$, if it is constant along all solutions of the differential equations, i.e. if $I'(y)f(y) = 0$ vanishes identically.

Example 3.3. Consider the Kepler problem (1.11). Besides the Hamiltonian (2.17), it also has the *angular momentum*

$$L(p_1, p_2, q_1, q_2) = q_1 p_2 - q_2 p_1 \tag{3.12}$$

and the so-called *Runge-Lenz-Pauli vector*

$$A(p, q) = \begin{pmatrix} p_1 \\ p_2 \\ 0 \end{pmatrix} \times \begin{pmatrix} 0 \\ 0 \\ q_1 p_2 - q_2 p_1 \end{pmatrix} - \frac{1}{\sqrt{q_1^2 + q_2^2}} \begin{pmatrix} q_1 \\ q_2 \\ 0 \end{pmatrix} \tag{3.13}$$

as first integrals. We take the numerical scheme to be the Störmer/Verlet method. We apply it to the Kepler problem with initial values as in Section 2.5, and we use the step size $h = 0.02$. Figure 3.3 shows the values of $H(p_n, q_n) - H(p_0, q_0)$ and of the first two components of $A(p_n, q_n) - A(p_0, q_0)$ along the numerical solution. The angular momentum $L(p, q)$ is exactly preserved by the method and therefore not visible in the figure. We see that, in agreement with (3.10), the error in the Hamiltonian is bounded by $\mathcal{O}(h^2)$ on the whole interval of integration. However, the Runge-Lenz-Pauli vector (3.13) is not preserved. The lower picture of Figure 3.3, where the errors obtained with step size $h = 0.04$ are included in grey, indicates that they behave like $e(h^2 t) + \mathcal{O}(h^2)$ with some smooth function $e(\tau)$.

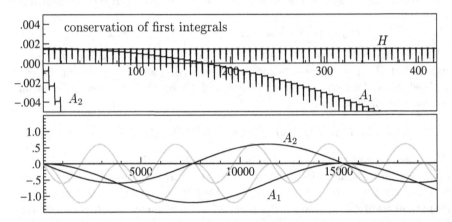

Fig. 3.3. Kepler problem: the Hamiltonian and the first two components of the Runge-Lenz-Pauli vector along the numerical solution of the Störmer/Verlet method with step sizes $h = 0.02$ (black) and $h = 0.04$ (grey).

Completely Integrable Systems The example above demonstrated that it is difficult to predict the conservation of general first integrals by numerical methods (even when they are symplectic). However, there is an important special case for which more information can be obtained. We mention some facts and refer the reader to Chapters X and XI of [12].

We call a Hamiltonian system (1.1) *completely integrable*, if there exists a symplectic transformation

$$(p, q) = \psi(a, \theta), \qquad 2\pi\text{-periodic in } \theta, \qquad (3.14)$$

such that the Hamiltonian becomes

$$H(p, q) = H(\psi(a, \theta)) = K(a). \qquad (3.15)$$

The new variables (a, θ) are called *action-angle variables*. Suppose we know explicitly the transformation ψ. Since it is symplectic, the Hamiltonian system (1.1) becomes in the new variables

$$\dot{a}_i = 0, \quad \dot{\theta} = \omega_i(a), \quad i = 1, \ldots, d$$

with $\omega_i(a) = \partial K / \partial a_i(a)$. This system can be readily solved, and gives $a_i(t) = a_{i0}$, $\theta_i(t) = \theta_{i0} + \omega_i(a_0)t$, so that

$$(p(t), q(t)) = \psi(a_0, \theta_0 + \omega(a_0)t).$$

This gives a periodic or quasi-periodic flow on the torus defined by $a = \text{const}$. Among the problems seen in this survey article, Hamiltonian systems with one degree of freedom (harmonic oscillator, pendulum) and the Kepler problem are completely integrable. Under some additional technical assumptions (see the general reference [12]), *symplectic numerical integrators* applied to such completely integrable Hamiltonian systems have the following interesting properties:

- the global error grows at most linearly with time, more precisely, for $t = nh$ we have

$$p_n - p(t) = \mathcal{O}(h^r t), \quad q_n - q(t) = \mathcal{O}(h^r t);$$

- first integrals that only depend on the action variables are well preserved on exponentially long time intervals; i.e. if $I(p, q)$ is such that $I(\psi(a, \theta))$ is independent of θ, then

$$I(p_n, q_n) - I(p_0, q_0) = \mathcal{O}(h^r) \quad \text{for} \quad nh \leq T$$

with T as in (3.10).

This result has to be seen in contrast to general methods, where the global error increases typically quadratically with time, and where the error in first integrals drifts linearly from the correct value.

We finally mention that the notion of complete integrability can be reinterpreted for ρ-reversible differential equations (not necessarily Hamiltonian), and the same results (linear error growth, conservation of action variables) hold for ρ-reversible integrators applied to such systems. Let us illustrate this with an interesting example.

Example 3.4 (Toda lattice). Let us consider particles on a line interacting pairwise with exponential forces, and suppose periodic boundary conditions $q_{d+1} = q_1$. The Hamiltonian is given by

$$H(p, q) = \sum_{k=1}^{d} \left(\frac{1}{2} p_k^2 + \exp(q_k - q_{k+1}) \right).$$

The corresponding Hamiltonian system has the interesting property that the d eigenvalues of the matrix

$$L = \begin{pmatrix} a_1 & b_1 & & & b_d \\ b_1 & a_2 & b_2 & 0 & \\ & b_2 & \ddots & \ddots & \\ & 0 & \ddots & a_{d-1} & b_{d-1} \\ b_d & & & b_{d-1} & a_d \end{pmatrix}, \qquad \begin{aligned} a_k &= -\tfrac{1}{2} p_k \\ b_k &= \tfrac{1}{2} \exp\left(\tfrac{1}{2}(q_k - q_{k+1}) \right) \end{aligned}$$

are first integrals. This Hamiltonian system is completely integrable within the class of Hamiltonian systems and also within the class of ρ-reversible systems, and the action variables are related to the eigenvalues of the matrix L.

We consider the case $d = 3$, and we apply an implicit Runge-Kutta method (Lobatto IIIB, $s = 3$, order $r = 4$) which is symmetric but not symplectic. The upper picture of Figure 3.4 shows the Euclidean norm of the vector of errors in the eigenvalues of L. It is of size $\mathcal{O}(h^4)$ on the whole interval of integration. The lower picture shows the norm of the global error, and we nicely observe the linear error growth. This confirms the statement about integrable systems of this section.

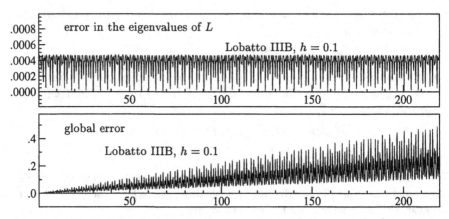

Fig. 3.4. The Toda lattice: error in the eigenvalues of L (upper picture), and global error of the numerical solution for an implicit Runge-Kutta method.

4   Matlab Programs of 'GniCodes'

We explain a few Matlab programs that implement the most important geometric integrators of the previous sections. They are collected in the Matlab package GniCodes which is available (together with short installation instructions) on the web at the address

http://www.unige.ch/math/folks/hairer

Fortran 77 versions of the programs are also available at the same address. Another Matlab package related to geometric integration is DiffMan of [7]. However, the philosophy of our package is completely different and it is closer to the standard Matlab ODE suite (ode45, ode23, etc.) of Shampine and Reichelt [34].

4.1   Standard Call of Integrators

We give an overview of how to use the three classes of geometric integrators that are implemented in the package GniCodes. For the solution of second order initial value problems

$$\ddot{q} = g(q), \quad q(0) = q_0, \quad \dot{q}(0) = \dot{q}_0 \qquad (4.1)$$

all these methods have the same syntax, and are usually called as

---

```
[T,P,Q] = gni_meth('g',tspan,y0,options,...);
```

---

where gni_meth has to be replaced by gni_irk2 for the implicit Runge-Kutta method, by gni_lmm2 for the linear multistep method, and by gni_comp for the composition method based on the Störmer/Verlet scheme. The '2' in irk2 and in lmm2 expresses the fact that these programs are applicable only to second order differential equations (4.1). For the composition method gni_comp there is the possibility of defining the basic integrator by altering the options structure, so that the method can be used for the solution of any differential equation. The relevant syntax will be explained in Section 4.6 below.

The meaning of the arguments in a call of gni_meth is as follows:

'g' This argument must be a string containing the name of a Matlab file describing the problem. The syntax for such a file is described in Section 4.2 below.

tspan should contain the time span over which the problem is to be solved. It has to be given in the form [t0,tf].

y0 This is a vector containing the initial values for q and $\dot{q}$. The initial values for q are given by the first d components and those for $\dot{q}$ by the remaining d components of y0.

options This argument should contain a GNI options structure created by gniset (the syntax of gniset is the same as for the standard odeset function). This option structure contains additional instructions for the integrator.

... After options, an arbitrary number of optional arguments can be given. These arguments are passed over to the function g.

The list of available options differs slightly from the standard ODE suite. Some of these options are also function-dependent and will be explained in the sequel. The following options are available for every integrator:

'OutputFcn' This is a string containing the name of an output function. The format for the output functions is the same as for the standard ODE suite, in particular the standard odeplot output function can be used (and will be used as the default output function). The vector passed to the output function contains in its d first components the value of the solution and in its d remaining components the values for its time-derivatives. If this parameter is set to phaseplot, the solution is drawn in the phase space corresponding to the first components of $(q, \dot{q})$.

'OutputSel' As for the standard ODE suite, this contains a vector of indices determining which components of the solution will be passed to the output function. By default, all the indices (including those corresponding to the time-derivative of the solution) are passed through.

'OutputSteps' tells the integrator which steps to take into account for the output. For example, if OutputSteps is equal to 10, only every 10th step generates some output. Putting OutputSteps equal to 0, output is made available only at the beginning and the end of the integration interval.

'Vectorized' has the same meaning as for the standard ODE suite.

'Events' If this option is set to 'on', event location is enabled. See Section 4.3 for an explanation of how to use event location.

'StepSize' Size h of one integration step. It is slightly altered by the code, if the length of the integration interval is not an integer multiple of h.

'NumSteps' is the number of integration steps. This option is only used when StepSize is not specified. If neither is specified, a warning is issued and the default step size $h = 0.01$ is used.

'Method' allows the user to select the type of method to use. The list of available methods depends on the integrator and is listed in the corresponding sections below.

Note that (like for ode45 for example) the arguments tspan, y0, and options are optional and can be defined in the file g.m instead.

On output, gni_meth returns three vectors [T,Q,P], containing the times at which the solution was evaluated, as well as the values of q and $\dot{q}$ at these times. If event location is turned on, additional return values are given as described in Section 4.3 below.

4.2 Problem Description

The problem to be solved should be described in a .m file. In the most simple case, this file only returns the right-hand side of the second-order differential equation. For example, in order to solve the equation $\ddot{q} = -q^3$, one may create a file trivial.m containing the following:

```
function out = trivial(t,q)
   out = -q^3;
```

From the command line, one would then use it for example as

```
options = gniset('StepSize', 0.1);
gni_meth('trivial', [0 10], [0 2.5], options);
```

An additional parameter flags can be used by the integrator to retrieve default parameters for the problem. Assume we want to solve the previous problem between $t = 0$ and $t = 10$, using a step size of 0.1 and with initial values $q(0) = 0$ and $\dot{q}(0) = 2.5$. We could then define the file trivial.m as

```
function [out,out2,out3] = trivial(t,q,flags)
if (nargin < 3) | isempty(flags)
   out = -q^3;
else
   switch flags
   case 'init',
      out = [0 10];
      out2 = [0 2.5];
      out3 = gniset('StepSize', 0.1);
   end
end
```

and call it from the command line in the most simple possible way as

```
gni_meth('trivial');
```

For a system of differential equations, out and q are column vectors. If the option 'Vectorized' is set to 'on' in the GNI options structure, the integrator may request an evaluation of the right-hand side of the problem for several values of t and q in one call. If the problem is of dimension d and the integrator requests m values, t is a line vector of size m and q is a $d \times m$ matrix. The right-hand side is also expected to be a $d \times m$ matrix. If 'Vectorized' is set to 'off', one can safely assume that $m = 1$.

When vectorized correctly, the .m file for the Kepler problem with initial values as in Section 2.5 looks like

```
function [out,out2,out3] = kepler(t,q,flags,ecc)
if (nargin < 3) | isempty(flags)
   rad=q(1,:).*q(1,:)+q(2,:).*q(2,:);
   rad=rad.*sqrt(rad);
   out(1,:)=-q(1,:)./rad;
   out(2,:)=-q(2,:)./rad;
else
   switch flags
   case 'init',
      if (ecc < 0) | (ecc >= 1)
         error('The␣eccentricity␣must␣lie␣between␣0␣and␣1');
      end
      out = [0 2*pi];
      out2 = [1-ecc,0,0,sqrt((1+ecc)/(1-ecc))];
      out3 = gniset('NumSteps',50,'Vectorized','on','Events','off',...
         'OutputFcn','phaseplot','OutputSel',[1,2]);
   end
end
```

Notice that this problem depends on the eccentricity ecc, which has been appended to the end of the parameter list. To solve this problem with ecc= 0.6, just type

```
gni_meth('kepler',[ ],[ ],[ ],0.6);
```

Entering [ ] in the parameter list tells the integrator to use the default values of the problem definition file instead. The parameter ecc is again simply appended at the end of the parameter list.

4.3   Event Location

In many situations (for example the computation of Poincaré sections), it is useful to know at which times some *event function* $g(t, q(t), p(t))$ vanishes. This is usually referred to as *event location*.

Event location is implemented in the GNI suite in a way that is again very similar to the standard ODE suite implementation. It can be enabled by specifying the value 'on' for the 'Events' selector of the GNI options structure.

When event location is turned on, the integrator can be called as

```
[T,P,Q,TE,PE,QE,IE] = gni_meth('g',tspan,y0,options,...);
```

The output vector TE contains the times at which events occurred. The vectors PE and QE contain the values of the solution and its derivative at these times. If more than one event function is defined, the vector IE contains the index of the event function that triggered the event.

When event location is turned on, the problem description file is expected to respond to the **flags** set to 'events' by returning in the first output argument a vector of event functions. Furthermore, it is supposed to return in the second and third output arguments vectors telling the integrator whether the corresponding event is terminal or not and which types of zero-crossings to consider. When a terminal event is encountered, the integration stops, whether the end of the integration interval has been reached or not. The following example shows how to define a problem description file that allows the user to retrieve the times at which the solution either crosses 1 upwards or 0 in any direction. The integration stops whenever the solution crosses -2 downwards.

```
function [out,out2,out3] = trivial(t,q,flags)
if (nargin < 3) | isempty(flags)
   out = -q^3;
else
   switch flags
   case 'init',
      out = [0 10];
      out2 = [-1 5];
      out3 = gniset('StepSize', 0.1,'Events','on');
   case 'events',
      out = [q(1)-1,q(1),q(1)+2];
      out2 = [0 0 1];
      out3 = [1 0 -1];
   end
end
```

4.4   Program gni_irk2

The program gni_irk2 uses an implicit Runge-Kutta scheme to solve second order differential equations $\ddot{q} = g(q)$. The following selectors of the GNI options structure are specific to gni_irk2:

'Method' This selector allows the user to specify which scheme is to be used. The accepted values for 'Method' are 'G4', 'G8', and 'G12'. The letter 'G' refers to the fact that all of these methods are Gauss methods (cf. Example 2.2), and the number that follows indicates the order of the corresponding method. If no method is specified, 'G12' is used.

'MaxIter' Since the schemes are implicit, a non-linear system of equations has to be solved at every integration step. This is achieved through a fixed point iteration. This selector allows the user to specify the maximal number of iterations that are performed. The default value is 50.

The coefficients of the different methods are contained in the file coeff_irk2. New methods can easily be incorporated.

Let us briefly explain the meaning of the required coefficients. The arrays C, B, BC correspond to the vectors with coefficients c_i, $\hat{b}_i$, $\bar{b}_i$ of (2.22), the 2-dimensional array AA of the matrix $\widetilde{A}$. Further coefficients are needed for an efficient solution of the nonlinear Runge-Kutta equations of (2.22), which are equivalent to

$$Q_i = q_n + hc_i v_n + h^2 \sum_{j=1}^{s} \widetilde{a}_{i,j}\, g(Q_j), \quad i = 1, \ldots, s. \qquad (4.2)$$

We solve this system by fixed point iteration and we use

$$Q_i^0 = q_n + hc_i v_n + h^2 \sum_{j=1}^{s+3} e_{i,j}\, g(Q_{j,n-1}) \qquad (4.3)$$

as the starting guess, where $Q_{1,n-1}, \ldots, Q_{s,n-1}$ are the internal stage values of the previously computed step, $Q_{s+1,n-1} = q_{n-1}$, $Q_{s+2,n-1} = q_n$, and

$$Q_{s+3,n-1} = q_n + \sum_{i=1}^{s} \mu_i \big(Q_{i,n-1} - q_{n-1} \big) + h\mu_{s+1} v_{n-1} + h\mu_{s+2} v_n$$

is an approximation to the solution at $t = t_{n-1} + \mu h$. The coefficients μ, μ_i, $e_{i,j}$ (stored in the arrays SM, AM, and E) are determined such that (4.3) coincides as far as possible with the Taylor series of the solution of (4.2). We refer to [12, Section VIII.6.1] and [20] for more details.

4.5   Program gni_lmm2

Linear multistep methods (2.25) for second order differential equations are
implemented in the code gni_lmm2. Since these methods are not self-starting,
we have to provide starting approximations. This is done by a call to gni_irk2
with 'Method' set to 'G12'. For the moment we have implemented the three
methods of Table 4.1, and the 'Method' options are '801', '802', and '803',
respectively.

	method 801		method 802		method 803	
i	C_{i-1}	$12096\,B_i$	C_{i-1}	$120960\,B_i$	C_{i-1}	$8640\,B_i$
1	1	17671	1	192481	1	13207
2	0	−23622	2	6582	1	−8934
3	1	61449	3	816783	1	42873
4	1	−50516	3.5	−156812	1	−33812

Table 4.1. Symmetric multistep methods for second order problems.

The coefficients of the methods are stored in the separate file coeff_lmm2
as follows. The generating polynomial $R(\zeta)$ has $\zeta = 1$ as a double zero, and
therefore it can be written as

$$R(\zeta) = (\zeta - 1)^2(C_0 + C_1\zeta + C_2\zeta^2 + \ldots + C_{K-2}\zeta^{K-2}).$$

Since the coefficients for explicit symmetric methods satisfy $B_{K-i} = B_i$ (with
$B_0 = 0$) and $C_{K-2-i} = C_i$, only those given in Table 4.1 have to be specified.
The coefficients C_i and B_i uniquely determine the method (2.25).

4.6   Program gni_comp

This program allows one to easily implement general composition methods. A
composition method (2.27) is characterized by the set of coefficients $\{\gamma_i\}$ and
by the basic method Φ_h. They are controlled by the following two options:

'Method' This option allows the user to choose between several predefined
sets of coefficients γ_i. The available methods are '21', '43', '45', '67',
'69', '815', '817', and '1033'. These methods are of order 2, 4, 6, 8,
and 10 respectively.

'PartialFlow' This option allows the user to specify the name of a Matlab
file defining the basic method Φ_h. The default method is the Störmer/Ver-
let method.

For reasons of efficiency we assume the basic method to be of the form

$$\Phi_h = \omega_{h/2} \circ \beta_{h/2,h,h/2} \circ \alpha_{h/2}\,, \tag{4.4}$$

where the following simplification formula holds:

$$\beta_{*,*,ha} \circ \alpha_{ha} \circ \omega_{hc} \circ \beta_{hc,*,*} = \beta_{*,*,ha+hc} \circ \beta_{ha+hc,*,*} \,. \qquad (4.5)$$

Every one-step method can be written in the form (4.4) by choosing $\omega_{h/2} = \alpha_{h/2} = id$ (the identity), and $\beta_{hc,hb,ha} = \Phi_{hb}$. But this is not the reason for writing Φ_h in this apparently complicated form. The advantage of the representation (4.4) is that in many important situations a large part of the work for evaluating Φ_h can be put into $\alpha_{h/2}$ and $\omega_{h/2}$ and, by the simplification formula (4.5), this part can be avoided unless one is at grid points where an output of the solution is required.

The code for the basic method must have the following structure:

```
function [outP,outQ] = basic(t,P,Q,ode,ha,hb,hc,first,last,flags,args)
if isempty(flags)
    if (first)
        apply αha to the vectors P and Q
    end
    apply βhc,hb,ha
    if (last)
        apply ωhc
    end
else
    switch flags
    case 'init',
        perform some initialization
    case 'done',
        perform some cleanup
    end
end
```

For example, the Störmer/Verlet method (2.12) for $\ddot{q} = g(q)$, considered as a splitting method (2.16), can be written as (4.4) with $\omega_{h/2} = id$, $\beta_{hc,hb,ha} = \varphi_{hc}^{[1]} \circ \varphi_{hb}^{[2]}$, and $\alpha_{ha} = \varphi_{ha}^{[1]}$. This presentation satisfies condition (4.5), because $\varphi_t^{[1]}$ has the group property. The Matlab program for this basic method is

```
function [outP,outQ] = stverl(t,P,Q,ode,ha,hb,hc,first,...
    last,flags,varargin)
if isempty(flags)
    if (first)
        Q = Q + ha*P;
    end
    F = feval(ode,t,Q,varargin{:});
    outP = P + hb*F;
    outQ = Q + hc*outP;
end
```

The actual implementation uses compensated summation (to reduce round-off error) and it is the default method used by gni_comp.

5 Some Typical Applications

Let us finally illustrate the use of our programs on some typical examples, where geometric integrators are recommended. We start with a comparison of geometric integrators for second order ordinary differential equations. We then show how Poincaré sections can be computed, and we finish with a slightly more sophisticated use of composition methods.

5.1 Comparison of Geometric Integrators

Often it is difficult to decide which integrator is the best for a given problem. The implicit Runge-Kutta and composition methods have a sound theoretical basis, but they typically need more function evaluations per step than linear multistep methods. On the other hand, multistep methods have larger local error, so that smaller step sizes are required. The best choice is in general problem dependent.

Consider first the Kepler problem with eccentricity ecc= 0.6. The file kepler.m, containing the problem description, is explained in Section 4.2. We compute the solution over the interval $[0, 400\pi]$ with many different step sizes. As we have seen in Figure 2.3 the efficiency of all three classes of integrators (implicit Runge-Kutta, multistep, composition) is about the same for this problem. This need not always be the case.

As another example consider the 6-body problem (sun and the five outer planets) with data and initial values as in Chapter I of [12] on a relatively short time interval $[0, 500\,000]$. Similar to Figure 2.3 we show in Figure 5.1 the work precision diagram of the different methods. It is somewhat surprising that for this problem (with orbits of very small eccentricity) the linear

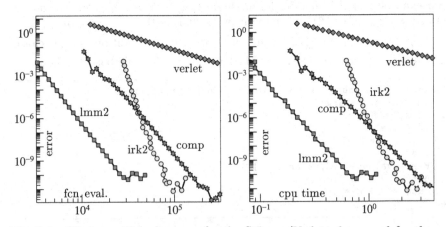

Fig. 5.1. Work precision diagrams for the Störmer/Verlet scheme and for three methods of order eight; implicit Runge-Kutta method (irk2), composition method (comp), and linear multistep method (lmm2), applied to the outer solar system.

multistep method is the most efficient integrator. The cpu times in Figure 2.3 and in Figure 5.1 are obtained with Fortran implementations of the codes.

5.2 Computation of Poincaré Sections

Consider the *Hénon-Heiles* Hamiltonian

$$H(p_1, p_2, q_1, q_2) = \frac{1}{2}\left(p_1^2 + p_2^2\right) + \frac{1}{2}\left(q_1^2 + q_2^2\right) + q_1^2 q_2 - \frac{1}{3} q_2^3. \tag{5.1}$$

The corresponding Hamiltonian system is integrable for sufficiently small energy, e.g. for the initial values $p_1(0) = p_2(0) = q_1(0) = q_2(0) = 0.18$, which we take for our computations. This means that the solution stays on a two-dimensional torus in the four-dimensional phase space. Its intersection with the hyperplane $q_1 = 0$ (Poincaré section) thus gives a closed curve in the phase space. We study the projection of this curve onto the (q_2, p_2)-plane.

The left picture of Figure 5.2 shows the Poincaré section for the numerical solution obtained by dop853 (an explicit Runge-Kutta method of order eight with step size control, see [14, Appendix]) and tolerance $Atol = Rtol = 10^{-5}$ on the interval $[0, 100\,000]$. The picture clearly demonstrates that the numerical solution is qualitatively wrong as it does not remain on a closed curve. The same experiment with the three geometric integrators gni_lmm2, gni_irk2, and gni_comp gives a correct simulation of the system, and it cannot be distinguished from a picture for the exact solution (right picture of Figure 5.2). If we use step sizes such that the error of the Hamiltonian remains below 10^{-5}, the code gni_lmm2 (with $h = 0.22$) requires $454\,716$ function evaluations, gni_irk2 (with $h = 1.5$) needs $3\,731\,867$ function evaluations, and gni_comp (with $h = 1.2$) $1\,416\,661$ function evaluations. For comparison,

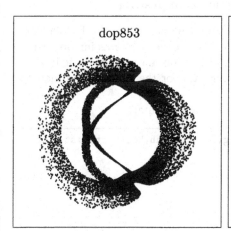
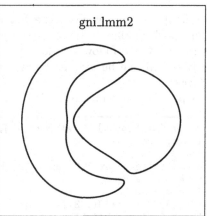

Fig. 5.2. Poincaré section for the Hénon-Heiles problem: dop853 with tolerance $Atol = Rtol = 10^{-5}$ (left picture) and gni_lmm2 with step size $h = 0.22$ (right picture); integration interval $[0, 100\,000]$.

the code dop853 requires 1 216 680 evaluations of the vector field, but the error in the Hamiltonian increases linearly with time. The high number of function evaluations for gni_irk2 is due to the fact that for low accuracy requirements (large step size) the convergence of the fixed point iterations for solving the nonlinear Runge-Kutta equations is rather slow.

For the computation of the Poincaré section (Figure 5.2) we have used the following program:

```
function [out,out2,out3] = henon(t,q,flags)
if (nargin < 3) | isempty(flags)
   out(1,:)=-q(1,:).*(1+2*q(2,:));
   out(2,:)=-q(2,:).*(1-q(2,:)) - q(1,:).^2;
else
   switch flags
   case 'init',
      out = [0 100000];
      out2 = [0.18 0.18 0.18 0.18];
      out3 = gniset('StepSize',0.22,'Vectorized','on',...
         'Events','on','OutputSteps',0);
   case 'events',
      out = [q(1)];
      out2 = [0];
      out3 = [0];
   end
end
```

The plot of the Poincaré section is then obtained with

```
[T,Q,P,TE,QE,PE]=gni_lmm2('henon');
plot(QE(:,2),PE(:,2),'.');
```

5.3   'Rattle' as a Basic Integrator for Composition

As a final example, we present a Matlab implementation of the Rattle algorithm (2.28) applied to the *two-body problem on the sphere* as introduced in Example 1.4. We follow the description of Section 4.6 and we do it in such a way that it can be used as a basic integrator for the composition method gni_comp.

A possible implementation is the following program:

```
function [outP,outQ] = rattwo(t,P,Q,gradpot,ha,hb,hc,first,last,...
   flags,varargin)
if isempty(flags)
   F = feval(gradpot,t,Q,varargin{:});
   EP = P - ha*F;
   EQ = Q + hb*EP;
   EE1 = EQ(1:3)'*EQ(1:3);
   EQ1 = EQ(1:3)'*Q(1:3);
   EE2 = EQ(4:6)'*EQ(4:6);
   EQ2 = EQ(4:6)'*Q(4:6);
   BET1 = 1 - EE1;
```

```
ALAM1 = -BET1/(hb*(EQ1+sqrt(BET1+EQ1^2)));
BET2 = 1 - EE2;
ALAM2 = -BET2/(hb*(EQ2+sqrt(BET2+EQ2^2)));
outP = EP - [ALAM1*Q(1:3);ALAM2*Q(4:6)];
outQ = Q + hb*outP;
if (last)
  F = feval(gradpot,t,outQ,varargin{:});
  outP = outP - hc*F;
  AMU1 = sum(outP(1:3).*outQ(1:3));
  AMU2 = sum(outP(4:6).*outQ(4:6));
  outP = outP - [AMU1*outQ(1:3);AMU2*outQ(4:6)];
end
end
```

We remark that, due to the simple structure of the Hamiltonian, the method is explicit in $p_{n+1/2}$, q_{n+1} and p_{n+1}, and it is implicit only in the Lagrange multipliers. Since the constraints are quadratic, we are only concerned with the solution of a scalar quadratic equation for each of the components of λ_n. This is why no iterations are involved in the above program. Since $\nabla_q H(p,q)$ does not depend on p, the first equation of (2.28) can be combined with the fourth equation of the preceding step into one formula to give $p_{n+1/2} = p_{n-1/2} + \dots$. This is the reason for putting the computation of p_{n+1} into $\omega_{h/2}$ of the decomposition (4.4).

The argument gradpot in the function rattwo is a function that computes the gradient of the potential (i.e. $\nabla_q U(q) = \nabla_q H(p,q)$). For the two-body problem on the sphere it is given by

```
function [out,out2,out3] = twobodysphere(t,q,flags)
if (nargin < 3) | isempty(flags)
  prod = q(1:3)'*q(4:6);
  out = -q([4:6,1:3])/(1-prod^2)^(3/2);
else
  switch flags
  case 'init',
    out = [0 10];
    phi = [1.3 -2.1];
    theta = [2.1 -1.1];
    out2([1 4]) = cos(phi).*sin(theta);
    out2([2 5]) = sin(phi).*sin(theta);
    out2([3 6]) = cos(theta);
    dphi = [1.2 0.1];
    dtheta = [0.1 -0.5];
    out2([7 10]) = -dphi.*sin(phi).*sin(theta) ...
       + dtheta.*cos(phi).*cos(theta);
    out2([8 11]) = dphi.*cos(phi).*sin(theta) ...
       + dtheta.*sin(phi).*cos(theta);
    out2([9 12]) = -dtheta.*sin(theta);
    out3 = gniset('StepSize',0.02,'Vectorized','off',...
       'Events','off','PartialFlow','rattwo','OutputFcn',...
       'sphereplot','OutputSteps',5,'Method','817');
  end
end
```

Here, the option `sphereplot` allows us to get a 3-dimensional plot of the solution. The problem is then simply solved by calling

```
gni_comp('twobodysphere');
```

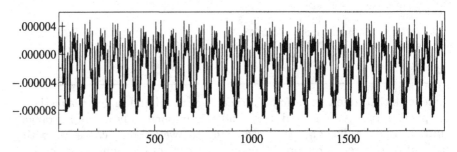

Fig. 5.3. Error in the Hamiltonian of an 8th order composition method with Rattle as a basic integrator; step size $h = 0.15$.

The experiment of Figure 5.3 confirms the statement of Theorem 3.3 for the Rattle algorithm applied to constrained Hamiltonian systems. We have plotted the error of the Hamiltonian for the composition method of Example 2.4, applied with step size $h = 0.15$ to the interval $[0, 2000]$. As expected for a symplectic integrator, there is no drift in the error of the Hamiltonian. This is also confirmed by integrations over much longer time intervals.

Hints for the implementation of the *rigid body* problem of Example 1.5 can be found in [10], where the extension of Theorem 3.3 to numerical methods for Hamiltonian systems on manifolds (including the Rattle algorithm) is also proved.

Acknowledgement

E.H. is grateful to the organizers of the 2002 Durham summer school, in particular to James F. Blowey, to Alan W. Craig, and to the local expert Sebastian Reich, for providing a sympathical and stimulating atmosphere. We also wish to thank Christian Lubich, Gerhard Wanner and numerous other colleagues for their useful discussions on the topic of this work.

This work was partially supported by the Fonds National Suisse.

References

1. H.C. Andersen (1983), Rattle: a "velocity" version of the Shake algorithm for molecular dynamics calculations, *J. Comput. Phys.* 52:24–34.
2. V.I. Arnold (1989), *Mathematical Methods of Classical Mechanics*, Second Edition, Springer-Verlag.

3. F. Bashforth (1883), *An attempt to test the theories of capillary action by comparing the theoretical and measured forms of drops of fluid. With an explanation of the method of integration employed in constructing the tables which give the theoretical form of such drops, by J.C.Adams.* Cambridge Univ. Press.

4. G. Benettin & A. Giorgilli (1994), On the Hamiltonian interpolation of near to the identity symplectic mappings with application to symplectic integration algorithms, *J. Statist. Phys.* 74:1117–1143.

5. J.C. Butcher (1963), Coefficients for the study of Runge-Kutta integration processes, *J. Austral. Math. Soc.* 3:185–201.

6. J.C. Butcher (1964), Implicit Rung-Kutta processes, *Math. Comput.* 18:50-64.

7. K. Engø, A. Marthinsen and H.Z. Munthe-Kaas (2001), DiffMan: an object-oriented MATLAB toolbox for solving differential equations on manifolds. Special issue: Themes in geometric integration. *Appl. Numer. Math.* 39:349–365.

8. K. Feng (1991), Formal power series and numerical algorithms for dynamical systems. In *Proceedings of international conference on scientific computation, Hangzhou, China, Eds. Tony Chan & Zhong-Ci Shi, Series on Appl. Math.* 1:28–35.

9. E. Hairer (1994), Backward analysis of numerical integrators and symplectic methods, *Annals of Numerical Mathematics* 1:107–132.

10. E. Hairer (2002), Global modified Hamiltonian for constrained symplectic integrators. *Numer. Math.* to appear.

11. E. Hairer and Ch. Lubich (1997), The life-span of backward error analysis for numerical integrators, *Numer. Math.* 76:441–462.

12. E. Hairer, C. Lubich and G. Wanner (2002), *Geometric Numerical Integration. Structure-Preserving Algorithms for Ordinary Differential Equations.* Springer Series in Computational Mathematics 31, Springer, Berlin.

13. E. Hairer, C. Lubich and G. Wanner (2003), Geometric numerical integration illustrated by the Störmer/Verlet method. *Acta Numerica* to appear.

14. E. Hairer, S.P. Nørsett and G. Wanner (1993), *Solving Ordinary Differential Equations I. Nonstiff Problems, 2nd edition.* Springer Series in Computational Mathematics 8, Springer, Berlin.

15. W. Kahan and R.-C. Li (1997), Composition constants for raising the orders of unconventional schemes for ordinary differential equations, *Math. Comput.* 66:1089–1099.

16. L. Jay (1994), *Runge-Kutta type methods for index three differential-algebraic equations with applications to Hamiltonian systems,* Thesis No. 2658, 1994, Univ. Genève.

17. U. Kirchgraber (1986), Multi-step methods are essentially one-step methods, *Numer. Math.* 48:85–90.

18. V.V. Kozlov and A.O. Harin (1992), *Kepler's problem in constant curvature spaces,* Celestial Mech. Dynam. Astronom. 54:393–399.

19. W. Kutta (1901), Beitrag zur näherungsweisen Integration totaler Differentialgleichungen, *Zeitschr. für Math. u. Phys.* 46:435–453.

20. M.P. Laburta (1998), Construction of starting algorithms for the RK-Gauss methods, *J. Comput. Appl. Math.* 90:239–261.

21. J.D. Lambert and I.A. Watson (1976), Symmetric multistep methods for periodic initial value problems, *J. Inst. Maths. Applics.* 18:189–202.

22. F.M. Lasagni (1988), Canonical Runge-Kutta methods, *ZAMP* 39:952-953.

23. B.J. Leimkuhler and R.D. Skeel (1994), Symplectic numerical integrators in constrained Hamiltonian systems, *J. Comput. Phys.* 112:117–125.

24. R.I. McLachlan (1995), On the numerical integration of ordinary differential equations by symmetric composition methods, *SIAM J. Sci. Comput.* 16:151–168.

25. R.I. McLachlan and G.R.W. Quispel (2002), Splitting methods. *Acta Numerica* 2002:341–434.

26. J. Moser (1968), Lectures on Hamiltonian systems, *Mem. Am. Math. Soc.* 81:1–60.

27. G.D. Quinlan and S. Tremaine (1990), Symmetric multistep methods for the numerical integration of planetary orbits, *Astron. J.* 100:1694–1700.

28. S. Reich (1994), Momentum conserving symplectic integrators, *Phys.* D 76:375–383.

29. S. Reich (1999), Backward error analysis for numerical integrators, *SIAM J. Numer. Anal.* 36:1549–1570.

30. C. Runge (1895), Ueber die numerische Auflösung von Differentialgleichungen, *Math. Ann.* 46:167–178.

31. J.-P. Ryckaert, G. Ciccotti and H.J.C. Berendsen (1977), Numerical integration of the cartesian equations of motion of a system with constraints: molecular dynamics of n-alkanes, *J. Comput. Phys.* 23:327–341.

32. J.M. Sanz-Serna (1988), Runge-Kutta schemes for Hamiltonian systems, *BIT* 28:877–883.

33. J.M. Sanz-Serna (1992), Symplectic integrators for Hamiltonian problems: an overview, *Acta Numerica* 1:243–286.

34. L.F. Shampine and M.W. Reichelt (1997), The MATLAB ODE suite. *SIAM J. Sci. Comput.* 18:1–22.

35. H.J. Stetter (1973), *Analysis of Discretization Methods for Ordinary Differential Equations*, Springer-Verlag, Berlin.

36. C. Störmer (1907), Sur les trajectoires des corpuscules électrisés. *Arch. sci. Phys. nat. Genève* 24:5–18, 113–158, 221–247.

37. Y.B. Suris (1988), On the conservation of the symplectic structure in the numerical solution of Hamiltonian systems (in Russian), In: *Numerical Solution of Ordinary Differential Equations*, ed. S.S. Filippov, Keldysh Institute of Applied Mathematics, USSR Academy of Sciences, Moscow, 1988, 148–160.

38. M. Suzuki (1990), Fractal decomposition of exponential operators with applications to many-body theories and Monte Carlo simulations, *Phys. Lett.* A 146:319–323.

39. Y.-F. Tang (1993), The symplecticity of multi-step methods, *Computers Math. Applic.* 25:83–90.

40. Y.-F. Tang (1994), Formal energy of a symplectic scheme for Hamiltonian systems and its applications (I), *Computers Math. Applic.* 27:31–39.

41. L. Verlet (1967), Computer "experiments" on classical fluids. I. Thermodynamical properties of Lennard-Jones molecules. *Phyiscal Review* 159:98–103.

42. R. de Vogelaere (1956), Methods of integration which preserve the contact transformation property of the Hamiltonian equations. Report Nr. 4, Dept. Math. Univ. of Notre Dame, Notre Dame, Ind.

43. G. Wanner (1973), Runge-Kutta-methods with expansion in even powers of h, *Computing* 11:81–85.

44. H. Yoshida (1990), Construction of higher order symplectic integrators, *Phys. Lett.* A 150:262–268.

45. H. Yoshida (1993), Recent progress in the theory and application of symplectic integrators, *Celestial Mech. Dynam. Astronom.* 56:27–43.

Numerical Approximations to Multiscale Solutions in Partial Differential Equations

Thomas Y. Hou*

Applied Mathematics, 217-50, Caltech, Pasadena, CA 91125, USA. Email: hou@ama.caltech.edu.

Abstract. Many problems of fundamental and practical importance have multiple scale solutions. The direct numerical solution of multiple scale problems is difficult to obtain even with modern supercomputers. The major difficulty of direct solutions is the scale of computation. The ratio between the largest scale and the smallest scale could be as large as 10^5 in each space dimension. From an engineering perspective, it is often sufficient to predict the macroscopic properties of the multiple-scale systems, such as the effective conductivity, elastic moduli, permeability, and eddy diffusivity. Therefore, it is desirable to develop a method that captures the small scale effect on the large scales, but does not require resolving all the small scale features. This paper reviews some of the recent advances in developing systematic multiscale methods such as homogenization, numerical samplings, multiscale finite element methods, variational multiscale methods, and wavelets based homogenization. Applications of these multiscale methods to transport through heterogeneous porous media and incompressible flows will be discussed. This paper is not intended to be a detailed survey and the discussion is limited by both the taste and expertise of the author.

1 Introduction

Many problems of fundamental and practical importance have multiple scale solutions. Composite materials, porous media, and turbulent transport in high Reynolds number flows are examples of this type. A complete analysis of these problems is extremely difficult. For example, the difficulty in analyzing groundwater transport is mainly caused by the heterogeneity of subsurface formations spanning over many scales. This heterogeneity is often represented by the multiscale fluctuations in the permeability of media. For composite materials, the dispersed phases (particles or fibers), which may be randomly distributed in the matrix, give rise to fluctuations in the thermal or electrical conductivity; moreover, the conductivity is usually discontinuous across the phase boundaries. In turbulent transport problems, the convective velocity field fluctuates randomly and contains many scales depending on the Reynolds number of the flow.

The direct numerical solution of multiple scale problems is difficult even with the advent of supercomputers. The major difficulty of direct solutions

* Research was in part supported by a grant DMS-0073916 from the National Science Foundation

is the scale of computation. For groundwater simulations, it is common that millions of grid blocks are involved, with each block having a dimension of tens of meters, whereas the permeability measured from cores is at a scale of several centimeters. This gives more than 10^5 degrees of freedom per spatial dimension in the computation. Therefore, a tremendous amount of computer memory and CPU time are required, and this can easily exceed the limit of today's computing resources. The situation can be relieved to some degree by parallel computing; however, the size of the discrete problem is *not* reduced. The load is merely shared by more processors with more memory. Whenever one can afford to resolve all the small scale features of a physical problem, direct solutions provide quantitative information of the physical processes at all scales. On the other hand, from an engineering perspective, it is often sufficient to predict the macroscopic properties of the multiscale systems, such as the effective conductivity, elastic moduli, permeability, and eddy diffusivity. Therefore, it is desirable to develop a method that captures the small scale effect on the large scales, but does not require resolving all the small scale features.

The purpose of these lecture notes is to review some recent advances in developing multiscale numerical methods that capture the small scale effect on the large scales, but do not require resolving all the small scale features. The ultimate goal is to develop a general method that works for problems with continuous spectrum of scales. Substantial progress has been made in recent years by combining modern mathematical techniques such as homogenization, numerical samplings, and multiresolution. My lectures can be roughly divided into five parts. In Section 2, I will review some homogenization theory for elliptic and hyperbolic equations as well as for incompressible flows. This homogenization theory provides the critical guideline for designing effective multiscale methods. Section 3 is devoted to numerical homogenization for semilinear hyperbolic systems using particle methods and sampling techniques. For hyperbolic systems, it is important to compute the advection of small scale information accurately and account for the nonlinear interaction properly. We also need to avoid certain resonant sampling of the grid in order to obtain convergence. In Section 4, we focus on some recent developments of numerical homogenization based on the multiscale finite element methods. We also discuss the issue of upscaling one-phase and two-phase flows through heterogeneous porous media. In Sections 5 and 6, I review the main ideas behind the wavelet-based numerical homogenization method and the variational multiscale method. There are many other multiscale methods which we will not cover due to the limited scope of these lectures. The above methods are chosen because they are similar philosophically and the materials complement each other very well. This paper is not intended to be a detailed survey of all available multiscale methods. The discussion is limited by scope of the lectures and expertise of the author.

2 Review of Homogenization Theory

In this section, we will review some classical homogenization theory for elliptic and hyperbolic PDEs. This homogenization theory will play an essential role in designing effective multiscale numerical methods for partial differential equations with multiscale solutions.

2.1 Homogenization Theory for Elliptic Problems

Consider the second order elliptic equation

$$\mathcal{L}(u_\varepsilon) \equiv -\frac{\partial}{\partial x_i}\left(a_{ij}\left(x/\varepsilon\right)\frac{\partial}{\partial x_j}\right)u_\varepsilon + a_0(x/\varepsilon)u_\varepsilon = f, \ u_\varepsilon|_{\partial\Omega} = 0, \qquad (2.1)$$

where $a_{ij}(y)$ and $a_0(y)$ are 1-periodic in both variables of y, and satisfy $a_{ij}(y)\xi_i\xi_j \geq \alpha\xi_i\xi_i$, with $\alpha > 0$, and $a_0 > \alpha_0 > 0$. Here we have used the Einstein summation notation, i.e. repeated index means summation with respect to that index.

This model equation represents a common difficulty shared by several physical problems. For porous media, it is the pressure equation through Darcy's law, the coefficient a_ε representing the permeability tensor. For composite materials, it is the steady heat conduction equation and the coefficient a_ε represents the thermal conductivity. For steady transport problems, it is a symmetrized form of the governing equation. In this case, the coefficient a_ε is a combination of transport velocity and viscosity tensor.

Homogenization theory is to study the limiting behavior $u_\varepsilon \to u$ as $\varepsilon \to 0$. The main task is to find the homogenized coefficients, a_{ij}^* and a_0^*, and the *homogenized equation* for the limiting solution u

$$-\frac{\partial}{\partial x_i}\left(a_{ij}^*\frac{\partial}{\partial x_j}\right)u + a_0^*u = f, \quad u|_{\partial\Omega} = 0. \qquad (2.2)$$

Define the L^2 and H^1 norms over Ω as follows

$$\|v\|_0^2 = \int_\Omega |v|^2\,dx, \quad \|v\|_1^2 = \|v\|_0^2 + \|\nabla v\|_0^2. \qquad (2.3)$$

Further, we define the bilinear form

$$a^\varepsilon(u,v) = \int_\Omega a_{i,j}^\varepsilon(x)\frac{\partial u}{\partial x_j}\frac{\partial v}{\partial x_i}\,dx + \int_\Omega a_0^\varepsilon uv\,dx. \qquad (2.4)$$

It is easy to show that

$$c_1\|u\|_1^2 \leq a^\varepsilon(u,u) \leq c_2\|u\|_1^2, \qquad (2.5)$$

with $c_1 = \min(\alpha,\alpha_0)$, $c_2 = \max(\|a_{ij}\|_\infty, \|a_0\|_\infty)$.

The elliptic problem can also be formulated as a variational problem: find $u_\varepsilon \in H_0^1$

$$a^\varepsilon(u_\varepsilon, v) = (f, v), \quad \text{for all} \quad v \in H_0^1(\Omega), \qquad (2.6)$$

where (f, v) is the usual L^2 inner product, $\int_\Omega fv\,dx$.

Special Case: One-Dimensional Problem Let $\Omega = (x_0, x_1)$ and take $a_0 = 0$. We have

$$-\frac{d}{dx}\left(a(x/\varepsilon)\frac{du_\varepsilon}{dx}\right) = f, \quad \text{in } \Omega, \tag{2.7}$$

where $u_\varepsilon(x_0) = u_\varepsilon(x_1) = 0$, and $a(y) > \alpha_0 > 0$ is y-periodic with period y_0.

By taking $v = u_\varepsilon$ in the bilinear form, we have

$$\|u_\varepsilon\|_1 \le c.$$

Therefore one can extract a subsequence, still denoted by u_ε, such that

$$u_\varepsilon \rightharpoonup u \quad \text{in } H_0^1(\Omega) \quad \text{weakly.} \tag{2.8}$$

On the other hand, we notice that

$$a^\varepsilon \rightharpoonup m(a) = \frac{1}{y_0}\int_0^{y_0} a(y)\,dy \quad \text{in } L^\infty(\Omega) \quad \text{weak star.} \tag{2.9}$$

It is tempting to conclude that u satisfies:

$$-\frac{d}{dx}\left(m(a)\frac{du}{dx}\right) = f,$$

where $m(a) = \frac{1}{y_0}\int_0^{y_0} a(y)\,dy$ is the arithmetic mean of a. However, this is *not* true. To derive the correct answer, we introduce

$$\xi^\varepsilon = a^\varepsilon \frac{du^\varepsilon}{dx}.$$

Since a^ε is bounded, and u_x^ε is bounded in $L^2(\Omega)$, so ξ^ε is bounded in $L^2(\Omega)$. Moreover, since $-\frac{d\xi^\varepsilon}{dx} = f$, we have $\xi^\varepsilon \in H^1(\Omega)$. Thus we get

$$\xi^\varepsilon \to \xi \quad \text{in } L^2(\Omega) \quad \text{strongly,}$$

so that

$$\frac{1}{a^\varepsilon}\xi^\varepsilon \to m(1/a)\xi \quad \text{in } L^2(\Omega) \quad \text{weakly.}$$

Further, we note that $\frac{1}{a^\varepsilon}\xi^\varepsilon = \frac{du^\varepsilon}{dx}$. Therefore, we arrive at

$$\frac{du}{dx} = m(1/a)\xi.$$

On the other hand, $-\frac{d\xi^\varepsilon}{dx} = f$ implies $-\frac{d\xi}{dx} = f$. This gives

$$-\frac{d}{dx}\left(\frac{1}{m(1/a)}\frac{du}{dx}\right) = f. \tag{2.10}$$

This is the correct homogenized equation for u. Note that $a^* = \frac{1}{m(1/a)}$ is the harmonic average of a^ε. It is in general not equal to the arithmetic average $\overline{a^\varepsilon} = m(a)$.

Multiscale Asymptotic Expansions The above analysis does not generalize to multi-dimensions. In this subsection, we introduce the multiscale expansion technique in deriving homogenized equations. This technique is very effective and can be used in a number of applications.

We shall look for $u_\varepsilon(x)$ in the form of asymptotic expansion

$$u_\varepsilon(x) = u_0(x, x/\varepsilon) + \varepsilon u_1(x, x/\varepsilon) + \varepsilon^2 u_2(x, x/\varepsilon) + \cdots, \qquad (2.11)$$

where the functions $u_j(x, y)$ are double periodic in y with period 1.

Denote by A^ε the second order elliptic operator

$$A^\varepsilon = -\frac{\partial}{\partial x_i}\left(a_{ij}(x/\varepsilon)\frac{\partial}{\partial x_j}\right). \qquad (2.12)$$

When differentiating a function $\phi(x, x/\varepsilon)$ with respect to x, we have

$$\frac{\partial}{\partial x_j} = \frac{\partial}{\partial x_j} + \frac{1}{\varepsilon}\frac{\partial}{\partial y_j},$$

where y is evaluated at $y = x/\varepsilon$. With this notation, we can expand A^ε as follows

$$A^\varepsilon = \varepsilon^{-2}A_1 + \varepsilon^{-1}A_2 + \varepsilon^0 A_3, \qquad (2.13)$$

where

$$A_1 = -\frac{\partial}{\partial y_i}\left(a_{ij}(y)\frac{\partial}{\partial y_j}\right), \qquad (2.14)$$

$$A_2 = -\frac{\partial}{\partial y_i}\left(a_{ij}(y)\frac{\partial}{\partial x_j}\right) - \frac{\partial}{\partial x_i}\left(a_{ij}(y)\frac{\partial}{\partial y_j}\right), \qquad (2.15)$$

$$A_3 = -\frac{\partial}{\partial x_i}\left(a_{ij}(y)\frac{\partial}{\partial x_j}\right) + a_0. \qquad (2.16)$$

Substituting the expansions for u_ε and A^ε into $A^\varepsilon u_\varepsilon = f$, and equating the terms of the same power, we get

$$A_1 u_0 = 0, \qquad (2.17)$$

$$A_1 u_1 + A_2 u_0 = 0, \qquad (2.18)$$

$$A_1 u_2 + A_2 u_1 + A_3 u_0 = f. \qquad (2.19)$$

Equation (2.17) can be written as

$$-\frac{\partial}{\partial y_i}\left(a_{ij}(y)\frac{\partial}{\partial y_j}\right)u_0(x, y) = 0, \qquad (2.20)$$

where u_0 is periodic in y. The theory of second order elliptic PDEs [35] implies that $u_0(x, y)$ is independent of y, i.e. $u_0(x, y) = u_0(x)$. This simplifies equation (2.18) for u_1,

$$-\frac{\partial}{\partial y_i}\left(a_{ij}(y)\frac{\partial}{\partial y_j}\right)u_1 = \left(\frac{\partial}{\partial y_i}a_{ij}(y)\right)\frac{\partial u}{\partial x_j}(x).$$

Define $\chi^j = \chi^j(y)$ as the solution to the following *cell problem*

$$\frac{\partial}{\partial y_i}\left(a_{ij}(y)\frac{\partial}{\partial y_j}\right)\chi^j = \frac{\partial}{\partial y_i}a_{ij}(y),\qquad(2.21)$$

where χ^j is double periodic in y. The general solution of equation (2.18) for u_1 is then given by

$$u_1(x,y) = -\chi^j(y)\frac{\partial u}{\partial x_j}(x) + \tilde{u}_1(x).\qquad(2.22)$$

Finally, we note that the equation for u_2 is given by

$$\frac{\partial}{\partial y_i}\left(a_{ij}(y)\frac{\partial}{\partial y_j}\right)u_2 = A_2 u_1 + A_3 u_0 - f.\qquad(2.23)$$

The solvability condition implies that the right hand side of (2.23) must have mean zero in y over one periodic cell $Y = [0,1]\times[0,1]$, i.e.

$$\int_Y (A_2 u_1 + A_3 u_0 - f)\,dy = 0.$$

This solvability condition for second order elliptic PDEs with periodic boundary condition [35] requires that the right hand side of equation (2.23) have mean zero with respect to the fast variable y. This solvability condition gives rise to the homogenized equation for u:

$$-\frac{\partial}{\partial x_i}\left(a_{ij}^*\frac{\partial}{\partial x_j}\right)u + m(a_0)u = f,\qquad(2.24)$$

where $m(a_0) = \frac{1}{|Y|}\int_Y a_0(y)\,dy$ and

$$a_{ij}^* = \frac{1}{|Y|}\left(\int_Y (a_{ij} - a_{ik}\frac{\partial\chi^j}{\partial y_k})\,dy\right).\qquad(2.25)$$

Justification of Formal Expansions The above multiscale expansion is based on a formal asymptotic analysis. However, we can justify its convergence rigorously.

Let $z_\varepsilon = u_\varepsilon - (u + \varepsilon u_1 + \varepsilon^2 u_2)$. Applying A^ε to z_ε, we get

$$A^\varepsilon z_\varepsilon = -\varepsilon r_\varepsilon,$$

where $r_\varepsilon = A_2 u_2 + A_3 u_1 + \varepsilon A_3 u_2$. If f is smooth enough, so is u_2. Thus we have $\|r_\varepsilon\|_\infty \le c$.

On the other hand, we have

$$z_\varepsilon|_{\partial\Omega} = -(\varepsilon u_1 + \varepsilon^2 u_2)|_{\partial\Omega}.$$

Thus, we obtain

$$\|z_\varepsilon\|_{L^\infty(\partial\Omega)} \le c\varepsilon.$$

It follows from the maximum principle [35] that

$$\|z_\varepsilon\|_{L^\infty(\Omega)} \le c\varepsilon$$

and therefore we conclude that

$$\|u_\varepsilon - u\|_{L^\infty(\Omega)} \le c\varepsilon.$$

Boundary Corrections The above asymptotic expansion does not take into account the boundary condition of the original elliptic PDEs. If we add a boundary correction, we can obtain higher order approximations.

Let $\theta_\varepsilon \in H^1(\Omega)$ denote the solution to

$$\nabla_x \cdot a^\varepsilon \nabla_x \theta_\varepsilon = 0 \text{ in } \Omega, \quad \theta_\varepsilon = u_1(x, x/\varepsilon) \text{ on } \partial\Omega.$$

Then we have

$$(u_\varepsilon - (u + \varepsilon u_1(x, x/\varepsilon) - \varepsilon\theta_\varepsilon))|_{\partial\Omega} = 0.$$

Moskow and Vogelius [52] have shown that

$$\|u_\varepsilon - u - \varepsilon u_1(x, x/\varepsilon) + \varepsilon\theta_\varepsilon\|_0 \le C_\omega \varepsilon^{1+\omega}\|u\|_{2+\omega}, \qquad (2.26)$$

$$\|u_\varepsilon - u - \varepsilon u_1(x, x/\varepsilon) + \varepsilon\theta_\varepsilon\|_1 \le C\varepsilon\|u\|_2, \qquad (2.27)$$

where we assume $u \in H^{2+\omega}(\Omega)$ with $0 \le \omega \le 1$, and Ω is assumed to be a bounded, convex curvilinear polygon of class C^∞. This improved estimate will be used in the convergence analysis of the multiscale finite element method to be presented in Section 4.

2.2   Homogenization for Hyperbolic Problems

In this subsection, we will review some homogenization theory for semilinear hyperbolic systems. As we will see below, homogenization for hyperbolic problems is very different from that for elliptic problems. The phenomena are also very rich.

Consider the semilinear Carleman equations [14]:

$$\frac{\partial u_\varepsilon}{\partial t} + \frac{\partial u_\varepsilon}{\partial x} = v_\varepsilon^2 - u_\varepsilon^2,$$

$$\frac{\partial v_\varepsilon}{\partial t} - \frac{\partial v_\varepsilon}{\partial x} = u_\varepsilon^2 - v_\varepsilon^2,$$

with oscillatory initial data, $u_\varepsilon(x, 0) = u_0^\varepsilon(x)$, $v_\varepsilon(x, 0) = v_0^\varepsilon(x)$.

Assume that the initial conditions are positive and bounded. Then it can be shown that there exists a unique *bounded* solution for all times. Thus we can extract a subsequence of u_ε and v_ε such that $u_\varepsilon \rightharpoonup u$ and $v_\varepsilon \rightharpoonup v$ as $\varepsilon \to 0$.

Denote u_m as the weak limit of u_ε^m, and v_m as the weak limit of v_ε^m. By taking the weak limit of both sides of the equations, we get

$$\frac{\partial u_1}{\partial t} + \frac{\partial u_1}{\partial x} = v_2 - u_2,$$

$$\frac{\partial v_1}{\partial t} - \frac{\partial v_1}{\partial x} = u_2 - v_2.$$

By multiplying the Carleman equations by u_ε and v_ε respectively, we get

$$\frac{\partial u_\varepsilon^2}{\partial t} + \frac{\partial u_\varepsilon^2}{\partial x} = 2u_\varepsilon v_\varepsilon^2 - 2u_\varepsilon^3,$$

$$\frac{\partial v_\varepsilon^2}{\partial t} + \frac{\partial v_\varepsilon^2}{\partial x} = 2v_\varepsilon u_\varepsilon^2 - 2v_\varepsilon^3.$$

Thus the weak limit of u_ε^2 depends on the weak limit of u_ε^3 and the weak limit of $u_\varepsilon v_\varepsilon^2$.

Denote by $\overline{w_\varepsilon}$ as the weak limit of w_ε. To obtain a closure, we would like to express $\overline{u_\varepsilon v_\varepsilon^2}$ in terms of the product $\overline{u_\varepsilon}$ and $\overline{v_\varepsilon^2}$. This is *not* possible in general. In this particular case, we can use the Div-Curl Lemma [53,54,58] to obtain a closure.

The Div-Curl Lemma. *Let Ω be an open set of $\mathbb{R}^N$ and u_ε and v_ε be two sequences such that*

$$u_\varepsilon \rightharpoonup u, \quad in \quad \left(L^2(\Omega)\right)^N \quad weakly,$$
$$v_\varepsilon \rightharpoonup v, \quad in \quad \left(L^2(\Omega)\right)^N \quad weakly.$$

Further, we assume that

$\operatorname{div} u_\varepsilon$ *is bounded in $L^2(\Omega)($ or compact in $H^{-1}(\Omega))$,*

$\operatorname{curl} v_\varepsilon$ *is bounded in $\left(L^2(\Omega)\right)^{N^2}$ (or compact in $\left(H^{-1}(\Omega)\right)^{N^2}$).*

Let $\langle \cdot, \cdot \rangle$ denote the inner product in $\mathbb{R}^N$, i.e.

$$\langle u, v \rangle = \sum_{i=1}^N u_i v_i.$$

Then we have

$$\langle u_\varepsilon \cdot v_\varepsilon \rangle \rightharpoonup \langle u \cdot v \rangle \quad weakly. \tag{2.28}$$

Remark 2.1. We remark that the Div-Curl Lemma is the simplest form of the more general Compensated Compactness Theory developed by Tartar [58] and Murat [53,54].

Applying the Div-Curl Lemma to $(u_\varepsilon, u_\varepsilon)$ and $(v_\varepsilon^2, v_\varepsilon^2)$ in the space-time domain, one can show that $\overline{u_\varepsilon v_\varepsilon^2} = \overline{u_\varepsilon}\,\overline{v_\varepsilon^2}$. Similarly, one can show that $\overline{u_\varepsilon^2 v_\varepsilon} = \overline{u_\varepsilon^2}\,\overline{v_\varepsilon}$. Using this fact, Tartar [59] obtained the following infinite hyperbolic system for u_m and v_m [59]:

$$\frac{\partial u_m}{\partial t} + \frac{\partial u_m}{\partial x} = m u_{m-1} v_2 - m u_{m+1},$$

$$\frac{\partial v_m}{\partial t} - \frac{\partial v_m}{\partial x} = m v_{m-1} u_2 - m v_{m+1}.$$

Note that the weak limit of u_ε^m, u_m, depends on the weak limit of u_ε^{m+1}, u_{m+1}. Similarly, v_m depends on v_{m+1}. Thus one cannot obtain a closed system for the weak limits u_ε and v_ε by a finite system. This is a generic phenomenon for nonlinear partial differential equations with microstructure. It is often referred to as the closure problem. On the other hand, for the Carleman equations, Tartar showed that the infinite system is hyperbolic and the system is well-posed.

The situation is very different for a 3×3 system of Broadwell type [13]:

$$\frac{\partial u_\varepsilon}{\partial t} + \frac{\partial u_\varepsilon}{\partial x} = w_\varepsilon^2 - u_\varepsilon v_\varepsilon, \tag{2.29}$$

$$\frac{\partial v_\varepsilon}{\partial t} - \frac{\partial v_\varepsilon}{\partial x} = w_\varepsilon^2 - u_\varepsilon v_\varepsilon, \tag{2.30}$$

$$\frac{\partial w_\varepsilon}{\partial t} + \alpha \frac{\partial w_\varepsilon}{\partial x} = u_\varepsilon v_\varepsilon - w_\varepsilon^2, \tag{2.31}$$

with oscillatory initial data, $u_\varepsilon(x,0) = u_0^\varepsilon(x)$, $v_\varepsilon(x,0) = v_0^\varepsilon(x)$ and $w_\varepsilon(x,0) = w_0^\varepsilon(x)$. When $\alpha = 0$, the above system reduces to the original Broadwell model. We will refer to the above system as the generalized Broadwell model.

Note that in the generalized Broadwell model, the right hand side of the w-equation depends on the product of uv. If we try to obtain an evolution equation for w_ε^2, it will depend on the triple product $u_\varepsilon v_\varepsilon w_\varepsilon$. The Div-Curl Lemma cannot be used here to characterize the weak limit of this triple product in terms of the weak limits of u_ε, v_ε and w_ε.

Assume the initial oscillations are periodic, i.e.

$$u_0^\varepsilon = u_0(x, x/\varepsilon), \quad v_0^\varepsilon = v_0(x, x/\varepsilon), \quad w_0^\varepsilon = w_0(x, x/\varepsilon).$$

where $u_0(x,y)$, $v_0(x,y)$, $w_0(x,y)$ are 1-periodic in y.

There are two cases to consider.

Case 2.1. $\alpha = m/n$ is a rational number. Let $\{U(x,y,t), V(x,y,t), W(x,y,t)\}$ be the homogenized solution which satisfies

$$U_t + U_x = \int_0^1 W^2 \, dy - U \int_0^1 V \, dy,$$

$$V_t - V_x = \int_0^1 W^2 \, dy - U \int_0^1 V \, dy,$$

$$W_t + \alpha W_x = -W^2 + \frac{1}{n} \int_0^n U(x, y + (\alpha - 1)z, t) V(x, y + (\alpha + 1)z, t) \, dz,$$

where $U|_{t=0} = u_0(x,y)$, $V|_{t=0} = v_0(x,y)$ and $W|_{t=0} = w_0(x,y)$. Then we have

$$\|u_\varepsilon(x,t) - U(x, \tfrac{x-t}{\varepsilon}, t)\|_{L^\infty} \le C\varepsilon,$$

$$\|v_\varepsilon(x,t) - V(x, \tfrac{x+t}{\varepsilon}, t)\|_{L^\infty} \le C\varepsilon,$$

$$\|w_\varepsilon(x,t) - W(x, \tfrac{x-\alpha t}{\varepsilon}, t)\|_{L^\infty} \le C\varepsilon.$$

Case 2.2. α is an irrational number. Let $\{U(x,y,t), V(x,y,t), W(x,y,t)\}$ be the homogenized solution which satisfies

$$U_t + U_x = \int_0^1 W^2 \, dy - U \int_0^1 V \, dy,$$

$$V_t - V_x = \int_0^1 W^2 \, dy - U \int_0^1 V \, dy,$$

$$W_t + \alpha W_x = -W^2 + \left(\int_0^1 U \, dy \right) \left(\int_0^1 V \, dy \right).$$

where $U|_{t=0} = u_0(x,y)$, $V|_{t=0} = v_0(x,y)$ and $W|_{t=0} = w_0(x,y)$. Then we have

$$\|u_\varepsilon(x,t) - U(x, \tfrac{x-t}{\varepsilon}, t)\|_{L^\infty} \leq C\varepsilon,$$

$$\|v_\varepsilon(x,t) - V(x, \tfrac{x+t}{\varepsilon}, t)\|_{L^\infty} \leq C\varepsilon,$$

$$\|w_\varepsilon(x,t) - W(x, \tfrac{x-\alpha t}{\varepsilon}, t)\|_{L^\infty} \leq C\varepsilon.$$

We refer the reader to [37] for the proof of the above results.

Note that when α is a rational number, the interaction of u_ε and v_ε can generate a high frequency contribution to w_ε. This is *not* the case when α is an irrational number. The rational α case corresponds to a resonance interaction.

The derivation and analysis of the above results rely on the following two Lemmas:

Lemma 2.1. *Let* $f(x), g(x,y) \in C^1$. *Assume that* $g(x,y)$ *is* n-periodic in y, *then we have*

$$\int_a^b f(x) g(x, x/\varepsilon) \, dx = \int_a^b f(x) \left(\frac{1}{n} \int_0^n g(x,y) \, dy \right) dx + O(\varepsilon).$$

Lemma 2.2. *Let* $f(x,y,z) \in C^1$. *Assume that* $f(x,y,z)$ *is* 1-periodic in y *and* z. *If* γ_2/γ_1 *is an irrational number, then we have*

$$\int_a^b f\left(x, \tfrac{x_1+\gamma_1 x}{\varepsilon}, \tfrac{x_2+\gamma_1 x}{\varepsilon}\right) dx = \int_a^b \left(\int_0^1 \int_0^1 f(x,y,z) \, dy \, dz \right) dx + O(\varepsilon).$$

The proof uses some basic ergodic theory. It can be seen easily by expanding in Fourier series in the periodic variables [37]. For the sake of completeness, we present a simple proof of the above homogenization result for the case of $\alpha = 0$ in the next subsection.

Homogenization of the Broadwell Model In this subsection, we give a simple proof of the homogenization result in the special case of $\alpha = 0$. The homogenized equations can be derived by multiscale asymptotic expansions [50].

Consider the Broadwell model

$$\partial_t u + \partial_x u = w^2 - uv \text{ in } \mathbb{R} \times (0,T), \qquad (2.32)$$

$$\partial_t v - \partial_x v = w^2 - uv \text{ in } \mathbb{R} \times (0,T), \qquad (2.33)$$

$$\partial_t w = uv - w^2 \text{ in } \mathbb{R} \times (0,T), \qquad (2.34)$$

with oscillatory initial values

$$u(x,0) = u_0(x,\tfrac{x}{\varepsilon}), \quad v(x,0) = v_0(x,\tfrac{x}{\varepsilon}), \quad w(x,0) = w_0(x,\tfrac{x}{\varepsilon}), \qquad (2.35)$$

where $u_0(x,y), v_0(x,y), w_0(x,y)$ are 1-periodic in y. We introduce an extra variable, y, to describe the fast variable, x/ε. Let the solution of the homogenized equation be $\{U(x,y,t), V(x,y,t), W(x,y,t)\}$ which satisfies

$$\partial_t U + \partial_x U + U \int_0^1 V\, dy - \int_0^1 W^2\, dy = 0 \text{ in } \mathbb{R} \times (0,T), (2.36)$$

$$\partial_t V - \partial_x V + V \int_0^1 U\, dy - \int_0^1 W^2\, dy = 0 \text{ in } \mathbb{R} \times (0,T), (2.37)$$

$$\partial_t W + W^2 - \int_0^1 U(x,y-z,t)V(x,y+z,t)\, dz = 0 \text{ in } \mathbb{R} \times (0,T), (2.38)$$

with initial values given by

$$U(x,y,0) = u_0(x,y), \quad V(x,y,0) = v_0(x,y), \quad W(x,y,0) = w_0(x,y).(2.39)$$

Note that $U(x,y,t), V(x,y,t), W(x,y,t)$ are 1-periodic in y and the system (2.36)-(2.39) is a set of partial differential equations in (x,t) with $y \in [0,1]$ as a parameter. The global existence of the systems (2.32)-(2.35) and (2.36)-(2.39) has been established, see the references cited in [32].

Theorem 2.1. *Let (u,v,w) and (U,V,W) be the solutions of the systems (2.32)-(2.35) and (2.36)-(2.39), respectively. Then we have the following error estimate*

$$\max_{0 \le t \le T} E(t) \le \Big[5(M(T)^2 + 2TK(T)M(T))\exp(6M(T)T)\Big]\, \varepsilon := C_1(T)\varepsilon,(2.40)$$

where the error function $E(t)$ is given by

$$E(t) = \max_{x \in \mathbb{R}} \Big\{ \Big|u(x,t) - U(x,\tfrac{x-t}{\varepsilon},t)\Big| + \Big|v(x,t) - V(x,\tfrac{x+t}{\varepsilon},t)\Big|$$

$$+ \Big|w(x,t) - W(x,\tfrac{x}{\varepsilon},t)\Big| \Big\}$$

and the constants $M(T)$ and $K(T)$ are given by

$$M(T) = \max_{(x,y,t)\in\mathbb{R}\times[0,1]\times[0,T]} \Big(|u|,|v|,|w|,|U|,|V|,|W|\Big), \qquad (2.41)$$

$$N(T) = \max_{(x,y,t)\in\mathbb{R}\times[0,1]\times[0,T]} \Big(|\partial_x U|,|\partial_t U|,|\partial_x V|,|\partial_t V|,|\partial_x W|,|\partial_t W|\Big). \quad (2.42)$$

This homogenization result was first obtained by McLaughlin, Papanico-laou and Tartar using an L^p norm estimate ($0 < p < \infty$) [50]. Since we need an L^∞ norm estimate in the convergence analysis of our particle method, we give another proof of this result in L^∞ norm. As a first step, we prove the following lemma.

Lemma 2.3. *Let $g(x,y) \in C^1(\mathbb{R} \times [0,1])$ be 1-periodic in y and satisfy the relation $\int_0^1 g(x,y)\,dy = 0$. Then for any $\varepsilon > 0$ and for any constants a and b, the following estimate holds*

$$\left| \int_a^b g(x, \tfrac{x}{\varepsilon})\,dx \right| \leq B(g)\varepsilon + |b - a|B(\partial_x g)\varepsilon, \tag{2.43}$$

where $B(\zeta) = \max_{(x,y)\in\mathbb{R}\times[0,1]} |\zeta(x,y)|$ for any function ζ defined on $\mathbb{R}\times[0,1]$.

Proof. The estimate (2.43) is a direct consequence of the identity

$$g(x, \tfrac{x}{\varepsilon}) = \frac{d}{dx}\int_a^x g(x, \tfrac{s}{\varepsilon})\,ds - \int_a^x \frac{\partial g}{\partial x}(x, \tfrac{s}{\varepsilon})\,ds$$

and the estimates

$$\left| \int_a^b g(x, \tfrac{s}{\varepsilon})\,ds \right| \leq B(g)\varepsilon, \quad \left| \int_a^x \frac{\partial g}{\partial x}(x, \tfrac{s}{\varepsilon})\,ds \right| \leq B(\partial_x g)\varepsilon,$$

which follow from the 1-periodicity of $g(x,y)$ in y and that $\int_0^1 g(x,y)\,dy = 0$. This completes the proof. $\square$

Proof (of Theorem 2.1). Subtracting (2.36) from (2.32) and integrating the resulting equation along the characteristics from 0 to t, we get

$$u(x,t) - U(x, \tfrac{x-t}{\varepsilon}, t)$$

$$= \int_0^t \left[w(x-t+s, s)^2 - W(x-t+s, \tfrac{x-t+s}{\varepsilon}, s)^2 \right] ds$$

$$+ \int_0^t \left[W(x-t+s, \tfrac{x-t+s}{\varepsilon}, s)^2 - \int_0^1 W(x-t+s, y, s)^2\,dy \right] ds$$

$$- \int_0^t \left[u(x-t+s, s)v(x-t+s, s) \right.$$

$$\left. -U(x-t+s, \tfrac{x-t}{\varepsilon}, s)V(x-t+s, \tfrac{x-t+2s}{\varepsilon}, s) \right] ds$$

$$- \int_0^t U(x-t+s, \tfrac{x-t}{\varepsilon}, s)\left[V(x-t+s, \tfrac{x-t+2s}{\varepsilon}, s) \right.$$

$$\left. - \int_0^1 V(x-t+s, y, s)\,dy \right] ds$$

$$:= (\mathrm{I})_1 + \cdots + (\mathrm{I})_4. \tag{2.44}$$

It is clear from the definition of $E(t)$ and $M(T)$ that

$$|(\mathrm{I})_1 + (\mathrm{I})_3| \leq 2M(T) \int_0^t E(s) \, ds.$$

To estimate $(\mathrm{I})_2$, we define for fixed $(x, t) \in \mathbb{R} \times [0, T]$,

$$g_{(x,t)}(s, y) = W(x - t + s, \tfrac{x-t}{\varepsilon} + y, s)^2.$$

Since the 1-periodicity of $W(x, y, t)$ in y implies

$$\int_0^1 W(x - t + s, y, s)^2 \, dy = \int_0^1 W(x - t + s, \tfrac{x-t}{\varepsilon} + y, s)^2 \, dy,$$

we obtain by applying Lemma 2.1 that

$$|(\mathrm{I})_2| = \left| \int_0^t \left[g_{(x,t)}(s, \tfrac{s}{\varepsilon}) - \int_0^1 g_{(x,t)}(s, y) \, dy \right] ds \right|$$
$$\leq M(T)^2 \varepsilon + 2M(T) K(T) T \varepsilon.$$

Similarly, we have

$$(\mathrm{I})_4 \leq M(T)^2 \varepsilon + 2M(T) K(T) T \varepsilon.$$

Substituting these estimates into (2.44) we get

$$\left| u(x, t) - U(x, \tfrac{x-t}{\varepsilon}, t) \right| \leq 2M(T) \int_0^t E(s) \, ds + 2M(T)^2 \varepsilon + 4M(T) K(T) T \varepsilon. \tag{2.45}$$

Similarly, we conclude from (2.37)-(2.38) and (2.33)-(2.34) that

$$\left| v(x, t) - V(x, \tfrac{x+t}{\varepsilon}, t) \right| \leq 2M(T) \int_0^t E(s) \, ds + 2M(T)^2 \varepsilon + 4M(T) K(T) T \varepsilon, \tag{2.46}$$

$$\left| w(x, t) - W(x, \tfrac{x}{\varepsilon}, t) \right| \leq 2M(T) \int_0^t E(s) \, ds + M(T)^2 \varepsilon + 2M(T) K(T) T \varepsilon. \tag{2.47}$$

Now the desired estimate (2.40) follows from summing (2.45)-(2.47) and using the Gronwall inequality. □

Remark 2.2. The homogenization theory tells us that the initial oscillatory solutions propagate along their characteristics. The nonlinear interaction can generate only low frequency contributions to the u and v components. On the other hand, the nonlinear interaction of u, v on w can generate both low and high frequency contribution to w. That is, even if w has no oscillatory component initially, the dynamical interaction of u, v and w can generate a high frequency contribution to w at later time. This is not the case for the u

and v components. Due to this resonant interaction of u, v and w, the weak limit of $u_\varepsilon v_\varepsilon w_\varepsilon$ is not equal to the product of the weak limits of u_ε, v_ε, w_ε. This explains why the Compensated Compactness result does not apply to this 3×3 system [59].

Although it is difficult to characterize the weak limit of the triple product, $u_\varepsilon v_\varepsilon w_\varepsilon$ for arbitrary oscillatory initial data, it is possible to say something about the weak limit of the triple product for oscillatory initial data that have periodic structure, such as the ones studies here. Depending on α being rational or irrational, the limiting behavior is very different. In fact, one can show that $\overline{u_\varepsilon v_\varepsilon w_\varepsilon} = \overline{u_\varepsilon}\,\overline{v_\varepsilon}\,\overline{w_\varepsilon}$ when α is equal to an irrational number. This is not true in general when α is a rational number.

2.3   Convection of Microstructure

It is most interesting to see if one can apply homogenization technique to obtain an averaged equation for the large scale quantity for incompressible Euler or Navier-Stokes equations. In 1985, McLaughlin, Papanicolaou and Pironneau [51] attempted to obtain a homogenized equation for the 3-D incompressible Euler equations with highly oscillatory velocity field. More specifically, they considered the following initial value problem:

$$u_t + (u \cdot \nabla)u = -\nabla p,$$

with $\nabla \cdot u = 0$ and highly oscillatory initial data

$$u(x,0) = U(x) + W(x, x/\varepsilon).$$

They then constructed multiscale expansions for both the velocity field and the pressure. In doing so, they made an important assumption that the microstructure is convected by the mean flow. Under this assumption, they constructed a multiscale expansion for the velocity field as follows:

$$u^\varepsilon(x,t) = u(x,t) + w\left(\tfrac{\theta(x,t)}{\varepsilon}, \tfrac{t}{\varepsilon}, x, t\right) + \varepsilon u_1\left(\tfrac{\theta(x,t)}{\varepsilon}, \tfrac{t}{\varepsilon}, x, t\right) + O(\varepsilon^2).$$

The pressure field p^ε is expanded similarly. From this ansatz, one can show that θ is convected by the mean velocity:

$$\theta_t + u \cdot \nabla \theta = 0 , \qquad \theta(x,0) = x .$$

It is a very challenging problems to develop a systematic approach to study the large scale solution in three dimensional Euler and Navier-Stokes equations. The work of McLaughlin, Papanicolaou and Pironneau provided some insightful understanding into how small scales interact with large scale and how to deal with the closure problem. However, the problem is still not completely resolved since the cell problem obtained this way does not have a unique solution. Additional constraints need to be enforced in order to derive a large scale averaged equation. With additional assumptions, they managed to derive a variant of the $k - \varepsilon$ model in turbulence modeling.

Remark 2.3. One possible way to improve the work of [51] is take into account the oscillation in the Lagrangian characteristics, θ_ε. The oscillatory part of θ_ε in general could have order one contribution to the mean velocity of the incompressible Euler equation. With Dr. Danping Yang [41], we have studied convection of microstructure of the 2-D and 3-D incompressible Euler equations using a new approach. We do not assume that the oscillation is propagated by the mean flow. In fact, we found that it is crucial to include the effect of oscillations in the characteristics on the mean flow. Using this new approach, we can derive a well-posed cell problem which can be used to obtain an effective large scale average equation.

More can be said for a passive scalar convection equation.

$$v_t + \frac{1}{\varepsilon}\nabla \cdot (u(x/\varepsilon)v) = \alpha\Delta v,$$

with $v(x, 0) = v_0(x)$. Here $u(y)$ is a known incompressible periodic (or stationary random) velocity field with zero mean. Assume that the initial condition is smooth.

Expand the solution v^ε in powers of ε

$$v^\varepsilon = v(t, x) + \varepsilon v_1(t, x, x/\varepsilon) + \varepsilon^2 v_2(t, x, x/\varepsilon) + \cdots.$$

The coefficients of ε^{-1} lead to

$$\alpha\Delta_y v_1 - u \cdot \nabla_y v_1 - u \cdot \nabla_x v = 0.$$

Let e_k, $k = 1, 2, 3$ be the unit vectors in the coordinate directions and let $\chi^k(y)$ satisfy the cell problem:

$$\alpha\Delta_y \chi^k - u \cdot \nabla_y \chi^k - u \cdot e_k = 0.$$

Then we have

$$v_1(t, x, y) = \sum_{k=1}^{3} \chi^k(y)\frac{v(t, x)}{\partial x_k}.$$

The coefficients of ε^0 give

$$\alpha\Delta_y v_2 - u \cdot \nabla_y v_2 = u \cdot \nabla_x v_1 - 2\alpha\nabla_x \cdot \nabla_y v_1 - \alpha\Delta_x v + v_t.$$

The solvability condition for v_2 requires that the right hand side has zero mean with respect to y. This gives rise to the equation for homogenized solution v

$$v_t = \alpha\Delta_x v - \overline{u \cdot \nabla_x v_1}.$$

Using the cell problem, McLaughlin, Papanicolaou, and Pironneau obtained [51]

$$v_t = \sum_{i,j=1}^{3} (\alpha\delta_{ij} + \alpha_{T_{ij}})\frac{\partial^2 v}{\partial x_i \partial x_j},$$

where $\alpha_{T_{ij}} = -\overline{u_i\chi^j}$.

Nonlocal Memory Effect of Homogenization It is interesting to note that for certain degenerate problem, the homogenized equation may have a nonlocal memory effect.

Consider the simple 2-D linear convection equation:

$$\frac{\partial u_\varepsilon(x,y,t)}{\partial t} + a_\varepsilon(y)\frac{\partial u_\varepsilon(x,y,t)}{\partial x} = 0,$$

with initial condition $u_\varepsilon(x,y,0) = u_0(x,y)$.

We assume that a_ε is bounded and u_0 has compact support. While it is easy to write down the solution explicitly,

$$u_\varepsilon(x,y,t) = u_0(x - a_\varepsilon(y)t, y),$$

it is not an easy task to derive the homogenized equation for the weak limit of u_ε.

Using Laplace Transform and measure theory, Luc Tartar [60] showed that the weak limit u of u_ε satisfies

$$\frac{\partial}{\partial t}u(x,y,t) + A_1(y)\frac{\partial}{\partial x}u(x,y,t) = \int_0^t \int \frac{\partial^2}{\partial x^2}u(x - \lambda(t-s),y,s)d\mu_y(\lambda)\,ds,$$

with $u(x,y,0) = u_0(x,y)$, where $A_1(y)$ is the weak limit of $a_\varepsilon(y)$, and μ_y is a probability measure of y and has support in $[\min(a_\varepsilon), \max(a_\varepsilon)]$.

As we can see, the degenerate convection induces a nonlocal history dependent diffusion term in the propagating direction (x). The homogenized equation is *not* amenable to computation since the measure μ_y cannot be expressed explicitly in terms of a_ε.

3   Numerical Homogenization Based on Sampling Techniques

Homogenization theory provides a critical guideline for us to design effective numerical methods to compute multiscale problems. Whenever homogenized equations are applicable they are very useful for computational purposes. There are, however, many situations for which we do not have well-posed effective equations or for which the solution contains different frequencies such that effective equations are not practical. In these cases we would like to approximate the original equations directly. In this part of my lectures, we will investigate the possibility of approximating multiscale problems using particle methods together with sampling technique. The classes of equations we consider here include semilinear hyperbolic systems and the incompressible Euler equation with oscillatory solutions.

When we talk about convergence of an approximation to an oscillatory solution, we need to introduce a new definition. The traditional convergence concept is too weak in practice and does not discriminate between solutions which are highly oscillatory and those which are smooth. We need the error to be small essentially independent of the wavelength in the oscillation when the computational grid size is small. On the other hand we cannot expect the approximation to be well behaved pointwise. It is enough if the continuous solution and its discrete approximation have similar local or moving averages.

Definition 3.1 (Engquist [30]). Let v^n be the numerical approximation to u at time $t_n(t_n = n\Delta t)$, ε represents the wave length of oscillation in the solution. The approximation v^n converges to u as $\Delta t \to 0$, essentially independent of ε, if for any $\delta > 0$ and $T > 0$ there exists a set $s(\varepsilon, \Delta t_0) \in (0, \Delta t_0)$ with measure $(s(\varepsilon, \Delta t_0)) \geq (1 - \delta)\Delta t_0$ such that

$$\|u(\cdot, t_n) - v^n\| \leq \delta, \qquad 0 \leq t_n \leq T$$

is valid for all $\Delta t \in s(\varepsilon, \Delta t_0)$ and where Δt_0 is independent of ε.

The convergence concept of "essentially independent of ε" is strong enough to mimic the practical case where the high frequency oscillations are not well resolved on the grid. A small set of values of Δt has to be removed in order to avoid resonance between Δt and ε. Compare the almost always convergence for the Monte Carlo methods [55].

It is natural to compare our problem with the numerical approximation of discontinuous solutions of nonlinear conservation laws. Shock capturing methods do not produce the correct shock profiles but the overall solution may still be good. For this the scheme must satisfy certain conditions such as conservation form. We are here interested in analogous conditions on algorithms for oscillatory solutions. These conditions should ideally guarantee that the numerical approximation in some sense is close to the solution of the corresponding effective equation when the wave length of the oscillation tends to zero.

There are three central sources of problems for discrete approximations of highly oscillatory solutions.

(i) The first one is the sampling of the computational mesh points $(x_j = j\Delta x, j = 0, 1, ...)$. There is the risk of resonance between the mesh points and the oscillation. For example, if Δx equals the wave length of the periodic oscillation, the discrete initial data may only get values from the peaks of a curve like the upper envelope of the oscillatory solution. We can never expect convergence in that case. Thus Δx cannot be completely independent of the wave length.

(ii) Another problem comes from the approximation of advection. The group velocity for the differential equation and the corresponding discretization are often very different [33]. This means that an oscillatory pulse which is not well resolved is not transported correctly even in average by the approximation. Furthermore, dissipative schemes do not advect oscillations correctly. The oscillations are damped out very fast in time.

(iii) Finally, the nonlinear interaction of different high frequency components in a solution must be modeled correctly. High frequency interactions may produce lower frequencies that influence the averaged solution. We can show that this nonlinear interaction is well approximated by certain particle methods applied to a class of semilinear differential equations. The problem is open for the approximation of more general nonlinear equations.

In [31,32], we studied a particle method approximation to the nonlinear discrete Boltzmann equations in kinetic theory of discrete velocity with multiscale initial data. In such equations, high frequency components can be transformed into lower frequencies through nonlinear interactions, thus affecting the average of solutions. We assume that the initial data are of the form $a(x, x/\varepsilon)$ with $a(x, y)$ 1-periodic in each component of y. As we see from the homogenization theory in the previous section, the behavior of oscillatory solutions for the generalized Broadwell model is very sensitive to the velocity coefficients. It depends on whether a certain ratio among the velocity components is a rational number or an irrational number.

It is interesting to note that the structure of oscillatory solutions for the generalized Broadwell model is quite stable when we perturb the velocity coefficient α around irrational numbers. In this case, the resonance effect of u and v on w vanishes in the limit of $\varepsilon \to 0$. However, the behavior of oscillatory solutions for the generalized Broadwell model becomes singular when perturbing around integer velocity coefficients. There is a strong interaction between the high frequency components of u and v, and the interaction in the uv term would create an oscillation of order $O(1)$ on the w component. In [59], Tartar showed that for the Carleman model the weak limit of all powers of the initial data will uniquely determine the weak limit of the oscillatory solutions at later times, using the Compensated Compactness Theorem. We found that this is no longer true for the generalized Broadwell model with integer-values velocity coefficients [37].

In [31,32], we showed that this subtle behavior for the generalized Broadwell model with oscillatory initial data can be captured correctly by a particle method even on a coarse grid. The particle method converges to the effective solution essentially independent of ε. For the Broadwell model, the hyperbolic part is solved exactly by the particle method. No averaging is therefore needed in the convergence result. We also analyze a numerical approximation of the Carleman equations with variable coefficients. The scheme is designed such that particle interaction can be accounted for without introducing interpolation. There are errors in the particle method approximation of the linear part of the system. As a result, the convergence can only be proved for moving averages. The convergence proofs for the Carleman and the Broadwell equations have one feature in common. The local truncation errors in both cases are of order $O(\Delta t)$. In order to show convergence, we need to take into account cancellation of the local errors at different time levels. This is very different from the conventional convergence analysis for finite difference methods. This is also the place where numerical sampling becomes crucial in order to obtain error cancellation at different time levels.

In the next two subsections, we present a careful study of the Broadwell model with highly oscillatory initial data in order to demonstrate the basic idea of the numerical homogenization based on sampling techniques.

3.1 Convergence of the Particle Method

Now we consider how to capture this oscillatory solution on a coarse grid using a particle method. Since the discrete velocity coefficients are integers for the Broadwell model, we can express a particle method in the form of a special finite difference method by choosing $\Delta x = \Delta t$. Denote by u_i^n, v_i^n, w_i^n the approximations of $u(x_i, t^n), v(x_i, t^n)$ and $w(x_i, t^n)$ respectively with $x_i = i\Delta x$ and $t^n = n\Delta t$. Our particle scheme is given by

$$u_i^n = u_{i-1}^{n-1} + \Delta t(w^2 - uv)_{i-1}^{n-1}, \tag{3.1}$$

$$v_i^n = v_{i+1}^{n-1} + \Delta t(w^2 - uv)_{i+1}^{n-1}, \tag{3.2}$$

$$w_i^n = w_i^{n-1} - \Delta t(w^2 - uv)_i^{n-1}, \tag{3.3}$$

with the initial conditions given by

$$u_i^0 = u(x_i, 0), \quad v_i^0 = v(x_i, 0), \quad w_i^0 = w(x_i, 0). \tag{3.4}$$

To study the convergence of the particle scheme (3.1)-(3.4) we need the following lemma, which is a discrete analogue of Lemma 2.1.

Lemma 3.1. *Let $g(x, y) \in C^3([0, T] \times [0, 1])$ be 1-periodic in y and satisfy the relation $\int_0^1 g(x, y)\, dy = 0$. Let $x_k = kh$ and $r = h/\varepsilon$. If $h \in S(\varepsilon, h_0)$ where*

$$S(\varepsilon, h_0) = \{0 < h \leq h_0 : \frac{kh}{\varepsilon} \notin \left(i - \frac{\tau}{|k|^{3/2}}, i + \frac{\tau}{|k|^{3/2}}\right),$$

$$for\ i = 1, 2, \cdots, \left[\frac{kh_0}{\varepsilon}\right] + 1, 0 \neq k \in Z, 0 < \varepsilon \leq 1\},$$

then we have

$$\left|\sum_{k=0}^{n-1} g(x_k, \frac{x_k}{\varepsilon})h\right| \leq \frac{C_0(1 + T)L(g)h}{\tau}, \quad \forall\, n = 1, 2, \cdots, \left[\frac{T}{h}\right],$$

where C_0 is a constant independent of h, ε, T, τ and g, and

$$L(g) = \max_{(x,y)\in[0,T]\times[0,1]} \left(|\partial_y^3 g(x, y)|, |\partial_x \partial_y^3 g(x, y)|\right).$$

Moreover, it is obvious that

$$|S(\varepsilon, h_0)| \geq h_0\left(1 - \tau\sum_{k=1}^{\infty} k^{-3/2}\right) \geq h(1 - 3\tau).$$

Proof. Since g is 1-periodic in y with mean zero, it can be expanded in a Fourier series

$$g(x,y) = \sum_{m \neq 0} a_m(x)e^{2\pi imy}, \text{ where } a_m(x) = \int_0^1 g(x,y)e^{-2\pi imy}\, dy.$$

Simple integration by parts yields that

$$|a_m(x)| \leq \frac{1}{(2\pi|m|)^3}L(g), \quad |a'_m(x)| \leq \frac{1}{(2\pi|m|)^3}L(g).$$

Thus we have

$$\left|\sum_{k=0}^{n-1} g(x_k, \frac{x_k}{\varepsilon})h\right| = \left|\sum_{k=0}^{n-1}\sum_{m \neq 0} a_m(x_k)e^{2\pi imx_k/\varepsilon}h\right|$$

$$= \left|\sum_{m \neq 0}\sum_{k=0}^{n-1} a_m(x_k)e^{2\pi imkh/\varepsilon}h\right|.$$

Summation by parts yields

$$\left|\sum_{k=0}^{n-1} a_m(x_k)e^{2\pi ikh/\varepsilon}\right| \leq \left|a_m(x_{n-1})\sum_{k=0}^{n-1} e^{2\pi ikh/\varepsilon}\right|$$

$$+ \left|\sum_{k=0}^{n-1}\left(\sum_{j=1}^{k} e^{2\pi imjh/\varepsilon}\right)(a_m(x_k) - a_m(x_{k+1}))\right|$$

$$\leq \frac{2(1+T)L(g)}{(2\pi|m|)^3|1 - e^{2\pi imh/\varepsilon}|}.$$

But for $h \in S(\varepsilon, h_0)$ we have

$$|1 - e^{2\pi imh/\varepsilon}| = 2|\sin(\pi mh/\varepsilon)| \geq \frac{2\pi\tau}{|m|^{3/2}}.$$

Hence, for $h \in S(\varepsilon, h_0)$,

$$\sum_{k=0}^{n-1}\left|g(x_k, \frac{x_k}{\varepsilon})h\right| \leq \frac{2(1+T)L(g)h}{(2\pi)^4\tau}\sum_{m \neq 0}\frac{1}{|m|^{3/2}} =: \frac{C_0 h(1+T)L(g)}{\tau}.$$

This completes the proof. □

Now we are ready to study the approximation property of the particle scheme (3.1)-(3.4). First denote by

$$E^n = \max_i \left(|u(x_i, t^n) - u_i^n|, |v(x_i, t^n) - v_i^n|, |w(x_i, t^n) - w_i^n|\right). \quad (3.5)$$

Integrating (2.29) from 0 to t^n along its characteristics, we get

$$u(x_i, t^n) = u(x_i - t^n, 0) + \int_0^{t^n} (w^2 - uv)(x_i - t^n + s, s)\, ds. \tag{3.6}$$

From (3.1) we know that

$$u_i^n = u_{i-n}^0 + \sum_{k=0}^{n-1} (w^2 - uv)_{i-k}^k \Delta t. \tag{3.7}$$

Subtracting (3.7) from (3.6) we obtain that

$$u(x_i, t^n) - u_i^n$$
$$= \int_0^{t^n} (w^2 - uv)(x_i - t^n + s, s)\, ds - \sum_{k=0}^{n-1} (w^2 - uv)(x_i - t^k, t^k)\Delta t$$
$$+ \sum_{k=0}^{n-1} \Delta t \left[(w^2 - uv)(x_i - t^k, t^k) - (w^2 - uv)_{i-k}^k \right]$$
$$:= (\mathrm{II}) + (\mathrm{III}). \tag{3.8}$$

Let $M(T)$ be defined as in (2.41) and $N(T)$ be given by

$$N(T) = \max \left\{ |u_i^k|, |v_i^k|, |w_i^k| : i \in Z, 0 \le k \le [T/\Delta t] \right\}. \tag{3.9}$$

It can be shown that $N(T)$ is bounded for finite time independent of ε, see [32]. Then it is clear that

$$(\mathrm{III}) \le (M(T) + N(T)) \sum_{k=0}^{n-1} \Delta t E^k. \tag{3.10}$$

It remains to estimate (II). For convenience, let $\theta = w^2 - uv$ and

$$\Theta(x, t) = W(x, \tfrac{x}{\varepsilon}, t)^2 - U(x, \tfrac{x-t}{\varepsilon}, t)V(x, \tfrac{x+t}{\varepsilon}, t).$$

Then we have

$$(\mathrm{II}) = \int_0^{t^n} \left[\theta(x_i - t^n + s, s)\, ds - \Theta(x_i - t^n + s, s) \right] ds$$
$$+ \left[\int_0^{t^n} \Theta(x_i - t^n + s, s)\, ds - \sum_{k=0}^{n-1} \Delta t \Theta(x_i - t^k, t^k) \right]$$
$$+ \sum_{k=0}^{n-1} \left[\Theta(x_i - t^k, t^k) - \theta(x_i - t^k, t^k) \right] \Delta t$$
$$:= (\mathrm{II})_1 + \cdots + (\mathrm{II})_3. \tag{3.11}$$

By Theorem 2.1 we get

$$|(II)_1 + (II)_3| \leq 2TM(T)C_1(T)\varepsilon. \tag{3.12}$$

To proceed further, let, for fixed (x_i, t^n),

$$g_i^n(s,y) = W(x_i - t^n + s, \tfrac{x_i - t^n}{\varepsilon} + y, s)^2.$$

It is clear that g_i^n is 1-periodic in y. Now by Lemmas 2.1-2.2 we have

$$
\int_0^{t^n} W(x_i - t^n + s, s)^2 \, ds - \sum_{k=0}^{n-1} W(x_i - t^k, t^k)^2 \Delta t
$$

$$
= \int_0^{t^n} \left[g_i^n(s, \tfrac{s}{\varepsilon}) - \int_0^1 g_i^n(s, y) \, dy \right] ds
$$

$$
+ \int_0^{t^n} \int_0^1 g_i^n(s, y) \, dy \, ds - \sum_{k=0}^{n-1} \Delta t \int_0^1 g_i^n(t^{n-k}, y) \, dy
$$

$$
- \sum_{k=0}^{n-1} \Delta t \left[g(t^{n-k}, \tfrac{t^{n-k}}{\varepsilon}) - \int_0^1 g_i^n(t^{n-k}, y) \, dy \right] \leq C(T)(\varepsilon + \Delta t), \tag{3.13}
$$

where we have used standard methods to estimate the second term, since the derivative of g_i^n with respect to s is independent of ε. Here and in the remainder of this section, we will always denote by $C(T)$ the various constants which are independent of ε and Δt. Now similar to the reasoning leading to (3.13) we can obtain

$$|(II)_3| \leq C(T)(\varepsilon + \Delta t). \tag{3.14}$$

From (3.8)-(3.12) and (3.14) we finally get

$$|u(x_i, t^n) - u_i^n| \leq C(T)(\varepsilon + \Delta t) + (M(T) + N(T)) \sum_{k=0}^{n-1} \Delta t E^k. \tag{3.15}$$

Similarly, we have

$$|v(x_i, t^n) - v_i^n| \leq C(T)(\varepsilon + \Delta t) + (M(T) + N(T)) \sum_{k=0}^{n-1} \Delta t E^k, \tag{3.16}$$

$$|w(x_i, t^n) - w_i^n| \leq C(T)(\varepsilon + \Delta t) + (M(T) + N(T)) \sum_{k=0}^{n-1} \Delta t E^k. \tag{3.17}$$

To summarize, we have the following theorem by summing (3.15)-(3.17) and applying the Gronwall inequality.

Theorem 3.1. *Let (u, v, w) be the solution of (2.32)-(2.35) and (u_i^n, v_i^n, w_i^n) be the solution of the particle scheme (3.1)-(3.4). Assume that $\Delta t \in S(\varepsilon, \Delta t_0)$ where $S(\varepsilon, \Delta t_0)$ is defined in Lemma 3.1. Then the following estimate holds*

$$\max_{1 \leq n \leq [T/\Delta t]} E^n \leq C(T)(\varepsilon + \Delta t),$$

where $C(T)$ is independent of ε and Δt, and E^n is defined as in (3.5).

Remark 3.1. It is important that we perform the error analysis globally in time in order to account for cancellation of local truncation errors at different time steps. As we can see from the analysis, the local truncation error is of order Δt in one time step. If we do not take into account the error cancellation in time, we would obtain an error bound of order $O(1)$ which is an overestimate. The error cancellation is closely related to the sampling we choose. This is the place where we can see the difference between a good sampling and a resonant sampling.

Remark 3.2. As we can see from the error analysis, error cancellation along Lagrangian characteristics is essential in obtaining convergence independent of the oscillation. This idea can be generalized to hyperbolic systems with variable coefficient velocity fields. In the special case of the Carleman model with variable coefficients, we have analyzed the convergence of a particle method in [31]. However, the particle method analyzed in [31] does not generalize to multi-dimensions or 3×3 systems. Together with a Ph.D. student, Razvan Fetecau, we have designed a modified Lagrangian particle method. In this method, each component of the solution is updated along its own characteristic. So there is no fixed grid. When we update one component of the solution, say u, we need values of the other components (say v and w) along the u characteristic. We obtain these values by using some high order interpolation scheme (such as Fourier interpolation or cubic spline). This modified Lagrangian particle method in principle works for any number of families of characteristics and for multi-dimensions. From our preliminary numerical experiments, it produces excellent results for both the Broadwell and Carleman models, even in the oscillatory coefficients case.

Below we describe briefly the results we obtain for the variable coefficient Carleman equations

$$u_t + a(x,t)u_x = v^2 - u^2 , \qquad (3.18)$$

$$v_t - b(x,t)v_x = u^2 - v^2 , \qquad (3.19)$$

with initial data $u(x,0) = u_0(x, x/\varepsilon)$, $v(x,0) = v_0(x, x/\varepsilon)$. In Figure 3.1, we illustrate the particle trajectories for the u and v components.

We choose the oscillatory coefficients as follows:

$$a(x,t) = 1 + 0.5 \sin\left(\tfrac{xt}{\varepsilon}\right) \text{ and } b(x,t) = 1 + 0.2 \cos\left(\tfrac{xt}{\varepsilon}\right).$$

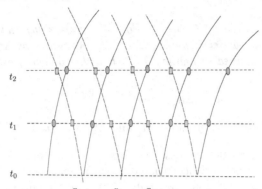

x_{i-1} $\qquad$ x_i $\qquad$ x_{i+1}

Fig. 3.1. Schematic particle trajectories for different components.

The initial conditions for u and v are chosen as

$$u_0(x, x/\varepsilon) = \begin{cases} 0.5\sin^4(\pi(x-3)/2)(1+\sin(2\pi(x-3)/\varepsilon)), & |x-4| < 1 \\ 0, & |x-4| \geq 1 \end{cases}$$

$$(3.20)$$

$$v_0(x, x/\varepsilon) = \begin{cases} 0.5\sin^4(\pi(x-4)/2)(1+\sin(2\pi(x-4)/\varepsilon)), & |x-5| < 1 \\ 0, & |x-5| \geq 1 \end{cases}$$

$$(3.21)$$

In our calculations, we choose $\Delta x = 0.01$, $\Delta t = \frac{\Delta x}{\sqrt{5}}$, and $\varepsilon = \Delta x \sqrt{2} \approx 0.014$. We plot the u-characteristic in Figure 3.2. The coarse grid solution for the u-component is plotted in Figure 3.3a. We can see that it captures very well the high frequency information. In Figure 3.3b, we put the coarse grid solution on top of the corresponding well-resolved solution. The agreement is very good. We also check the accuracy of the moving average [31] of the solution and the average of its second order moments. The results are plotted in Figure 3.4. Again, we observe excellent agreement between the coarse grid calculations and the well-resolved calculations.

We have also performed the same calculations for the 3×3 Broadwell model with rational or irrational coefficient α. The subtle homogenization behavior is captured correctly for both rational α and for irrational α. We do not present the results here.

3.2   Vortex Methods for Incompressible Flows

The generalization of the particle method to the incompressible flows is the vortex method. In [26], we have analyzed the convergence of the vortex method for 2-D incompressible Euler equations with oscillatory vorticity field. Our analysis relies on the observation that there are tremendous cancellations among the local errors at different space locations in the velocity approximation. Thus the local errors do not add up to $O(1)$ as predicted by the classical error estimate in the case where the grid size is large compared to the oscillatory wavelength.

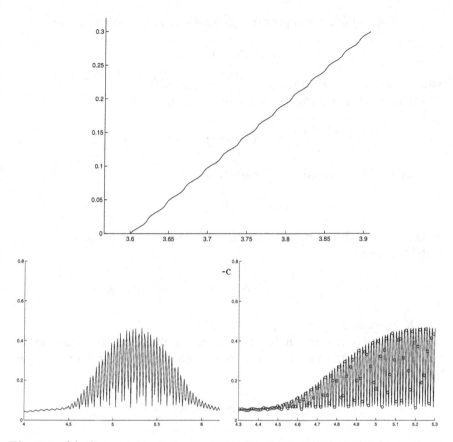

Fig. 3.3. (a): Coarse grid solution u at time $t = 1.28$. (b): Putting the coarse grid solution u on top of a well-resolved computation (solid line).

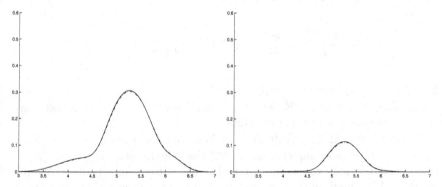

Fig. 3.4. (a): The averaged solution $\overline{u}$ (dashdot line); the solid line represents a very well resolved computation. (b): The averaged second order moment $\overline{u^2}$ (dashdot line); the solid line represents a very well resolved computation.

Consider the 2-D incompressible Euler equation in vorticity form:

$$\omega_t + (u \cdot \nabla)\omega = 0$$

with oscillatory initial vorticity $\omega(x,0) = \omega_0(x, x/\varepsilon)$.

Define the particle trajectory, denoted as $X(t, \alpha)$,

$$\frac{dX(t, \alpha)}{dt} = u(X(t, \alpha), t), \quad X(0, \alpha) = \alpha.$$

Vorticity is conserved along characteristics:

$$\omega(X(t, \alpha), t) = \omega_0(\alpha).$$

On the other hand, velocity can be expressed in terms of vorticity by the Biot-Savart law:

$$u(X(t, \alpha), t) = \int K(X(t, \alpha) - X(t, \alpha'))\omega_0(\alpha')d\alpha'$$

with K given by $K(x) = (-x_2, x_1)/(2\pi|x|^2)$.

The Biot-Savart kernel K has a singularity at the origin. To regularize the kernel, Chorin introduced the vortex blob method (see, e.g. [17], replacing K by $K_\delta = K * \zeta_\delta$,

$$\zeta_\delta = \frac{1}{\delta^2}\zeta\left(\frac{x}{\delta}\right), \quad \delta = h^\sigma, \text{ with } \sigma < 1.$$

ζ is typically chosen as a variant of Gaussian.

The vortex blob method is given by

$$\frac{dX_i^h(t)}{dt} = \sum_j K_\delta(X_i^h(t) - X_j^h(t))\omega_j h^2,$$

where $X_i^h(0) = \alpha_i$, and $w_j = w_0(\alpha_j, \alpha_j/\varepsilon)$.

Together with Weinan E, we have proved that the vortex method converges essentially independent of ε [26].

The case studied in [26] deals with bounded oscillatory vorticity. This assumption leads to strong convergence of the velocity field. It is more physical to consider homogenization for highly oscillatory velocity field. Would the vortex blob method still capture the correct large scale solution with a relatively coarse grid (or small number of particles)? Together with a Ph.D. student, Razvan Fetecau, we have recently derived a modified vortex method for the coarse (or macro) particle system by combining a local subgrid correction with a model reduction technique.

4   Numerical Homogenization Based on Multiscale FEMs

It is natural to consider the possibility of generalizing the sampling technique to second order elliptic equations with highly oscillatory coefficients. In [3], we showed that finite difference approximations converge essentially independent of the small scale ε for one-dimensional elliptic problems. In several space dimensions we found that only in the case of rapidly oscillating periodic coefficients do the above results generalize, in a weaker form. In the case of almost periodic or random coefficients in several space dimensions we showed, both theoretically and with a simple counterexample, that numerical homogenization by sampling does not work efficiently. New ideas seem to be needed.

In order to overcome the difficulty we mentioned above for the sampling technique, we have introduced a multiscale finite element method (MsFEM) for solving partial differential equations with multiscale solutions, see [38,40,39,28,16,61,1]. The central goal of this approach is to obtain the large scale solutions accurately and efficiently without resolving the small scale details. The main idea is to construct finite element base functions which capture the small scale information within each element. The small scale information is then brought to the large scales through the coupling of the global stiffness matrix. Thus, the effect of small scales on the large scales is correctly captured. In our method, the base functions are constructed from the leading order homogeneous elliptic equation in each element. As a consequence, the base functions are adapted to the local microstructure of the differential operator. In the case of two-scale periodic structures, we have proved that the multiscale method indeed converges to the correct solution independent of the small scale in the homogenization limit [40].

In practical computations, a large amount of overhead time comes from constructing the base functions. In general, these multiscale base functions are constructed numerically, except for certain special cases. Since the base functions are independent of each other, they can be constructed independently and can be done perfectly in parallel. This greatly reduces the overhead time in constructing these bases. In many applications, it is important to obtain a scale-up equation from the fine grid equation. Our multiscale finite element method can be used for a similar purpose [61]. The advantage of deriving a scale-up equation is that one can perform many useful tests on the scale-up (coarse grid) model with different boundary conditions or source terms. This would be very expensive if we have to perform all these tests on a fine grid. For time dependent problems, the scaled-up equation also allows for larger time steps. This results in additional computational saving. Another advantage of the method is its ability to scale down the size of a large scale computation. This offers a big saving in computer memory.

It should be mentioned that many numerical methods have been developed with goals similar to ours. These include wavelet based numerical ho-

268   Thomas Y. Hou

mogenization methods [10,21,19,45], methods based on the homogenization theory (cf. [9,25,18,34]), variational multiscale methods [42,12,43], matrix-dependent multigrid based homogenization [45,19], generalized p-FEM in homogenization [47,48], and some upscaling methods based on simple physical and/or mathematical motivations (cf. [23,49]). The methods based on the homogenization theory have been successfully applied to determine the effective conductivity and permeability of certain composite materials and porous media. However, their range of applications is usually limited by restrictive assumptions on the media, such as scale separation and periodicity [7,44]. They are also expensive to use for solving problems with many separate scales since the cost of computation grows exponentially with the number of scales. But for the multiscale method, the number of scales does not increase the overall computational cost exponentially. The upscaling methods are more general and have been applied to problems with random coefficients with partial success (cf. [23,49]). But the design principle is strongly motivated by the homogenization theory for periodic structures. Their application to nonperiodic structures is not always guaranteed to work.

We remark that the idea of using base functions governed by the differential equations has been applied to convection-diffusion equation with boundary layers (see, e.g., [6] and references therein). With a motivation different from ours, Babuska et al. applied a similar idea to 1-D problems [5] and to a special class of 2-D problems with the coefficient varying locally in one direction [4]. However, most of these methods are based on the special property of the harmonic average in one-dimensional elliptic problems. As indicated by our convergence analysis, there is a fundamental difference between one-dimensional problems and genuinely multi-dimensional problems. Special complications such as the resonance between the mesh scale and the physical scale never occur in the corresponding 1-D problems.

4.1 Multiscale Finite Element Methods for Elliptic PDEs

In this section we consider the multiscale finite element method applied to the following problem

$$L_\varepsilon u := -\nabla \cdot (a(\tfrac{x}{\varepsilon})\nabla u) = f \text{ in } \Omega, \quad u = 0 \text{ on } \Gamma = \partial\Omega, \qquad (4.1)$$

where Ω is a convex polygon in $\mathbb{R}^2$. ε is assumed to be a small parameter, and $a(x) = (a_{ij}(x/\varepsilon))$ is symmetric and satisfies $\alpha|\xi|^2 \leq a_{ij}\xi_i\xi_j \leq \beta|\xi|^2$, for all $\xi \in \mathbb{R}^2$ and with $0 < \alpha < \beta$. Furthermore, $a_{ij}(y)$ are smooth periodic function in y in a unit cube Y. We will always assume that $f \in L^2(\Omega)$. In fact, the smoothness assumption on a_{ij} can be relaxed. In [27], Efendiev has proved convergence of the multiscale finite element method in the case where a_{ij} is only piecewise continuous. Efendiev has also obtained convergence of MsFEM in the case where a_{ij} is random [27].

Let u_0 be the solution of the homogenized equation

$$L_0 u_0 := -\nabla \cdot (a^*\nabla u_0) = f \text{ in } \Omega, \quad u_0 = 0 \text{ on } \Gamma, \qquad (4.2)$$

where $\Gamma = \partial\Omega$ and

$$a_{ij}^* = \frac{1}{|Y|} \int_Y a_{ik}(y)(\delta_{kj} - \frac{\partial\chi^j}{\partial y_k})\,dy,$$

and $\chi^j(y)$ is the periodic solution of the cell problem

$$\nabla_y \cdot (a(y)\nabla_y\chi^j) = \frac{\partial}{\partial y_i}a_{ij}(y) \ \text{ in } Y, \quad \int_Y \chi^j(y)\,dy = 0.$$

It is clear that $u_0 \in H^2(\Omega)$ since Ω is a convex polygon. Denote by $u_1(x,y) = -\chi^j(y)\frac{\partial u_0(x)}{\partial x_j}$ and let θ_ε be the solution of the problem

$$L_\varepsilon\theta_\varepsilon = 0 \ \text{ in } \Omega, \quad \theta_\varepsilon(x) = u_1(x, \tfrac{x}{\varepsilon}) \ \text{ on } \Gamma. \tag{4.3}$$

Our analysis of the multiscale finite element method relies on the following homogenization result obtained by Moskow and Vogelius [52].

Lemma 4.1. *Let $u_0 \in H^2(\Omega)$ be the solution of (4.2), $\theta_\varepsilon \in H^1(\Omega)$ be the solution to (4.3) and $u_1(x) = -\chi^j(x/\varepsilon)\partial u_0(x)/\partial x_j$. Then there exists a constant C independent of u_0, ε and Ω such that*

$$\| u - u_0 - \varepsilon(u_1 - \theta_\varepsilon) \|_{1,\Omega} \leq C\varepsilon(|u_0|_{2,\Omega} + \| f \|_{0,\Omega}).$$

Now we are going to introduce the multiscale finite element methods. Let $\mathcal{T}_h$ be a regular partition of Ω into triangles. Let $\{x_j\}_{j=1}^J$ be the interior nodes of the mesh $\mathcal{T}_h$ and $\{\psi_j\}_{j=1}^J$ be the nodal basis of the standard linear finite element space $W_h \subset H_0^1(\Omega)$. Denote by $S_i = \text{supp}(\psi_i)$ and define ϕ^i with support in S_i as follows:

$$L_\varepsilon\phi^i = 0 \ \text{ in } K, \quad \phi^i = \psi_i \ \text{ on } \partial K \ \forall \, K \in \mathcal{T}_h, K \subset S_i. \tag{4.4}$$

It is obvious that $\phi^i \in H_0^1(S_i) \subset H_0^1(\Omega)$. Finally, let $V_h \subset H_0^1(\Omega)$ be the finite element space spanned by $\{\phi^i\}_{i=1}^J$.

With above notation we can introduce the following discrete problem: find $u_h \in V_h$ such that

$$(a(\tfrac{x}{\varepsilon})\nabla u_h, \nabla v_h) = (f, v_h) \ \forall \, v_h \in V_h, \tag{4.5}$$

where and hereafter we denote by $(\cdot, \cdot)$ the L^2 inner product in $L^2(\Omega)$.

As we will see later, the choice of boundary conditions in defining the multiscale bases will play a crucial role in approximating the multiscale solution. Intuitively, the boundary condition for the multiscale base function should reflect the multiscale oscillation of the solution u across the boundary of the coarse grid element. By choosing a linear boundary condition for the base function, we will create a mismatch between the exact solution u and the finite element approximation across the element boundary. In the next section, we will discuss this issue further and introduce an over-sampling technique to alleviate this difficulty. The over-sampling technique plays an important role when we need to reconstruct the local fine grid velocity field from a coarse grid pressure computation for two-phase flows. This technique enables us to remove the artificial numerical boundary layer across the coarse grid boundary element.

We remark that the multiscale finite element method with linear boundary conditions for the multiscale base functions is similar in spirit to the residual-free bubbles finite element method [11] and the variational multiscale method [42,12]. In a recent paper [57], Dr. G. Sangalli derives a multiscale method based on the residual-free bubbles formulation in [11] and compares it with the multiscale finite element method described here. There are many striking similarities between the two approaches. In Section 6, we will discuss the variational multiscale method in some more detail and compare it with the multiscale finite element method.

To gain some insight into the multiscale finite element method, we next perform an error analysis for the multiscale finite element method in the simplest case, i.e. we use linear boundary conditions for the multiscale base functions.

4.2   Error Estimates ($h < \varepsilon$)

The starting point is the well-known Cea's lemma.

Lemma 4.2. *Let u be the solution of (4.1) and u_h be the solution of (4.5). Then we have*
$$\| u - u_h \|_{1,\Omega} \leq C \inf_{v_h \in V_h} \| u - v_h \|_{1,\Omega}.$$

Let $\Pi_h : C(\bar{\Omega}) \to W_h \subset H_0^1(\Omega)$ be the usual Lagrange interpolation operator:
$$\Pi_h u(x) = \sum_{j=1}^J u(x_j)\psi_j(x) \ \forall u \in C(\bar{\Omega})$$

and $I_h : C(\bar{\Omega}) \to V_h$ be the corresponding interpolation operator defined through the multiscale base function ϕ
$$I_h u(x) = \sum_{j=1}^J u(x_j)\phi^j(x) \ \forall u \in C(\bar{\Omega}).$$

From the definition of the basis function ϕ^i in (4.4) we have
$$L_\varepsilon(I_h u) = 0 \ \text{in} \ K, \quad I_h u = \Pi_h u \ \text{on} \ \partial K, \tag{4.6}$$

for any $K \in \mathcal{T}_h$.

Lemma 4.3. *Let $u \in H^2(\Omega)$ be the solution of (4.1). Then there exists a constant C independent of h, ε such that*

$$\| u - I_h u \|_{0,\Omega} + h \| u - I_h u \|_{1,\Omega} \leq C h^2 (| u |_{2,\Omega} + \| f \|_{0,\Omega}). \tag{4.7}$$

Proof. At first it is known from the standard finite element interpolation theory that

$$\| u - \Pi_h u \|_{0,\Omega} + h \| u - \Pi_h u \|_{1,\Omega} \leq C h^2 (| u |_{2,\Omega} + \| f \|_{0,\Omega}). \tag{4.8}$$

On the other hand, since $\Pi_h u - I_h u = 0$ on ∂K, the standard scaling argument yields

$$\| \Pi_h u - I_h u \|_{0,K} \leq C h | \Pi_h u - I_h u |_{1,K} \quad \forall\, K \in \mathcal{T}_h. \tag{4.9}$$

To estimate $|\Pi_h u - I_h u|_{1,K}$ we multiply the equation in (4.6) by $I_h u - \Pi_h u \in H_0^1(K)$ to get

$$(a(\tfrac{x}{\varepsilon}) \nabla I_h u, \nabla (I_h u - \Pi_h u))_K = 0,$$

where $(\cdot, \cdot)_K$ denotes the L^2 inner product of $L^2(K)$. Thus, upon using the equation in (4.1), we get

$$(a(\tfrac{x}{\varepsilon}) \nabla (I_h u - \Pi_h u), \nabla (I_h u - \Pi_h u))_K$$

$$= (a(\tfrac{x}{\varepsilon}) \nabla (u - \Pi_h u), \nabla (I_h u - \Pi_h u))_K - (a(\tfrac{x}{\varepsilon}) \nabla u, \nabla (I_h u - \Pi_h u))_K$$

$$= (a(\tfrac{x}{\varepsilon}) \nabla (u - \Pi_h u), \nabla (I_h u - \Pi_h u))_K - (f, I_h u - \Pi_h u)_K.$$

This implies that

$$|I_h u - \Pi_h u|_{1,K} \leq C h | u |_{2,K} + \| I_h u - \Pi_h u \|_{0,K} \| f \|_{0,K}.$$

Hence

$$|I_h u - \Pi_h u|_{1,K} \leq C h (| u |_{2,K} + \| f \|_{0,K}), \tag{4.10}$$

where we have used (4.9). Now the lemma follows from (4.8)-(4.10). □

In conclusion, we have the following estimate by using Lemmas 4.2-4.3.

Theorem 4.1. *Let $u \in H^2(\Omega)$ be the solution of (4.1) and $u_h \in V_h$ be the solution of (4.5). Then we have*

$$\| u - u_h \|_{1,\Omega} \leq C h (| u |_{2,\Omega} + \| f \|_{0,\Omega}). \tag{4.11}$$

Note that the estimate (4.11) blows up like h/ε as $\varepsilon \to 0$ since $| u |_{2,\Omega} = O(1/\varepsilon)$. This is insufficient for practical applications. In next subsection we derive an error estimate which is uniform as $\varepsilon \to 0$.

4.3 Error Estimates ($h > \varepsilon$)

In this section, we will show that the multiscale finite element method gives a convergence result uniform in ε as ε tends to zero. This is the main feature of this multiscale finite element method over the traditional finite element method. The main result in this subsection is the following theorem.

Theorem 4.2. *Let $u \in H^2(\Omega)$ be the solution of (4.1) and $u_h \in V_h$ be the solution of (4.5). Then we have*

$$\| u - u_h \|_{1,\Omega} \le C(h + \varepsilon) \| f \|_{0,\Omega} + C\left(\frac{\varepsilon}{h}\right)^{1/2} \| u_0 \|_{1,\infty,\Omega}, \qquad (4.12)$$

where $u_0 \in H^2(\Omega) \cap W^{1,\infty}(\Omega)$ is the solution of the homogenized equation (4.2).

To prove the theorem, we first denote by

$$u_I(x) = I_h u_0(x) = \sum_{j=1}^{J} u_0(x_j) \phi^j(x) \in V_h.$$

From (4.6) we know that $L_\varepsilon u_I = 0$ in K and $u_I = \Pi_h u_0$ on ∂K for any $K \in \mathcal{T}_h$. The homogenization theory in Lemma 3.1 implies that

$$\| u_I - u_{I0} - \varepsilon(u_{I1} - \theta_{I\varepsilon}) \|_{1,K} \le C\varepsilon(\| f \|_{0,K} + | u_{I0} |_{2,K}), \qquad (4.13)$$

where u_{I0} is the solution of the homogenized equation on K:

$$L_0 u_{I0} = 0 \text{ in } K, \quad u_{I0} = \Pi_h u_0 \text{ on } \partial K, \qquad (4.14)$$

u_{I1} is given by the relation

$$u_{I1}(x, y) = -\chi^j(y) \frac{\partial u_{I0}}{\partial x_j} \text{ in } K, \qquad (4.15)$$

and $\theta_{I\varepsilon} \in H^1(K)$ is the solution of the problem:

$$L_\varepsilon \theta_{I\varepsilon} = 0 \text{ in } K, \quad \theta_{I\varepsilon}(x) = u_{I1}(x, \tfrac{x}{\varepsilon}) \text{ on } \partial K. \qquad (4.16)$$

It is obvious from (4.14) that

$$u_{I0} = \Pi_h u_0 \text{ in } K, \qquad (4.17)$$

since $\Pi_h u_0$ is linear on K. From Lemma 3.1 and (4.13) we obtain that

$$\| u - u_I \|_{1,\Omega} \le \| u_0 - u_{I0} \|_{1,\Omega} + \| \varepsilon(u_1 - u_{I1}) \|_{1,\Omega}$$
$$+ \| \varepsilon(\theta_\varepsilon - \theta_{I\varepsilon}) \|_{1,\Omega} + C\varepsilon \| f \|_{0,\Omega}, \qquad (4.18)$$

where we have used the regularity estimate $\| u_0 \|_{2,\Omega} \le C \| f \|_{0,\Omega}$. Now it remains to estimate the terms at the right-hand side of (4.18).

Lemma 4.4. *We have*

$$\| u_0 - u_{I0} \|_{1,\Omega} \le Ch\| f \|_{0,\Omega}, \tag{4.19}$$

$$\| \varepsilon(u_1 - u_{I1}) \|_{1,\Omega} \le C(h+\varepsilon)\| f \|_{0,\Omega}. \tag{4.20}$$

Proof. The estimate (4.19) is a direct consequence of the standard finite element interpolation theory since $u_{I0} = \Pi_h u_0$ by (4.17). Next we note that $\chi^j(x/\varepsilon)$ satisfies

$$\| \chi^j \|_{0,\infty,\Omega} + \varepsilon\| \nabla\chi^j \|_{0,\infty,\Omega} \le C \tag{4.21}$$

for some constant C independent of h and ε. Thus we have, for any $K \in \mathcal{T}_h$,

$$\| \varepsilon(u_1 - u_{I1}) \|_{0,K} \le C\varepsilon\| \chi^j \frac{\partial}{\partial x_j}(u_0 - \Pi_h u_0) \|_{0,K} \le Ch\varepsilon| u_0 |_{2,K},$$

$$\| \varepsilon\nabla(u_1 - u_{I1}) \|_{0,K} = \varepsilon\| \nabla(\chi^j \frac{\partial(u_0 - \Pi_h u_0)}{\partial x_j}) \|_{0,K}$$

$$\le C\| \nabla(u_0 - \Pi_h u_0) \|_{0,K} + C\varepsilon| u_0 |_{2,K}$$

$$\le C(h+\varepsilon)| u_0 |_{2,K}.$$

This completes the proof. $\square$

Lemma 4.5. *We have*

$$\| \varepsilon\theta_\varepsilon \|_{1,\Omega} \le C\sqrt{\varepsilon}\| u_0 \|_{1,\infty,\Omega} + C\varepsilon| u_0 |_{2,\Omega}. \tag{4.22}$$

Proof. Let $\zeta \in C_0^\infty(\mathbb{R}^2)$ be the cut-off function which satisfies $\zeta \equiv 1$ in $\Omega\backslash\Omega_{\delta/2}$, $\zeta \equiv 0$ in Ω_δ, $0 \le \zeta \le 1$ in $\mathbb{R}^2$, and $|\nabla\zeta| \le C/\delta$ in Ω, where for any $\delta > 0$ sufficiently small, we denote by Ω_δ as

$$\Omega_\delta = \{x \in \Omega : \text{dist}(x, \partial\Omega) \ge \delta\}.$$

With this definition, it is clear that $\theta_\varepsilon - \zeta u_1 = \theta_\varepsilon + \zeta(\chi^j \partial u_0/\partial x_j) \in H_0^1(\Omega)$. Multiplying the equation in (4.3) by $\theta_\varepsilon - \zeta u_1$, we get

$$(a(\tfrac{x}{\varepsilon})\nabla\theta_\varepsilon, \nabla(\theta_\varepsilon + \zeta\chi^j \frac{\partial u_0}{\partial x_j})) = 0,$$

which yields, by using (4.21),

$$\| \nabla\theta_\varepsilon \|_{0,\Omega} \le C\| \nabla(\zeta\chi^j \partial u_0/\partial x_j) \|_{0,\Omega}$$

$$\le C\| \nabla\zeta \cdot \chi^j \partial u_0/\partial x_j \|_{0,\Omega} + C\| \zeta\nabla\chi^j \partial u_0/\partial x_j \|_{0,\Omega}$$

$$+C\| \zeta\chi^j \partial^2 u_0/\partial^2 x_j \|_{0,\Omega}$$

$$\le C\sqrt{|\partial\Omega| \cdot \delta}\frac{D}{\delta} + C\sqrt{|\partial\Omega| \cdot \delta}\frac{D}{\varepsilon} + C| u_0 |_{2,\Omega}, \tag{4.23}$$

where $D = \| u_0 \|_{1,\infty,\Omega}$ and the constant C is independent of the domain Ω. From (4.23) we have

$$\| \varepsilon\theta_\varepsilon \|_{0,\Omega} \le C(\frac{\varepsilon}{\sqrt{\delta}} + \sqrt{\delta})\| u_0 \|_{1,\infty,\Omega} + C\varepsilon| u_0 |_{2,\Omega}$$

$$\le C\sqrt{\varepsilon}\| u_0 \|_{1,\infty,\Omega} + C\varepsilon| u_0 |_{2,\Omega}. \tag{4.24}$$

Moreover, by applying the maximum principle to (4.3), we get

$$\| \theta_\varepsilon \|_{0,\infty,\Omega} \leq \| \chi^j \partial u_0/\partial x_j \|_{0,\infty,\partial\Omega} \leq C\| u_0 \|_{1,\infty,\Omega}. \qquad (4.25)$$

Combining (4.24) and (4.25) completes the proof. □

Lemma 4.6. *We have*

$$\| \varepsilon\theta_{I\varepsilon} \|_{1,\Omega} \leq C\Big(\frac{\varepsilon}{h}\Big)^{1/2} \| u_0 \|_{1,\infty,\Omega}. \qquad (4.26)$$

Proof. First we remember that for any $K \in \mathcal{T}_h$, $\theta_{I\varepsilon} \in H^1(K)$ satisfies

$$L_\varepsilon\theta_{I\varepsilon} = 0 \ \ \text{in } K, \quad \theta_{I\varepsilon} = -\chi^j\Big(\frac{x}{\varepsilon}\Big)\frac{\partial(\Pi_h u_0)}{\partial x_j} \ \ \text{on } \partial K. \qquad (4.27)$$

By applying maximum principle and (4.21) we get

$$\| \theta_{I\varepsilon} \|_{0,\infty,K} \leq \| \chi^j \partial(\Pi_h u_0)/\partial x_j \|_{0,\infty,\partial K} \leq C\| u_0 \|_{1,\infty,K}.$$

Thus we have

$$\| \varepsilon\theta_{I\varepsilon} \|_{0,\Omega} \leq C\varepsilon\| u_0 \|_{1,\infty,\Omega}. \qquad (4.28)$$

On the other hand, since the constant C in (4.23) is independent of Ω, we can apply the same argument leading to (4.23) to obtain

$$\begin{aligned}
\| \varepsilon\nabla\theta_{I\varepsilon} \|_{0,K} &\leq C\varepsilon\| \Pi_h u_0 \|_{1,\infty,K}(\sqrt{|\partial K|}/\sqrt{\delta} + \sqrt{|\partial K| \cdot \delta}/\varepsilon) + C\varepsilon| \Pi_h u_0 |_{2,K} \\
&\leq C\sqrt{h}\| u_0 \|_{1,\infty,K}\Big(\frac{\varepsilon}{\sqrt{\delta}} + \sqrt{\delta}\Big) \\
&\leq C\sqrt{h\varepsilon}\| u_0 \|_{1,\infty,K},
\end{aligned}$$

which implies that

$$\| \varepsilon\nabla\theta_{I\varepsilon} \|_{0,\Omega} \leq C\Big(\frac{\varepsilon}{h}\Big)^{1/2}\| u_0 \|_{1,\infty,\Omega}.$$

This completes the proof. □

Proof (of Theorem 3.2.). The theorem is now a direct consequence of (4.18) and the Lemmas 4.4-4.6 and the regularity estimate $\| u_0 \|_{2,\Omega} \leq C\| f \|_{0,\Omega}$. □

Remark 4.1. As we pointed out earlier, the multiscale FEM indeed gives correct homogenized result as ε tends to zero. This is in contrast with the traditional FEM which does not give the correct homogenized result as $\varepsilon \to 0$. The error would grow like $O(h^2/\varepsilon^2)$. On the other hand, we also observe that when $h \sim \varepsilon$, the multiscale method attains large error in both H^1 and L^2 norms. This is what we call the *resonance* effect between the grid scale (h) and the small scale (ε) of the problem. This estimate reflects the intrinsic scale interaction between the two scales in the *discrete* problem. Our extensive numerical experiments confirm that this estimate is indeed generic and sharp. From the viewpoint of practical applications, it is important to reduce or completely remove the resonance error for problems with many scales since the chance of hitting a resonance sampling is high. In the next subsection, we propose an over-sampling method to overcome this difficulty.

We illustrate the main idea by considering the following differential equation

$$-\frac{d}{dx}\left(a_\varepsilon(x)\frac{d}{dx}u_\varepsilon(x)\right) = f(x),$$

where $a_\varepsilon(x) = a(\frac{x}{\varepsilon})$. Discretizing the equation on a fine mesh h gives $L_{\varepsilon,h}u_h^\varepsilon = f_h$. We denote by L_{J+1} the discrete operator on the finest level. We have

$$L_{J+1}U = F.$$

If we use a centered difference approximation to the derivative operator, we have

$$L_{J+1} = h^{-2}\Delta_+ \operatorname{diag}(a_\varepsilon)\Delta_-$$

where $\Delta_+ u_k = u_{k+1} - u_k$ is the forward difference operator, and $\Delta_- u_k = u_k - u_{k-1}$ is the backward difference operator, which satisfies $\Delta_- = \Delta_+^T$. Here $\operatorname{diag}(a)$ is the diagonal matrix with diagonal entry given by $a(x_i)$.

We first determine the coefficient in the space of V_{J+1}. A linear operator L_{J+1} acting on the space V_{J+1} can be decomposed into four operators $L_{J+1} = A_J + B_J + C_J + L_J$ acting on the subspace W_J and V_J, where

$$\begin{aligned} A_J &= Q_J L_{J+1} Q_J : & W_J &\to W_J \\ B_J &= Q_J L_{J+1} P_J : & V_J &\to W_J \\ C_J &= P_J L_{J+1} Q_J : & W_J &\to V_J \\ L_J &= P_J L_{J+1} P_J : & V_J &\to V_J. \end{aligned}$$

Applying the transformation $\mathcal{W}_J$ on L_{J+1}, we have for $U \in V_{J+1}$,

$$\mathcal{W}_J L_{J+1} \mathcal{W}_J^T(\mathcal{W}_J U) = \begin{bmatrix} A_J & B_J \\ C_J & L_J \end{bmatrix}\begin{bmatrix} Q_J U \\ P_J U \end{bmatrix}, \tag{5.3}$$

or simply

$$\mathcal{W}_J L_{J+1} \mathcal{W}_J^T = \begin{bmatrix} A_J & B_J \\ C_J & L_J \end{bmatrix}. \tag{5.4}$$

Let us now consider

$$L_{J+1}U = F, \qquad U, F \in V_{J+1}.$$

This equation may originate from a finite difference, finite element or finite volume discretization of a given equation. We identify U as a piecewise constant approximation of $u(x)$, the solution to the continuous problem. After the same wavelet transformation as in (5.3), we have

$$\begin{pmatrix} A_J & B_J \\ C_J & L_J \end{pmatrix}\begin{pmatrix} U_h \\ U_l \end{pmatrix} = \begin{pmatrix} F_h \\ F_l \end{pmatrix}, \qquad U_h, F_h \in W_J, \qquad U_l, F_l \in V_J,$$

where $U_h = Q_J U$ and $U_l = P_J U$ and similarly for F. For the Haar basis this means that U_h is essentially the high frequency part and U_l is the low frequency part of U. The first equation is

$$U_h = A_J^{-1}(F_h - B_J U_l).$$

Eliminating U_h yields the equation for U_l

$$(L_J - C_J A_J^{-1} B_J)U_l = F_l - C_J A_J^{-1} F_h.$$

Our new "coarse grid operator" is the Schur complement

$$\bar{L}_J = L_J - C_J A_J^{-1} B_J, \tag{5.5}$$

which includes subgrid phenomena via $C_J A_J^{-1} B_J$. We also get the homogenized right hand side,

$$\bar{F}_J = F_l - C_J A_J^{-1} F_h.$$

Note that this is in fact a block Gaussian elimination procedure. Further note that the above procedure can be repeated on $\bar{L}_J$ to get $\bar{L}_{J-1}$ and so on. To make this efficient in real applications it is necessary to be able to approximate $\bar{L}_J$ with a sparse matrix. This sparse matrix can be seen as a discretization of a local differential operator.

There is a striking relation between the Schur complement $\bar{L}_j$ in (5.5) and the analytically homogenized operator (2.24)-(2.25) in Section 1, repeated here for convenience,

$$\bar{L}\bar{u} = -\frac{\partial}{\partial x_i}\left(a_{ij}^*\frac{\partial}{\partial x_i \partial x_j}\right)\bar{u} + \left(\frac{1}{|Y|}\int_Y a_0(y)\,dy\right)\bar{u}, \tag{5.6}$$

where

$$a_{ij}^* = \frac{1}{|Y|}\int_Y (a_{ij} - a_{ik}\frac{\partial \chi^j(y)}{\partial y_k})\,dy. \tag{5.7}$$

The first terms in (5.5) and (5.7) both represent averaged operators, L_J in a discrete sense and

$$-\left(\sum_{ij}\frac{1}{|Y|}\int_Y a_{ij}\,dy\right)\frac{\partial^2}{\partial x_i \partial x_j} \tag{5.8}$$

in an integral sense. In both formulations a correction term is subtracted from the average. Furthermore, in the correction term χ is the solution of an elliptic equation and A_j^{-1} is a discrete positive definite operator.

The above discussion on one-dimensional problems can be generalized to two-dimensional problems. In the two-dimensional case, the maps

$$\mathcal{W}_J : \mathbf{V}_{J+1} \to \mathbf{W}_J \oplus \mathbf{V}_J \tag{5.9}$$

can be written as a tensor product of one-dimensional transforms,

$$\mathcal{W}_J^{2d} = \mathcal{W}_J \otimes \mathcal{W}_J.$$

A linear operator L_{J+1} that acts on the space $\mathbf{V}_{J+1}$ can be decomposed in a way similar to the one-dimensional case. To get a convenient matrix representation we use $\widetilde{\mathcal{W}} = P\mathcal{W}$, instead of $\mathcal{W}$. The matrix P is a suitable permutation. The equation

$$L_{J+1}U = F, \qquad U, F \in \mathbf{V}_{J+1},$$

can then be transformed to

$$\begin{pmatrix} A_J & B_J \\ C_J & L_J \end{pmatrix} \begin{pmatrix} U_h \\ U_l \end{pmatrix} = \begin{pmatrix} F_h \\ F_l \end{pmatrix}, \qquad U_h, F_h \in \mathbf{W}_J, \qquad U_l, F_l \in \mathbf{V}_J,$$

and the coarse grid operator is again the Schur complement,

$$\bar{L}_J = L_J - C_J A_J^{-1} B_J.$$

Note that the high frequency part of U can be decomposed as

$$U_h = \begin{bmatrix} U_{hh} \\ U_{lh} \\ U_{hl} \end{bmatrix}, \qquad U_{hh} \in W_J \otimes W_J, \quad U_{lh} \in V_J \otimes W_J, \quad U_{hl} \in W_J \otimes V_J.$$

Similar tensor product extensions can be made also for high dimensions.

Unlike the homogenized operator $\bar{L}$ in the continuous case, the discrete "homogenized" operator, $\bar{L}_J$, is a nonlocal dense operator, since A_J^1 is dense. For elliptic operators, A_J is diagonally dominant. The compression property of wavelets makes it possible to approximate A_J^{-1} by a sparse matrix. This is an essential property that makes this numerical homogenization procedure efficient. This discrete homogenization procedure can be applied recursively to yield a coarse grid operator, $\bar{L}_I$ at a desired coarse level.

It turns out that it is more effective to write $\bar{L}_J$ in a conservative form. In the 1-D case, this amounts to expressing $\bar{L}_J$ as follows:

$$\bar{L}_J = \frac{1}{(2h)^2} \Delta_+ H \Delta_-,$$

where $\Delta_+ u_j = u_{j+1} - u_j$ and $\Delta_- u_j = u_j - u_{j-1}$, and H is a strongly diagonal dominant matrix.

We look at the extreme case when $a(x) = \bar{a} + \tilde{a}(x)$ is the sum of a constant and the highest frequency represented on the grid, i.e., $a(x_m) = \bar{a} + |\tilde{a}|(-1)^m$. We have that $\bar{a}$ and $\tilde{a}$ are represented as constant vectors in the bases of V_J and W_J. The fact that $a(x) > 0$ implies $|\tilde{a}| < |\bar{a}|$.

The following theorem shows that the wavelet homogenized operator $\frac{1}{h^2}\Delta_+ H \Delta_-$ equals the discrete form $\alpha \frac{1}{h^2}\Delta_+ \Delta_-$ of the classically homogenized equation, apart from a second order error term of order h^2.

Theorem 5.1. *[21] Let $a(x) = \bar{a} + \tilde{a} \in V_{J+1}$ be such that $\bar{a} \in V_0$ is a constant and the oscillatory part $\tilde{a} \in W_J$ has constant amplitude and satisfies the condition $|\tilde{a}| < \bar{a}$. Let $L_{J+1} = \frac{1}{(h/2)^2}\Delta_+ a\Delta_-$ and α be the harmonic average*

$$\alpha = \left(\frac{1}{2h} \int_0^{2h} \frac{1}{a(x)}\, dx \right)^{-1}.$$

Then there exists a function $v(x)$ with a continuous and bounded fourth derivative such that

$$\|\overline{L}_J v - \alpha \frac{1}{h^2}\Delta_+\Delta_- v\|_\infty \le Ch^2 \|v^{(4)}\|_{L^\infty}.$$

In practice, we want to approximate the homogenized operator $\overline{L}_J$ by a sparse approximation. Due to the decay in the off-diagonal entries, we can approximate $\overline{L}_J$ by a band-diagonal matrix $\overline{L}_{J,\nu}$ where ν is the band-width. Let us consider the operator band defined by

$$\mathrm{band}(M,\nu)_{i,j} = \begin{cases} M_{i,j}, & \text{if } 2|i-j| \le \nu - 1, \\ 0, & \text{otherwise.} \end{cases}$$

We have in fact two obvious strategies available for producing $\overline{L}_{J,\nu}$: We can set directly $\overline{L}_{J,\nu} = \mathrm{band}(\overline{L}_J, \nu)$ or use the homogenized coefficient form and build $\overline{L}_{J,\nu} = \frac{1}{h^2}\Delta_+ \mathrm{band}(H, \nu - 2)\Delta_-$. Both approaches produce small perturbations of $\overline{L}_J$. However, important properties, such as divergence form, are lost in the first approach and numerical experiments show that ν needs to be rather large to compensate for this. The second approach produces $\overline{L}_{J,\nu}$ in divergence form. Moreover, the approximation error can be estimated, as in the following result:

Theorem 5.2. *[21] If the conditions of Theorem 5.1 are valid, then we have*

$$\|H - \mathrm{band}(H,\nu)\| \le C\rho^\nu, \qquad \rho = \frac{2(\bar{a} + |\tilde{a}|)}{6\bar{a} - 2|\tilde{a}|} < 1.$$

If v is the discretization of a smooth function $v(x)$, then

$$\|(\overline{L}_J - \overline{L}_{J,\nu})v\|_\infty \le C\rho^\nu \|v''\|_{L^\infty}.$$

In the current approach, the numerical homogenization starts from a fine grid, and the operator is global. It would be nice to derive a local procedure to implement this idea. Further, it remains to study the decay rate of H_{ij} away from diagonal for more general a_ε and more general wavelet bases. One also needs to find an efficient way of computing the inverse of $\overline{L}_J$ (incomplete LU decomposition of A). Multigrid method and preconditional conjugate gradient method are good candidates for this purpose.

6   Variational Multiscale Method

In this section, we will briefly review the main idea of the variational multi-scale method introduced by Hughes and Brezzi et al in [42,12,43].

Consider an abstract variational problem: Find $u \in V$ such that

$$a(u,v) = F(v), \quad \text{for all} \quad v \in V, \tag{6.1}$$

where V is a Hilbert space, $a(\cdot, \cdot)$ is a continuous and coercive bilinear form on V, and $F(\cdot)$ is a continuous linear form on V.

A typical example is the elliptic problem (2.1) in Section 1. In this case, we have

$$a(u,v) = \int_\Omega a_\varepsilon(x)\nabla u \cdot \nabla v \, dx + \int_\Omega a_0(x)uv \, dx, \quad F(v) = \int_\Omega f(x)v(x) \, dx.$$

If we choose $V = H_0^1(\Omega)$, then the above variational problem is equivalent to the elliptic equation (2.1) in Section 1.

The classical Galerkin approximation of (6.1) consists of taking a finite dimensional subspace V_h of V and solving (6.1) in V_h, i.e. find $u_h \in V_h$ such that

$$a(u_h, v_h) = F(v_h), \quad \text{for all} \quad v_h \in V_h. \tag{6.2}$$

Let $\mathcal{T}_h = \{K\}$ be a triangulation of Ω, $h_K = \text{diam}\{K\}$, $h = \max_K h_K$. Typically V_h consists of continuous functions which are polynomials of some degrees on a triangular element K.

To be specific, we consider piecewise linear elements. We set

$$V_R^h = \left\{ v_R \in H_0^1(\Omega), \; v_R|_K \text{ is linear in each} \quad K \right\}.$$

The variational problem (6.2) can be written as follows: Find $u_R \in V_R^h$ such that

$$a(u_R, v_R) = F(v_R), \quad \text{for all } v_R \in V_R^h. \tag{6.3}$$

Here, u_R represents the resolvable part of the solution.

Let V_U^b be a closed subspace of $H_0^1(\Omega)$ such that $V_R^h \cap V_U = \{0\}$.

Further, we define

$$V_h = V_R^h \oplus V_U^b. \tag{6.4}$$

We can consider V_h as the augmented space of V_R^h.

Using the decomposition $V_h = V_R^h \oplus V_U^b$, we can express any $v_h \in V_h$ as the sum of a *resolvable part*, $v_R \in V_R^h$, and an *unresolvable part*, $v_U \in V_U^b$ in a unique way:

$$v_h = v_R + v_U \in V_R^h \oplus V_U^b.$$

In turn, the variational problem (6.2) can be expressed as follows: Find $u_h = u_R + u_U \in V_R^h \oplus V_U^b$ such that

$$a(u_R + u_U, v_R) = F(v_R), \quad \text{for all } v_R \in V_R^h \tag{6.5}$$

$$a(u_R + u_U, v_U) = F(v_U), \quad \text{for all } v_U \in V_U^b. \tag{6.6}$$

Using the bilinearity of $a(\cdot, \cdot)$, equation (6.6) can be written as

$$a(u_U, v_U) = -\left(a(u_R, \cdot) - F(\cdot)\right)(v_U), \quad \text{for all } v_U \in V_U^b. \tag{6.7}$$

Problem (6.7) can be "solved" for any $u_R \in V_R^h$, and the solution can be formally written as

$$u_U = M(\mathcal{L}u_R - f), \tag{6.8}$$

where the operator $\mathcal{L}$ is defined as in (2.1), M is a linear solution operator from $H^{-1}(\Omega)$ to $H_0^1(\Omega)$. One can also view M as the fine grid solution operator or the discrete Green function operator acting on the unresolvable scales.

Substituting the unresolvable part of the solution, $u_U = M(\mathcal{L}u_R - f)$ into equation (6.5) for the resolvable part, we get

$$a(u_R, v_R) + \underbrace{a(M(\mathcal{L}u_R - f), v_R)}_{\text{effect of the space } v_U^b} = (f, v_R), \quad \text{for all } v_R \in V_R^h. \tag{6.9}$$

The term $a(M(\mathcal{L}u_R - f), v_R)$ represents the contribution of small scales to large scales, which resembles the so-called "Reynolds stress" term in turbulence modeling.

Solving u_U exactly would be as expensive as solving the fine grid solution globally. In order to *localize* the computation of u_U, the authors in [42,12,43] made the following assumption:

$$V_U = \oplus_K H_0^1(K).$$

In other words, they take into account only those unresolvable scales that *vanish* on the boundaries of the coarse grid elements. In some sense, the multiscale finite element method with linear boundary conditions for the multiscale base functions is very similar to the variational multiscale method described here. As we see from the analysis of the multiscale finite element method in [38,28], by forcing the unresolvable bases to vanish on the boundaries of the coarse grid elements, the resulting multiscale method may introduce $O(1)$ errors when the physical small scale is of the same order as the coarse grid size.

Using the assumption $u_U|_K = 0$, u_U can be uniquely decomposed among each element, K:

$$u_U = \sum_K u_{U,K}, \quad u_{U,K} \in H_0^1(K).$$

The variational problem now becomes: Find $u_h = u_R + u_U = u_R + \sum_K u_{U,K} \in V_R^h \oplus V_U$ such that

$$a(u_R + u_U, v_R) = F(v_R), \quad \text{for all } v_R \in V_R^h \tag{6.10}$$

$$a(u_R + u_{U,K}, v_{U,K})_K = F(v_{U,K})_K, \quad \text{for all } v_{U,K} \in H_0^1(K), \ \forall \ K, \tag{6.11}$$

where $a(u_R + u_{U,K}, v_{U,K})_K$ and $F(v_{U,K})_K$ are the restrictions of $a(u_R + u_{U,K}, v_{U,K})$ and $F(v_{U,K})$ on K respectively. Again, we obtain an equation for u_R as

$$a(u_R, v_R) + \sum_K a(u_{U,K}, v_R)_K = F(v_R), \quad \text{for all } v_R \in V_R^h, \ \forall \ K. \tag{6.12}$$

The "local equation" for $u_{U,R}$ becomes

$$a(u_{U,K}, v_{U,K})_K = -[a(u_R, \cdot) - F(\cdot)]|_K(v_{U,K}), \quad \text{for } v_{U,K} \in H_0^1(K).$$

Equivalently, we have

$$\mathcal{L}u_{U,K} = -\underbrace{(\mathcal{L}u_R - f)}_{\text{residual}}, \quad \text{in } K, \quad \text{with } u_{U,K} = 0 \quad \text{on } \partial K.$$

In some sense, the un-resolvable bases, $\{u_{U,K}\}$, play the same role as the residual-free bubbles in the residual-free bubbles finite element method introduced by Brezzi and Russo in [11]. Let g_y^K be the Green function on K for operator $\mathcal{L}$, i.e.

$$\mathcal{L}g_y^K(x) = \delta_y(x), \quad \text{for } x \in K, \quad g_y^K = 0 \quad \text{on } \partial K.$$

Then we can write $u_{U,K}$ formally as

$$u_{U,K} = -\int_K g_y^K(x)(\mathcal{L}u_R - f)(y) \, dy.$$

This is not very practical since it is expensive to construct the Green function numerically in each element. In [42,12,43], various approximations to the discrete Green function are proposed to study the stabilizing residual-free bubble method. Hughes and his co-workers have also applied this idea to a number of interesting applications [43]. Recently, Todd Arbogast has introduced a subgrid upscaling method for two-phase flow in porous media using a similar approach [2]. From the analytical view point, the variational multi-scale method or the residual-free bubble approach provides a good framework to design multiscale methods in a systematic way.

References

1. J. Aarnes and T. Y. Hou *An Efficient Domain Decomposition Preconditioner for Multiscale Elliptic Problems with High Aspect Ratios*, Acta Mathematicae Applicatae Sinica, **18** (2002), 63-76.

298   Thomas Y. Hou

2. T. Arbogast, *Numerical Subgrid Upscaling of Two-Phase Flow in Porous Media*, in Numerical treatment of multiphase flows in porous media, Z. Chen et al., eds., Lecture Notes in Physics 552, Springer, Berlin, 2000, pp. 35-49.
3. M. Avellaneda, T. Y. Hou and G. Papanicolaou, *Finite Difference Approximations for Partial Differential Equations with Rapidly Oscillating Coefficients*, Mathematical Modelling and Numerical Analysis, **25** (1991), 693-710.
4. I. Babuska, G. Caloz, and E. Osborn, *Special Finite Element Methods for a Class of Second Order Elliptic Problems with Rough Coefficients*, SIAM J. Numer. Anal., **31** (1994), 945-981.
5. I. Babuska and E. Osborn, *Generalized Finite Element Methods: Their Performance and Their Relation to Mixed Methods*, SIAM J. Numer. Anal., **20** (1983), 510-536.
6. I. Babuska and W. G. Szymczak, *An Error Analysis for the Finite Element Method Applied to Convection-Diffusion Problems*, Comput. Methods Appl. Math. Engrg, **31** (1982), 19-42.
7. A. Bensoussan, J. L. Lions, and G. Papanicolaou, *Asymptotic Analysis for Periodic Structures*, Volume 5 of Studies in Mathematics and Its Applications, North-Holland Publ., 1978.
8. G. Beylkin, R. Coifman, and V. Rokhlin *Fast Wavelet Transforms and Numerical Algorithm I* Comm. Pure Appl. Math., **44** (1991), 141-183.
9. A. Bourgeat, *Homogenized Behavior of Two-Phase Flows in Naturally Fractured Reservoirs with Uniform Fractures Distribution*, Comp. Meth. Appl. Mech. Engrg, **47** (1984), 205-216.
10. M. Brewster and G. Beylkin, *A Multiresolution Strategy for Numerical Homogenization*, ACHA, **2**(1995), 327-349.
11. F. Brezzi and A. Russo, *Choosing Bubbles for Advection-Diffusion Problems*, Math. Models Methods Appl. Sci, **4** (1994), 571-587.
12. F. Brezzi, L. P. Franca, T. J. R. Hughes and A. Russo, $b = \int g$, Comput. Methods in Appl. Mech. and Engrg., **145** (1997), 329-339.
13. J. E. Broadwell, *Shock Structure in a Simple Discrete Velocity Gas*, Phys. Fluids, **7** (1964), 1243-1247.
14. T. Carleman, *Problèms Mathématiques dans la Théorie Cinétique de Gaz*, Publ. Sc. Inst. Mittag-Leffler, Uppsala, 1957.
15. H. Ceniceros and T. Y. Hou, *An Efficient Dynamically Adaptive Mesh for Potentially Singular Solutions.* J. Comput. Phys., **172** (2001), 609-639.
16. Z. Chen and T. Y. Hou, *A Mixed Finite Element Method for Elliptic Problems with Rapidly Oscillating Coefficients*, to appear in Math. Comput..
17. A. J. Chorin, *Vortex Models and Boundary Layer Instabilities*, SIAM J. Sci. Statist. Comput., **1** (1980), 1-21.
18. M. E. Cruz and A. Petera, *A Parallel Monte-Carlo Finite Element Procedure for the Analysis of Multicomponent Random Media*, Int. J. Numer. Methods Engrg, **38** (1995), 1087-1121.
19. J. E. Dendy, J. M. Hyman, and J. D. Moulton, *The Black Box Multigrid Numerical Homogenization Algorithm*, J. Comput. Phys., **142** (1998), 80-108.
20. I. Daubechies *Ten Lectures on Wavelets.* SIAM Publications, 1991.
21. M. Dorobantu and B. Engquist, *Wavelet-based Numerical Homogenization*, SIAM J.Numer. Anal., **35** (1998), 540-559.

22. J. Douglas, Jr. and T.F. Russell, *Numerical Methods for Convection-dominated Diffusion Problem Based on Combining the Method of Characteristics with Finite Element or Finite Difference Procedures*, SIAM J. Numer. Anal. **19** (1982), 871–885.

23. L. J. Durlofsky, *Numerical Calculation of Equivalent Grid Block Permeability Tensors for Heterogeneous Porous Media*, Water Resour. Res., **27** (1991), 699–708.

24. L.J. Durlofsky, R.C. Jones, and W.J. Milliken, *A Nonuniform Coarsening Approach for the Scale-up of Displacement Processes in Heterogeneous Porous Media*, Adv. Water Resources, **20** (1997), 335–347.

25. B. B. Dykaar and P. K. Kitanidis, *Determination of the Effective Hydraulic Conductivity for Heterogeneous Porous Media Using a Numerical Spectral Approach: 1. Method*, Water Resour. Res., **28** (1992), 1155-1166.

26. W. E and T. Y. Hou, *Homogenization and Convergence of the Vortex Method for 2-D Euler Equations with Oscillatory Vorticity Fields*, Comm. Pure and Appl. Math., **43** (1990), 821-855.

27. Y. R. Efendiev, *Multiscale Finite Element Method (MsFEM) and its Applications*, Ph. D. Thesis, Applied Mathematics, Caltech, 1999.

28. Y. R. Efendiev, T. Y. Hou, and X. H. Wu, *Convergence of A Nonconforming Multiscale Finite Element Method*, SIAM J. Numer. Anal., **37** (2000), 888-910.

29. Y. R. Efendiev, L. J. Durlofsky, S. H. Lee, *Modeling of Subgrid Effects in Coarse-scale Simulations of Transport in Heterogeneous Porous Media*, WATER RESOUR RES, **36** (2000), 2031-2041.

30. B. Engquist, *Computation of Oscillatory Solutions to Partial Differential Equations*, in Proc. Conference on Hyperbolic Partial Differential Equations, Carasso, Raviart, and Serre, eds, Lecture Notes in Mathematics No. 1270, Springer-Verlag, 10-22, 1987.

31. B. Engquist and T. Y. Hou, *Particle Method Approximation of Oscillatory Solutions to Hyperbolic Differential Equations*, SIAM J. Numer. Anal., **26** (1989), 289-319.

32. B. Engquist and T. Y. Hou, *Computation of Oscillatory Solutions to Hyperbolic Equations Using Particle Methods*, Lecture Notes in Mathematics No. 1360, Anderson and Greengard eds., Springer-Verlag, 68-82, 1988.

33. B. Engquist and H. O. Kreiss, *Difference and Finite Element Methods for Hyperbolic Differential Equations*, Comput. Methods Appl. Mech. Engrg., **17/18** (1979), 581-596.

34. B. Engquist and E.D. Luo, *Convergence of a Multigrid Method for Elliptic Equations with Highly Oscillatory Coefficients*, SIAM J. Numer. Anal., **34** (1997), 2254-2273.

35. D. Gilbarg and N. S. Trudinger, Elliptic Partial Differential Equations of Second Order. Springer, Berlin, New York, 2001.

36. J. Glimm, H. Kim, D. Sharp, and T. Wallstrom *A Stochastic Analysis of the Scale Up Problem for Flow in Porous Media*, Comput. Appl. Math., **17** (1998), 67-79.

37. T. Y. Hou, *Homogenization for Semilinear Hyperbolic Systems with Oscillatory Data*, Comm. Pure and Appl. Math., **41** (1988), 471-495.

38. T. Y. Hou and X. H. Wu, *A Multiscale Finite Element Method for Elliptic Problems in Composite Materials and Porous Media*, J. Comput. Phys., **134** (1997), 169-189.

39. T. Y. Hou and X. H. Wu, *A Multiscale Finite Element Method for PDEs with Oscillatory Coefficients*, Proceedings of 13th GAMM-Seminar Kiel on Numerical Treatment of Multi-Scale Problems, Jan 24-26, 1997, Notes on Numerical Fluid Mechanics, Vol. 70, ed. by W. Hackbusch and G. Wittum, Vieweg-Verlag, 58-69, 1999.

40. T. Y. Hou, X. H. Wu, and Z. Cai, *Convergence of a Multiscale Finite Element Method for Elliptic Problems With Rapidly Oscillating Coefficients*, Math. Comput., **68** (1999), 913-943.

41. T. Y. Hou and D.-P. Yang, *Convection of Microstructure in Two and Three Dimensional Incompressible Euler Equations*, in preparation, 2002.

42. T. J. R. Hughes, *Multiscale Phenomena: Green's Functions, the Dirichlet-to-Neumann Formulation, Subgrid Scale Models, Bubbles and the Origins of Stabilized Methods*, Comput. Methods Appl. Mech Engrg., **127** (1995), 387-401.

43. T. J. R. Hughes, G. R. Feijóo, L. Mazzei, J.-B. Quincy, *The Variational Multiscale Method – A Paradigm for Computational Mechanics*, Comput. Methods Appl. Mech Engrg., **166**(1998), 3-24.

44. V. V. Jikov, S. M. Kozlov, and O. A. Oleinik, *Homogenization of Differential Operators and Integral Functionals*, Springer-Verlag, 1994, Translated from Russian.

45. S. Knapek, *Matrix-Dependent Multigrid-Homogenization for Diffusion Problems*, in the Proceedings of the Copper Mountain Conference on Iterative Methods, edited by T. Manteuffal and S. McCormick, volume I, SIAM Special Interest Group on Linear Algebra, Cray Research , 1996.

46. P. Langlo and M.S. Espedal, *Macrodispersion for Two-phase, Immiscible Flow in Porous Media*, Adv. Water Resources **17** (1994), 297–316.

47. A. M. Matache, I. Babuska, and C. Schwab, *Generalized p-FEM in Homogenization*, Numer. Math. **86**(2000), 319-375.

48. A. M. Matache and C. Schwab, *Homogenization via p-FEM for Problems with Microstructure*, Appl. Numer. Math. **33** (2000), 43-59.

49. J. F. McCarthy, *Comparison of Fast Algorithms for Estimating Large-Scale Permeabilities of Heterogeneous Media*, Transport in Porous Media, **19** (1995), 123-137.

50. D. W. McLaughlin, G. C. Papanicolaou, and L. Tartar, *Weak Limits of Semilinear Hyperbolic Systems with Oscillating Data*, Lecture Notes in Physics **230** (1985), 277-289, Springer-Verlag, Berlin, New York.

51. D. W. McLaughlin, G. C. Papanicolaou, and O. Pironneau, *Convection of Microstructure and Related Problems*, SIAM J. Applied Math, **45** (1985), 780-797.

52. S. Moskow and M. Vogelius, *First Order Corrections to the Homogenized Eigenvalues of a Periodic Composite Medium: A Convergence Proof*, Proc. Roy. Soc. Edinburgh, A, **127** (1997), 1263-1299.

53. F. Murat, *Compacité par Compensation*, Ann. Scuola Norm. Sup. Pisa, **5** (1978), 489-507.

54. F. Murat, *Compacité par compensation II,*, Proceedings of the International Meeting on Recent Methods in Nonlinear Analysis, Rome, May 8-12, 1978, ed. by E. De Giorgi, E. Magenes and U. Mosco, Pitagora Editrice, Bologna, 245-256, 1979.

55. H. Neiderriter, *Quasi-Monte Carlo Methods and Pseudo-Random Numbers*, Bull. Amer. Math. Soc., **84** (1978), 957-1041.

56. O. Pironneau, *On the Transport-diffusion Algorithm and its Application to the Navier-Stokes Equations*, Numer. Math. **38** (1982), 309–332.

57. G. Sangalli, *Capturing Small Scales in Elliptic Problems Using a Residual-Free Bubbles Finite Element Method*, to appear in Multiscale Modeling and Simulation.

58. L. Tartar, *Compensated Compactness and Applications to P.D.E.*, Nonlinear Analysis and Mechanics, Heriot-Watt Symposium, Vol. IV, ed. by R. J. Knops, Research Notes in Mathematics **39**, Pitman, Boston, 136-212, 1979.

59. L. Tartar, *Solutions oscillantes des équations de Carleman*, Seminaire Goulaouic-Meyer-Schwartz (1980-1981), exp. XII. Ecole Polytechnique (Palaiseau), 1981.

60. L. Tartar, *Nonlocal Effects Induced by Homogenization*, in PDE and Calculus of Variations, ed by F. Culumbini, et al, Birkhäuser, Boston, 925-938, 1989.

61. X.H. Wu, Y. Efendiev, and T. Y. Hou, *Analysis of Upscaling Absolute Permeability*, Discrete and Continuous Dynamical Systems, Series B, **2** (2002), 185-204.

62. P. M. De Zeeuw, *Matrix-dependent Prolongation and Restrictions in a Blackbox Multigrid Solver*, J. Comput. Applied Math, **33**(1990), 1-27.

63. S. Verdiere and M.H. Vignal, *Numerical and Theoretical Study of a Dual Mesh Method Using Finite Volume Schemes for Two-phase Flow Problems in Porous Media*, Numer. Math. **80** (1998), 601–639.

64. T. C. Wallstrom, M. A. Christie, L. J. Durlofsky, and D. H. Sharp, *Effective Flux Boundary Conditions for Upscaling Porous Media Equations*, Transport in Porous Media, **46** (2002), 139-153.

65. T. C. Wallstrom, M. A. Christie, L. J. Durlofsky, and D. H. Sharp, *Application of Effective Flux Boundary Conditions to Two-phase Upscaling in Porous Media*, Transport in Porous Media, **46** (2002), 155-178.

66. T. C. Wallstrom, S. L. Hou, M. A. Christie, L. J. Durlofsky, and D. H. Sharp, *Accurate Scale Up of Two Phase Flow Using Renormalization and Nonuniform Coarsening*, Comput. Geosci, **3** (1999), 69-87.

Numerical Methods for Eigenvalue and Control Problems

Volker Mehrmann

Technische Universität Berlin, Institut für Mathematik MA 4-5, Str. des 17. Juni 136, D–10623 Berlin, Germany

Abstract. We briefly survey some of the classical methods for the numerical solution of eigenvalue problems, including methods for large scale problems. We also briefly discuss some of the basics of linear control theory, including stabilization and optimal control and show how they lead to several types of eigenvalue problems. We then discuss how these problems from control theory can be solved via classical and also nonclassical eigenvalue methods. The latter include recently developed structure preserving methods for the solution of Hamiltonian eigenvalue problems. We also demonstrate how structured eigenvalue methods can be developed for large scale and polynomial eigenvalue problems.

1 Introduction

The numerical solution of linear and nonlinear eigenvalue problems is an important task in many applications such as vibration analysis, stability or sensitivity analysis [6,45,49,55], the solution of boundary value problems for linear differential equations as well as many areas of control theory [31,40].

Let us begin with a classical example from mechanics.

Example 1.1. [42,50] Consider the classical model of a robot with electric motors in the joints, given by the system

$$M(q)\ddot{q} + h(q,\dot{q}) + K(q-p) = 0, \qquad \text{(robot model)}$$
$$J\ddot{p} + D\dot{p} - V(q-p) = 0, \qquad \text{(motor mechanics)} \qquad (1.1)$$

with coefficient matrices satisfying $K = K^T, V = V^T$ positive definite, J diagonal and positive definite, D diagonal and positive semidefinite.

Linearization $(h(q,\dot{q}) = G\dot{q} + Cq)$ and simplification $(M(q) \equiv M_4)$ in the robot equations leads to an equation for the robot dynamics of the form

$$M_4\ddot{q} + G\dot{q} + (C+K)q - Kp = 0,$$

with $M_4 = M_4^T$ positive definite, $G = -G^T$ and $C = C^T$. Solving this equation for p and inserting in the second equation of (1.1) leads to the fourth order system of differential equations

$$M_4 q^{(4)} + M_3 q^{(3)} + M_2 q^{(2)} + M_1\dot{q} + M_0 q = 0, \qquad (1.2)$$

where

$$M_3 = G + KJ^{-1}DK^{-1}M_4,$$
$$M_2 = C + K + KJ^{-1}DK^{-1}G + KJ^{-1}VK^{-1}M_4,$$
$$M_1 = KJ^{-1}(DK^{-1}C + D + VK^{-1}G),$$
$$M_0 = KJ^{-1}VK^{-1}C.$$

Substituting $q = e^{\lambda t}v$ then yields a polynomial eigenvalue problem of the form

$$P(\lambda)v = \sum_{i=0}^{4} \lambda^i M_i v = 0.$$

By solving this eigenvalue problem we can represent the solution of (1.2) in terms of the eigenvalues and eigenvectors.

Turning the fourth order system into a first order system by introducing new variables $x_1 = q$, $x_2 = \dot{q}$, ..., $x_4 = q^{(3)}$ leads to linear system of differential-algebraic equations $E\dot{x} = Ax$ with

$$E = \begin{bmatrix} I & 0 & 0 & 0 \\ 0 & I & 0 & 0 \\ 0 & 0 & I & 0 \\ 0 & 0 & 0 & M_4 \end{bmatrix}, \quad A = \begin{bmatrix} 0 & I & 0 & 0 \\ 0 & 0 & I & 0 \\ 0 & 0 & 0 & I \\ -M_0 & -M_1 & -M_2 & -M_3 \end{bmatrix} x, \quad x := \begin{bmatrix} x_1 \\ x_2 \\ x_3 \\ x_4 \end{bmatrix}.$$

We can solve this system via the generalized eigenvalue problem $\lambda Ey = Ay$. Furthermore, if M_4 is invertible, then we may obtain an ordinary differential equation $\dot{x} = E^{-1}Ax$ which can be solved via the standard eigenvalue problem $E^{-1}Ay = \lambda y$ with

$$E^{-1}A = \begin{bmatrix} 0 & I & 0 & 0 \\ 0 & 0 & I & 0 \\ 0 & 0 & 0 & I \\ -M_4^{-1}M_0 & -M_4^{-1}M_1 & -M_4^{-1}M_2 & -M_4^{-1}M_3 \end{bmatrix}.$$

Thus we have several alternatives to relate the solution of the fourth order system of differential equations to eigenvalue problems. We can also use the eigenvalues and eigenvectors to detect stability of the system, carry out a sensitivity analysis or to determine resonance frequencies.

These approaches are tried and trusted classical methods, but there are several subtleties that have to be considered and that are still an important issue in current research on the solution of eigenvalue problems. Let us discuss some of these difficulties ignoring the fact that we have linearized the system, which may be problematic in itself. We have taken the coupled system of second order equations and turned it into a system of fourth order equations with coefficients that have to be computed numerically and thus may be corrupted by round-off or cancellation errors. Then we have turned the system into a system of first order equations by introducing new variables. It is known that this may lead to an increased sensitivity of the solution to small changes in the data, see [54]. Then we have inverted M_4 and formed the matrix $E^{-1}A$ which may lead to large inaccuracies in the matrix $E^{-1}A$ if M_4 is ill-conditioned with respect to inversion. Finally, and this is an important point, originally our coefficient matrices had significant structure, for example symmetry, and this is destroyed when forming $E^{-1}A$.

This simple example illustrates some of the effects that may occur when we turn a problem that is complicated into one that we know how to solve and where a very well established software package like LAPACK or MATLAB [3,39] is available.

The basic philosophy that we would like to promote is that one should *try to solve the numerical problem in terms of original data and make use of the existing structures as much as possible*. Algebraic structures typically represent some physical properties of the system and ignoring the structure may produce physically meaningless results. We shall see examples of this kind in the course of this paper and also discuss some approaches that follow this philosophy. But it should also be noted that many of the important questions in the use of the underlying structures of the model and the design of numerical methods that make maximal use of these structures remain unsolved. See for example the recent survey articles [43,55].

In this paper we will first (in Section 2) give a brief survey of basic eigenvalue methods such as the classical QR- and QZ-algorithm for the computation of the Schur form, generalized Schur-form and the singular value decomposition. These methods are numerically backward stable and particularly appropriate for the solution of small and medium size problems. They work with dense matrices and require typically $\mathcal{O}(n^3)$ *floating point operations (flops)*, where n is the size of the matrix. For large scale problems, where one typically is only interested in part of the spectrum and associated eigenvectors, different methods are used. We will briefly discuss, as an example, the Arnoldi method.

A major source for eigenvalue problems is the area of control theory. In Section 3 we will briefly discuss some basic elements of linear control theory such as controllability and stabilizability and we discuss how these properties can be checked numerically via the solution of eigenvalue problems. We then discuss the linear quadratic optimal control problem that can be used in the stabilization via feedback and is also a basic task in many algorithms for more general optimal control problems. The solution of linear quadratic optimal problems leads to generalized eigenvalue problems that have a special (Hamiltonian) structure that is associated with the variational problem that is solved. We will make a detailed analysis of this structure in Section 4. Classical eigenvalue methods like the QZ-algorithm are not well suited for this problem, since they do not respect the structure and therefore sometimes return inappropriate results. In order to overcome these problems, structure preserving methods have to be designed. We discuss current techniques that exploit the structure in Section 5. Finally, we then demonstrate how these methods can be extended to large scale problems in Section 6.

In this paper we do not give the proofs for all the results that we present. In particular we omit proofs that are easily accessible from standard text books or journals.

We use the following notation. Real and complex $n \times k$ matrices are denoted by $\mathbb{R}^{n,k}$ and $\mathbb{C}^{n,k}$, respectively. The j-th unit vector of the appropriate dimension is denoted e_j. The identity matrix of appropriate dimension is denoted I, the Euclidean norm in $\mathbb{R}^n$ or $\mathbb{C}^n$ is denoted $\| \cdot \|_2$ and the corresponding spectral norm in $\mathbb{R}^{n,n}$ or $\mathbb{C}^{n,n}$, while $\| \cdot \|_F$ denotes the Frobenius norm in $\mathbb{R}^{n,n}$ or $\mathbb{C}^{n,n}$. We denote the transpose and conjugate transpose of a matrix A by A^T and A^H. We denote the groups of real orthogonal and unitary matrices in $\mathbb{R}^{n,n}$ and $\mathbb{C}^{n,n}$ by $\mathcal{U}_n(\mathbb{R})$ and $\mathcal{U}_n(\mathbb{C})$, respectively . The spectrum of the matrix A is denoted $\sigma(A)$ and the sets of eigenvalues of A in the left half plane, right half plane and on the imaginary axis are denoted by $\sigma_-(A)$, $\sigma_+(A)$ and $\sigma_0(A)$, respectively. Analogously the associated invariant subspaces are denoted by $\mathrm{inv}_-(A)$, $\mathrm{inv}_+(A)$ and $\mathrm{inv}_0(A)$, respectively. The machine precision is eps.

2   Classical Techniques for Eigenvalue Problems

In this section we briefly review some of the basic techniques for the numerical solution of eigenvalue problems. These result can be found in many classical books e.g., [28,45,49,59] and for this reason we not discuss any details. The algorithms that we present here are included in standard software packages like LAPACK [3], ARPACK [35] and MATLAB [39]. For a survey on the state of the art in computational eigenvalue methods see [6].

2.1   The Schur Form and the QR-Algorithm

The fundamental result for the development of numerically stable eigenvalue methods is the Schur Theorem.

Theorem 2.1.

(i) If $A \in \mathbb{C}^{n,n}$, then there exists $Q \in \mathcal{U}_n(\mathbb{C})$ such that

$$Q^H A Q = T,$$

  is upper triangular. The matrix Q can be chosen so that the eigenvalues λ_j occur in any arbitrary order on the diagonal of T.

(ii) If $A \in \mathbb{R}^{n,n}$, then there exists $Q \in \mathcal{U}_n(\mathbb{R})$ such that

$$Q^T A Q = T,$$

  is quasi-upper triangular, i.e. $T = D + N$ with N strictly upper triangular and $D = \mathrm{diag}(D_1, \cdots, D_k)$, where D_i is either a 1×1 or a 2×2 matrix with a pair of complex conjugate eigenvalues. This form is called the real Schur form.

The Schur form allows us to read off the eigenvalues from the diagonal or block diagonal. However, it only gives partial information about the eigenvectors or other invariants under similarity transformations, such as the Jordan canonical form. The columns q_i of $Q = [q_1, \cdots, q_n]$ are called *Schur vectors* and for any k the columns of $S_k = [q_1, \cdots, q_k]$ span a k-dimensional invariant subspace of A. Note that only q_1 is an eigenvector.

The best (and most widely used) method for the computation of the Schur–form is the QR–Algorithm of Francis/Kublanovskaya.

This method is based on a sequence of QR-*decompositions*, i.e. factorizations into a product of an orthogonal and an upper triangular matrix. To compute a QR-decomposition of a matrix A, one generates a sequence of *Householder transformations*

$$P_i = I - 2\frac{uu^T}{u^T u}$$

to eliminate the elements below the diagonal of A via transformations from the left, see [28]. Schematically the QR-decomposition looks as follows:

$$A = \begin{bmatrix} x\ x\ x\ x\ x\ x \\ x\ x\ x\ x\ x\ x \\ x\ x\ x\ x\ x\ x \\ x\ x\ x\ x\ x\ x \\ x\ x\ x\ x\ x\ x \\ x\ x\ x\ x\ x\ x \end{bmatrix} \rightarrow P_1^T A = \begin{bmatrix} x\ x\ x\ x\ x\ x \\ 0\ x\ x\ x\ x\ x \\ 0\ x\ x\ x\ x\ x \\ 0\ x\ x\ x\ x\ x \\ 0\ x\ x\ x\ x\ x \\ 0\ x\ x\ x\ x\ x \end{bmatrix}$$

$$\rightarrow P_2^T P_1^T A = \begin{bmatrix} x\ x\ x\ x\ x\ x \\ 0\ x\ x\ x\ x\ x \\ 0\ 0\ x\ x\ x\ x \\ 0\ 0\ x\ x\ x\ x \\ 0\ 0\ x\ x\ x\ x \\ 0\ 0\ x\ x\ x\ x \end{bmatrix}$$

$$\rightarrow P_{n-1}^T \cdots P_2^T P_1^T A = \begin{bmatrix} x\ x\ x\ x\ x\ x \\ 0\ x\ x\ x\ x\ x \\ 0\ 0\ x\ x\ x\ x \\ 0\ 0\ 0\ x\ x\ x \\ 0\ 0\ 0\ 0\ x\ x \\ 0\ 0\ 0\ 0\ 0\ x \end{bmatrix}$$

This decomposition costs $\mathcal{O}(n^3)$ flops and is *backward stable*, i.e. the computed factors satisfy $\tilde{R} = Q^T(A+F)$ with $\|F\|_2 \approx \text{eps}\,\|A\|_2$ and the computed matrix (in finite precision arithmetic) $\tilde{Q}$ is almost orthogonal.

The QR-algorithm for a real matrix A then follows the following basic principle:

Let $H_0 = Q_0^T A Q_0$ for some $Q_0 \in \mathcal{U}_n(\mathbb{R})$.
FOR $k = 1, 2, \cdots$

      Compute a QR–decomposition
      $H_{k-1} = Q_k R_k$ and set
        $H_k = R_k Q_k.$

END

This iteration converges (under some further assumptions, see [28]) to the real Schur form. The cost is $\mathcal{O}(n^3)$ flops per step and the method needs $\mathcal{O}(n)$ iterations. To reduce this complexity to a total of $\mathcal{O}(n^3)$ flops, one uses Q_0 to transform A to upper Hessenberg form. A matrix is called *upper Hessenberg* if all elements below the first subdiagonal vanish.

The basic procedure to reduce a matrix to Hessenberg form is a sequence of orthogonal similarity transformations with Householder matrices to eliminate the elements below the subdiagonal, see [28]. Schematically we have the following procedure:

$$A = \begin{bmatrix} x & x & x & x & x & x \\ x & x & x & x & x & x \\ x & x & x & x & x & x \\ x & x & x & x & x & x \\ x & x & x & x & x & x \\ x & x & x & x & x & x \end{bmatrix} \rightarrow P_1^T A P_1 = \begin{bmatrix} x & x & x & x & x & x \\ x & x & x & x & x & x \\ 0 & x & x & x & x & x \\ 0 & x & x & x & x & x \\ 0 & x & x & x & x & x \\ 0 & x & x & x & x & x \end{bmatrix}$$

$$\rightarrow P_2^T P_1^T A P_1 P_2 = \begin{bmatrix} x & x & x & x & x & x \\ x & x & x & x & x & x \\ 0 & x & x & x & x & x \\ 0 & 0 & x & x & x & x \\ 0 & 0 & x & x & x & x \\ 0 & 0 & x & x & x & x \end{bmatrix}$$

$$\rightarrow P_{n-2}^T \cdots P_2^T P_1^T A P_1 P_2 \cdots P_{n-2} = \begin{bmatrix} x & x & x & x & x & x \\ x & x & x & x & x & x \\ 0 & x & x & x & x & x \\ 0 & 0 & x & x & x & x \\ 0 & 0 & 0 & x & x & x \\ 0 & 0 & 0 & 0 & x & x \end{bmatrix}$$

The costs for this finite procedure are again $\mathcal{O}(n^3)$ flops and it is backward stable.

The Hessenberg reduction is in general not unique. But if it is *unreduced*, i.e. if all elements in the lower subdiagonal are nonzero, then the reduction is uniquely defined by fixing the first column of Q.

Theorem 2.2 (Implicit Q–Theorem). *For $A \in \mathbb{R}^{n,n}$, let $Q = [q_1, \ldots, q_n]$, $V = [v_1, \cdots, v_n] \in \mathcal{U}_n(\mathbb{R})$ be such that both $Q^T A Q = H, V^T A V = G$ are upper Hessenberg matrices. Let k be the smallest index, so that $h_{k+1,k} = 0$ ($k = n$, if H is unreduced). If $v_1 = q_1$, then $v_i = \pm q_i$ and $|h_{i,i-1}| = |q_{i,i-1}|$ for $i = 2, \cdots, k$. If $k < n$, then $g_{k+1,k} = 0$.*

This theorem shows that if we start the basic QR-iteration with a Hessenberg matrix $H_0 = Q_0^T A Q_0$ then all iterates remain Hessenberg. This theorem also forms the basis for the Arnoldi method that we discuss later.

In the following we may assume that $h_{i+1,i} \neq 0$, otherwise the problem may be decomposed into subproblems by setting subdiagonal elements that satisfy

$$|h_{p+1,p}| \leq c \text{ eps}(|h_{p,p}| + |h_{p+1,p+1}|),$$

to 0. This yields the block upper-triangular matrix

$$H = \left[\begin{array}{c|c} H_{1,1} & H_{1,2} \\ \hline 0 & H_{2,2} \end{array}\right],$$

and the solution of the eigenvalue problem is then reduced to eigenvalue problems for $H_{1,1}, H_{2,2}$.

To accelerate the convergence one uses shifts. The basis for this is the following observation.

Lemma 2.1. *If μ is an eigenvalue of an unreduced upper Hessenberg-Matrix H and $H - \mu I = QR$ is a QR-decomposition, then for $\tilde{H} = RQ + \mu I$ it follows that $\tilde{h}_{n,n-1} = 0$ and $\tilde{h}_{n,n} = \mu$.*

Proof. Since H is unreduced, so is $H - \mu I$. Since $U^T(H - \mu I) = R$ is singular and since it follows from the QR-decomposition that $|r_{i,i}| \geq |h_{i+1,i}|$ for $i = 1, \ldots, n-1$, one has $r_{m,m} = 0$ and hence the last column of $\tilde{H}$ vanishes. $\square$

In general one does not know the exact eigenvalues and therefore one uses as shifts the eigenvalues of the trailing 2×2 block (Wilkinson shift)

$$\left[\begin{array}{cc} h_{n-1,n-1} & h_{n-1,n} \\ h_{n,n-1} & h_{n,n} \end{array}\right].$$

To avoid complex arithmetic in the real case one may combine two steps with complex conjugate shifts, since from

$$A_i - \lambda I = Q_1 R_1,$$
$$A_{i+1} = R_1 Q_1 + \lambda I,$$
$$A_{i+1} - \bar{\lambda} I = Q_2 R_2,$$
$$A_{i+2} = R_2 Q_2 + \bar{\lambda} I$$

it follows that $(A_i - \lambda I)(A_i - \bar{\lambda} I) = (Q_1 Q_2)(R_2 R_1)$. This is a QR-decomposition of the real matrix $(A_i - \lambda I)(A_i - \bar{\lambda} I) = A_i^2 - 2\Re(\lambda)A_i + |\lambda|^2$.

To improve the implementation one works out the iteration in an implicit way. Instead of forming $A_i^2 - 2\Re(\lambda)A_i + |\lambda_1|^2$ one determines $z = (A_i - \lambda_i I)(A_i - \bar{\lambda}_i I)e_1$ and uses first the implicit Q–theorem to determine an orthogonal matrix P_0 so that $P_0 z = \alpha e_1$. After this the first column of P_0 is kept fixed and one computes orthogonal matrices $P_1, P_2, \cdots, P_{n-2}$ so that $P_{n-2}^T \cdots P_0^T A_i P_0 \cdots P_{n-2}$ is again an upper Hessenberg matrix. This is done by exploiting as much of the zero-nonzero structure of the matrix as possible. As we have discussed, we need to deflate the problem into smaller subproblems in order to get fast convergence. The costs for the complete QR-algorithm are $\mathcal{O}(n^3)$ flops and the algorithm is backward stable.

If one wants the eigenvalues in a specific order on the diagonal, then the reordering can be performed after the Schur form has been computed by interchanging successively the eigenvalues on the diagonal, see [28].

For symmetric matrices the Hessenberg form is a tridiagonal form, and the algorithm can be implemented to become a lot faster, i.e. only $\mathcal{O}(n)$ per iterative step and the convergence is cubic [45].

2.2 The Generalized Schur Form and the QZ-Algorithm

Linear differential-algebraic systems $E\dot{x} = Ax$ lead to generalized eigenvalue problems. Given matrices $E, A \in \mathbb{R}^{n,n}$ such that the pencil $\lambda E - A$ is *regular*, i.e. $\det(\lambda E - A)$ does not vanish identically for all $\lambda \in \mathbb{C}$, then *the generalized eigenvalue problem* is to find eigenvectors x and eigenvalues λ so that

$$\lambda Ex = Ax.$$

Note that if E is invertible, then this problem is mathematically equivalent of solving the standard eigenvalue problem for $B = E^{-1}A$, but numerically this may be a bad idea, since if E is ill-conditioned with respect to inversion, then forming $E^{-1}A$ may lead to highly inaccurate elements in B. We have the following generalized Schur form.

Theorem 2.3.

(i) *Let $E, A \in \mathbb{C}^{n,n}$ be such that $\lambda E - A$ is a regular. Then there exist $Q, Z \in \mathcal{U}_n(\mathbb{C})$ such that*

$$Q^H(\lambda E - A)Z = \lambda S - T,$$

where S, T are upper triangular. The eigenvalues of $\lambda E - A$ are then $\lambda_j = t_{j,j}/s_{j,j}$ if $s_{j,j} \neq 0$ or ∞ otherwise. Again the ordering of the eigenvalues is arbitrary.

(ii) *If $E, A \in \mathbb{R}^{n,n}$, then there exist $Q, Z \in \mathcal{U}_n(\mathbb{R})$ such that*

$$Q^T(\lambda E - A)Z = \lambda S - T, \qquad (2.1)$$

with S upper triangular and T quasi upper triangular. This form is called the real generalized Schur form.

The algorithm to compute this form is called the QZ-algorithm. It has the following basic steps. First a reduction is carried out that reduces A to upper Hessenberg form and E to triangular form. Note that we use different unitary (or orthogonal) matrices from the left and right to achieve this form.

After this, the method implicitly works with the QR-algorithm for $E^{-1}A$ by carrying out analogous transformations as in the QR-algorithm but using different transformations from the right to keep the Hessenberg-triangular form, see [28].

The costs for the complete QZ-algorithm are $\mathcal{O}(n^3)$ flops and the algorithm is backward stable. It should be noted though that if the pencil is singular or close to singular then the computation of the generalized Schur form requires a more sophisticated procedure, see [23].

2.3   The Singular Value Decomposition (SVD)

Another important basic tool of numerical linear algebra that is used in many applications such as data compression, image compression and control is the singular value decomposition.

Theorem 2.4. *If $A \in \mathbb{R}^{m,n}$, then there exist $U = [u_1, \cdots, u_m] \in \mathcal{U}_m(\mathbb{R})$ and $V = [v_1, \cdots, v_n] \in \mathcal{U}_n(\mathbb{R})$ so that*

$$U^T A V = \operatorname{diag}(\sigma_1, \cdots, \sigma_p) \in \mathbb{R}^{m,n}, \ p = \min(m,n)$$

where $\sigma_1 \geq \sigma_2 \geq \cdots \geq \sigma_p \geq 0$.

The σ_i are called *singular values* and the u_i, v_i *left and right singular vectors*. They satisfy

$$A v_i = \sigma_i u_i, \ A^T u_i = \sigma_i v_i, \ i = 1, \cdots, \min(m,n).$$

If $\sigma_1 \geq \cdots \geq \sigma_r > \sigma_{r+1} = \cdots = \sigma_p = 0$, then

$$\operatorname{rank}(A) = r; \ \operatorname{Ker}(A) = \operatorname{span}\{v_{r+1}, \cdots, v_n\};$$
$$\operatorname{range}(A) = \operatorname{span}\{u_1, \cdots, u_r\};$$
$$\|A\|_F^2 = \sigma_1^2 + \cdots + \sigma_p^2, \ p = \min\{m,n\}; \ \|A\|_2 = \sigma_1.$$

In principle we could compute the singular value decomposition by solving a symmetric eigenvalue problem, since from $A = U \Sigma V^T \in \mathbb{R}^{m,n}$, $m \geq n$, it is clear that

$$AA^T = U \Sigma V^T V \Sigma^T U^T = U \Sigma^2 U^T = U \operatorname{diag}(\sigma_1^2, \cdots, \sigma_n^2) U^T$$

and

$$A^T A = V \Sigma^2 V^T = V \operatorname{diag}(\sigma_1^2, \cdots, \sigma_n^2, 0, \cdots, 0) V^T.$$

But forming these products may lead to a squaring of the condition number, i.e. to an increased sensitivity of the problem.

In order to avoid this, one first transforms the matrix to bidiagonal form using orthogonal transformations U_i from the left and V_i from the right. Schematically this is the following procedure.

$$
\begin{bmatrix} x & x & x & x \\ x & x & x & x \\ x & x & x & x \\ x & x & x & x \\ x & x & x & x \end{bmatrix}
\xrightarrow{U_1}
\begin{bmatrix} x & x & x & x \\ 0 & x & x & x \\ 0 & x & x & x \\ 0 & x & x & x \\ 0 & x & x & x \end{bmatrix}
\xrightarrow{V_1}
\begin{bmatrix} x & x & 0 & 0 \\ 0 & x & x & x \\ 0 & x & x & x \\ 0 & x & x & x \\ 0 & x & x & x \end{bmatrix}
\xrightarrow{U_2}
\begin{bmatrix} x & x & 0 & 0 \\ 0 & x & x & x \\ 0 & 0 & x & x \\ 0 & 0 & x & x \\ 0 & 0 & x & x \end{bmatrix}
\xrightarrow{V_2}
$$

$$
\begin{bmatrix} x & x & 0 & 0 \\ 0 & x & x & 0 \\ 0 & 0 & x & x \\ 0 & 0 & x & x \\ 0 & 0 & x & x \end{bmatrix}
\xrightarrow{U_3}
\begin{bmatrix} x & x & 0 & 0 \\ 0 & x & x & 0 \\ 0 & 0 & x & x \\ 0 & 0 & 0 & x \\ 0 & 0 & 0 & x \end{bmatrix}
\xrightarrow{U_4}
\begin{bmatrix} x & x & 0 & 0 \\ 0 & x & x & 0 \\ 0 & 0 & x & x \\ 0 & 0 & 0 & x \\ 0 & 0 & 0 & 0 \end{bmatrix} .
$$

After that one applies the symmetric version of the QR-algorithm implicitly to the tridiagonal matrix $B^T B$ without explicitly forming the product. The costs for this procedure are $\mathcal{O}(mn^2)$ flops and the method is backward stable.

2.4   The Arnoldi Algorithm

In many applications associated with quantum mechanics, molecular dynamics or structural analysis, see for example [6], the eigenvalue problem is large and sparse. Often a matrix is not explicitly available but only a subroutine, which performs multiplication of the matrix times a vector. Then methods that are based on factorizations are not applicable. In this case the methods of choice are *Krylov subspace methods* and their variants. The most well-known representatives are the *Lanczos method*, *Arnoldi method* and the *Jacobi-Davidson method* or their variations, see [6].

We will discuss only the Arnoldi method, which consists in its basic form in forming a *Krylov basis*

$$[q_1, Aq_1, A^2 q_1, \ldots, A^\ell q_1]$$

followed by an orthogonalization of this basis using a recursive Gram-Schmidt orthogonalization. If this process is carried out until $\ell = n$ then it corresponds, in exact arithmetic, to a Hessenberg reduction. However, one only creates a much smaller basis, $\ell \ll n$, corresponding to a partial Hessenberg reduction. The associated orthogonal basis $Q_\ell = [q_1, q_2, \ldots, q_\ell]$ is determined by the ordinary Arnoldi process that is given by the recursion

$$q_{j+1} = Aq_j - \sum_{i=1}^{j} q_i h_{ij}, \tag{2.2}$$

where $h_{ij} = q_i^T A q_j$. This can be written as

$$AQ_\ell = Q_\ell H_\ell + f_\ell e_\ell^T, \tag{2.3}$$

where f_ℓ would be zero if Q_ℓ represents an invariant subspace. One then uses the eigenvalues θ_j, $j = 1, \ldots, \ell$, called *Ritz values*, of the small Hessenberg matrix H_ℓ as approximations to the eigenvalues of the matrix A.

A detailed analysis [45] of this method shows several limitations of this approach. First of all the basis Q_ℓ is growing with each step, so the storage requirements grow rapidly. This limits the use of this method for very large scale problems. Furthermore, in finite precision arithmetic the recursive orthogonalization process (2.2) deteriorates after some steps, leading to a loss of orthogonality in the columns of Q_ℓ. This has the effect that spurious eigenvalues are computed that need to be removed. To avoid these spurious eigenvalues, the process could be combined with a reorthogonalization of the vectors. But this creates a lot of extra complexity. A very successful way out of both these limitations is the idea of implicit restarts [52]. This implicitly restarted Arnoldi IRA-algorithm has the following basic form.

Step 1. Build a length l Arnoldi process as in (2.3).
Step 2. For $i = 1, 2, \ldots$

(i) Compute the eigenvalues of H_ℓ using the QR-algorithm and split them into a wanted set $\lambda_1, \ldots, \lambda_k$ and an unwanted set $\lambda_{k+1}, \ldots, \lambda_\ell$.
(ii) Perform $p = \ell - k$ steps of the QR-iteration with the unwanted eigenvalues of H_ℓ as exact shifts and obtain $H_\ell V_\ell = V_\ell \tilde{H}_\ell$.
(iii) Restart: Postmultiply by the matrix V_k consisting of the k leading columns of V_ℓ.

$$AQ_\ell V_k = Q_\ell V_k \tilde{H}_k + \tilde{f}_k e_k^T,$$

where $\tilde{H}_k$ is the leading $k \times k$ principal submatrix of $\tilde{H}_\ell$.
(iv) Set $Q_k = Q_\ell V_k$ and extend the Arnoldi factorization again to length ℓ.

Typically one gets fast convergence for eigenvalues in the exterior part of the spectrum. To get interior eigenvalues near a focal point τ, we can use shift-and-invert, i.e. we apply the algorithm to $(A - \tau I)^{-1}$ for some value τ. This requires a solution method for linear systems with $A - \tau I$ to carry out the product $(A - \tau I)^{-1}x$. Another possibility is to use a polynomial in A to suppress part of the spectrum. More details can be found in [6,28,45,49].

Having briefly discussed some of the basic methods for the solution of eigenvalue problems, in the next section we give a brief overview of basic concepts of control theory.

3   Basics of Linear Control Theory

One of the major application areas of eigenvalue methods is control theory, see the classical monographs [5,31,32] for a detailed study of the basic theory and [46,40,51] for recent books on numerical methods in control. Basic software for control theory has been implemented in the SLICOT package [12] and can also be found in several MATLAB toolboxes [26,37,38].

Let us begin our brief tour through linear control theory with a simple example.

Example 3.1. Consider the control of a train modelled as a mass point. Denote the position at time t by $s(t)$ and the velocity by $v(t)$ and introduce the state

$$\begin{bmatrix} s(t) \\ v(t) \end{bmatrix} = x(t).$$

For a given initial time t_0 and initial state

$$\begin{bmatrix} s(t_0) \\ v(t_0) \end{bmatrix} = x(t_0)$$

the control is the accelerating or braking force

$$m\ddot{s}(t) = F(t) = u(t).$$

We can formulate the problem

$$\begin{bmatrix} \dot{s}(t) \\ \dot{v}(t) \end{bmatrix} = \begin{bmatrix} 0 & 1 \\ 0 & 0 \end{bmatrix} \begin{bmatrix} s(t) \\ v(t) \end{bmatrix} + \begin{bmatrix} 0 \\ \frac{1}{m} \end{bmatrix} u(t),$$

to control the movement of the train in a time optimal or energy optimal way or to follow a reference trajectory. Also we may want to consider constraints such as maximal braking force, maximal velocities, time tables, etc.

A general linear control problem has the form

$$\dot{x} = Ax(t) + Bu(t), \quad x(t_0) = x^0. \tag{3.1}$$

Here $x(t)$ is the *state* with $x \in \mathcal{X}$, the *state space*, and $u(t)$ is the input with $u(t) \in \mathcal{U}$, the *input space*. The spaces $\mathcal{X}$ and $\mathcal{U}$ are typically sets of (piecewise continuous) functions defined on $[t_0, \infty)$ with dimensions

$$x : [t_0, \infty) \longrightarrow \mathbb{R}^n,$$
$$u : [t_0, \infty) \longrightarrow \mathbb{R}^m.$$

The classical solution theory for ordinary differential equations [22] allows us to give an explicit representation of the *transfer function*

$$(t, u(t)) \longrightarrow x(t) \in \mathcal{X},$$

where $x(t)$ is the solution of (3.1) and $u \in \mathcal{U}$.

Theorem 3.1. *The transfer function for (3.1) is*

$$x(t) = \Phi(t, t_0)x^0 + \int_{t_0}^{t} \Phi(t, s)B(s)u(s)ds,$$

where $\Phi(t, s) = e^{A(t-s)}$ is the fundamental solution of the homogeneous matrix differential equation

$$\frac{\partial \Phi}{\partial t}(t, s) = A\Phi(t, s), \quad \Phi(s, s) = I.$$

One of the major tasks of control theory is the *stabilization* of a system.

Definition 3.1.   System (3.1) is called *stable*, if any solution of the homogeneous system $\dot{x} = Ax$ is bounded on (t_0, ∞). It is called *asymptotically stable* if $\lim_{t \to \infty} \|x(t)\| = 0$ for some norm.

It is well-known that stability can be characterized via the spectrum $\sigma(A)$.

Theorem 3.2. *[22]*

(i) *System (3.1) is asymptotically stable if and only if all eigenvalues of A have negative real part.*

(ii) *System (3.1) is stable if and only if all eigenvalues of A have nonpositive real part and all eigenvalues with real part zero have equal algebraic and geometric multiplicity.*

This theorem gives us a first application of eigenvalue methods in control theory. To check stability it suffices to compute the eigenvalues of A. A classical task of control theory is to influence an unstable system in such a way that it becomes stable. But this may not always be possible.

3.1   Controllability and Stabilizability

In this subsection we introduce the concepts of *controllability* and *stabilizability*. We will restrict ourself to constant coefficient systems, but the definition and some of the characterizations hold for variable coefficients and also locally for nonlinear systems.

Definition 3.2. Consider system (3.1) and a desired state x^1. A pair (t_0, x^0) is called *controllable to x^1 in time* $t_1 > t_0$, if there exists a control $u \in \mathcal{U}$ so that the solution of (3.1) with this u satisfies $x(t_1) = x^1$.

A pair (t_0, x^0) is called *controllable to x^1*, if it is controllable to x^1 in any time t_1, $t_0 < t_1 < \infty$. If for all (t_0, x_0) and for all x^1, the pair (t_0, x^0) is controllable to x^1, then (3.1) is called *(completely) controllable*.

Very often one chooses $u(t)$ in dependence of the current state by performing a feedback control. In our case the feedback will typically be linear and have the form $u(t) = Fx(t)$, so that the *closed loop system* is

$$\dot{x} = (A + BF)\, x(t).$$

In order to determine whether a system is controllable, one first studies the following question: Given an interval $[t_0, t_1]$, determine all x^0 so that (t_0, x^0) is controllable to $(t_1, 0)$. Denote the set of all these x^0 by $\mathcal{L}(t_0, t_1)$. By Theorem 3.1 the solution at t_1 satisfies

$$0 = x(t_1) = \Phi(t_1, t_0)x^0 + \int_{t_0}^{t_1} \Phi(t_1, s)Bu(s)\, ds$$

or

$$0 = x^0 + \int_{t_0}^{t_1} \Phi(t_0, s)Bu(s)\, ds, \tag{3.2}$$

since we have the *semigroup property* of the solution

$$\Phi(t_1, s) = \Phi(t_1, t_0)\Phi(t_0, s).$$

Hence $x^0 \in \mathcal{L}(t_0, t_1)$ if and only if (3.2) holds. Let $G(t)$ be a matrix valued, piecewise continuous function on $(-\infty, +\infty)$. Then for $t_0 < t_1$,

$$V = \int_{t_0}^{t_1} G(t)G(t)^T \, dt$$

is called the *Gramian* of $G(t)$ and we have the following Lemma.

Lemma 3.1. *We can write $x \in \mathbb{R}^n$ as*

$$x = \int_{t_0}^{t_1} G(t)u(t) \, dt \tag{3.3}$$

with piecewise continuous $u(t) \in \mathbb{R}^m$ if and only if x is in the range of V, i.e. there exists $z \in \mathbb{R}^n$ so that $x = Vz$.

Proof. The matrix V is symmetric, positive semidefinite, since $G(t)G(t)^T$ is positive semidefinite. The integral preserves this property, since for all $x \in \mathbb{R}^n$

$$x^T V x = \int_{t_0}^{t_1} x^T G(t)G(t)^T x \, dt = \int_{t_0}^{t_1} \|G(t)^T x\|_2^2 \, dt \geq 0.$$

Hence, $Vx = 0$ if and only if $G(t)^T x = 0$ for all $t \in [t_0, t_1]$. The set of all x for which there exist a $u(t)$ so that (3.3) holds, forms a linear space L that contains the range of V. This follows by definition with

$$u(t) = G(t)^T z, \quad z = \text{constant}.$$

We have to show that $L = \text{range}(V)$, i.e. $L \cap \text{Ker}(V) = \{0\}$. For $x \in L \cap \text{Ker}(V)$, one has

$$\|x\|_2^2 = x^T x = \int_{t_0}^{t_1} x^T(G(t)u(t)) \, dt = \int_{t_0}^{t_1} (G(t)^T x)^T u(t) \, dt = 0,$$

and hence $x = 0$. $\square$

With the Gramian

$$W(t_0, t_1) := \int_{t_0}^{t_1} \Phi(t_0, t)BB^T \Phi(t_0, t)^T \, dt \tag{3.4}$$

we then obtain the following result.

Theorem 3.3. *For W from (3.4) the following hold:*

(i) $\mathcal{L}(t_0, t_1) = \{x \mid$ there exists z with $x = W(t_0, t_1)z\}$.
(ii) $W(t_0, t_1)x = 0$ if and only if $x^T \Phi(t_0, t)B \equiv 0$ for all $t \in [t_0, t_1]$.

Proof. Use Lemma 3.1 with $G(t) = \Phi(t_0, t)B$. $\square$

Using the *adjoint equation*

$$\dot{z} = -A^T z \tag{3.5}$$

we obtain the following characterization of controllability.

Theorem 3.4. *The following are equivalent:*
1. *System (3.1) is completely controllable.*
2. *For every t_0 there exists t_1 so that the Gramian $W(t_0, t_1)$ from (3.4) is positive definite.*
3. *The controllability matrix*

$$K := [B, AB, A^2 B, \cdots, A^{n-1} B]$$

has full rank.
4. *If $p \neq 0$ is an eigenvector of A^T, then $p^T B \neq 0$.*
5. *We have* rank $([\lambda I - A, B]) = n$ *for all $\lambda \in \mathbb{C}$.*

Proof. For the proof, see for example [31,32]. Note that we have the following explicit expression for a control that moves the system from x^0 to x^1

$$u(t) = B^T \Phi(t_0, t)^T c.$$

Here c is the solution of the equation

$$x^0 = \Phi(t_0, t_1)x^1 + \int_{t_1}^{t_0} \Phi(t_0, t)B(t)B(t)^T \Phi(t_0, t)^T c \, dt$$

$$= \Phi(t_0, t_1)x^1 + W(t_0, t_1)c$$

which is unique, since $W(t_0, t_1)$ is nonsingular. $\square$

In many applications it is not important to control to an arbitrary position but to make the system asymptotically stable or stable.

Definition 3.3. A linear system is called *stabilizable*, if for every (t_0, x^0) there exists a piecewise continuous control $u(t)$, defined for all $t \geq t_0$, so that $\lim_{t \to \infty} x(t) = 0$ for the solution with this $u(t)$.

Theorem 3.5. *The following are equivalent.*

(i) The system $\dot{x} = Ax + Bu$ is stabilizable.
(ii) If λ with $\Re(\lambda) \geq 0$ is an eigenvalue of A with left eigenvector p, then $p^T B \neq 0$.
(iii) We have rank $([\lambda I - A, B]) = n$ *for all λ with $\Re(\lambda) \geq 0$.*

Proof. (i)$\Longrightarrow$(ii) If λ with $\Re(\lambda) \geq 0$ is an eigenvalue of A with left eigenvector p, then $\Re(e^{-\lambda t})p$ and $\Im(e^{-\lambda t})p$ solve (3.5) and are bounded. Theorem 3.4 implies $p^T B \neq 0$.

(ii)$\Longrightarrow$(iii) Suppose there exists $\lambda \in \mathbb{C}$, $\Re(\lambda) \geq 0$, so that rank $([\lambda I - A, B]) < n$. Then there exists $p \neq 0$, so that $p^T(\lambda I - A, B) = 0$. Hence p is a left eigenvector of A and $p^T B = 0$.

(iii)$\Longrightarrow$(i) rank $([\lambda I - A, B]) = n$ for all λ with $\Re(\lambda) \geq 0$. Then there exists $F \in \mathbb{R}^{m,n}$ so that all eigenvalues of $A - BF$ have negative real part. If this were not the case then there would exist λ with $\Re(\lambda) \geq 0$ and $p \neq 0$ so that $p^T(A - BG) = 0$ for all $G \in \mathbb{R}^{m,n}$. This contradicts (iii). $\square$

In this subsection we have given conditions for controllability and stabilizability of a system. In the following subsection we describe how these properties may be checked numerically.

3.2   System Equivalence

In order to explicitly compute controllers that perform certain control tasks and to check system properties it is essential to transform the system to simpler forms. The basic transformations to do this are

$$\begin{aligned} x &\longmapsto Px, & P \in \mathbb{R}^{n,n} & \quad \text{change of basis;} \\ u &\longmapsto Qu, & Q \in \mathbb{R}^{m,m} & \quad \text{change of basis;} \\ u &\longrightarrow -Fx + v, & F \in \mathbb{R}^{m,n} & \quad \text{linear state feedback.} \end{aligned} \quad (3.6)$$

Theorem 3.6. *If the system (3.1) is controllable (stabilizable) then the transformed system $\dot{\tilde{x}} = \tilde{A}\tilde{x} + \tilde{B}\tilde{u}$, as in (3.6) is controllable (stabilizable).*

Proof. This follows immediately, since

$$\begin{aligned} \text{rank}\,([\lambda I - A, B]) &= \text{rank}\left(P^{-1}[\lambda I - A, B]\begin{pmatrix} P & 0 \\ -F & Q \end{pmatrix}\right) \\ &= \text{rank}\left(\left[\lambda I - \tilde{A}, \tilde{B}\right]\right), \end{aligned}$$

for P, Q invertible. $\square$

In order to achieve backward stability of the numerical methods, we perform the changes of basis with orthogonal or unitary transformations. Note that feedback is not an orthogonal transformation.

To check the system properties, i.e. whether the system is controllable or stabilizable, the best method is the staircase algorithm of Van Dooren [56].

Lemma 3.2. *For $A \in \mathbb{R}^{n,n}$, $B \in \mathbb{R}^{n,m}$, there exist $P \in \mathcal{U}_n(\mathbb{R})$, $Q \in \mathcal{U}_m(\mathbb{R})$ so that*

$$PAP^T = \begin{bmatrix} A_{1,1} & \cdots & & A_{1,s-1} & A_{1,s} \\ A_{2,1} & \ddots & & \vdots & \vdots \\ & \ddots & \ddots & \vdots & \vdots \\ & & A_{s-1,s-2} & A_{s-1,s-1} & A_{s-1,s} \\ 0 & \cdots & 0 & 0 & A_{s,s} \end{bmatrix} \begin{matrix} n_1 \\ n_2 \\ \vdots \\ n_{s-1} \\ n_s \end{matrix} \quad , PBQ = \begin{bmatrix} B_1 & 0 \\ 0 & 0 \\ \vdots & \vdots \\ \vdots & \vdots \\ 0 & 0 \end{bmatrix} \begin{matrix} n_1 \\ n_2 \\ \vdots \\ \vdots \\ n_s \end{matrix}$$

$$ n_1 \ \cdots \quad n_{s-2} \quad n_{s-1} \quad n_s \qquad\qquad n_1$$

(3.7)

where $n_1 \geq n_2 \geq \cdots \geq n_{s-1} \geq n_s \geq 0, n_{s-1} > 0$,

$$A_{i,i-1} = \begin{bmatrix} \Sigma_{i,i-1} & 0 \end{bmatrix} \begin{matrix} n_i \\ \end{matrix} \quad i = 1, \cdots, s-1$$
$$\phantom{A_{i,i-1} = \begin{bmatrix} \Sigma \end{bmatrix}} n_i \quad n_{i-1} - n_i$$

$\Sigma_{i,i-1}$ *is a square nonsingular matrix, $\Sigma_{s-1,s-2}$ is a diagonal matrix and B_1 is a square nonsingular matrix.*

The proof is given by the following algorithm, which generates a sequence of singular value decompositions (SVD).

Algorithm 3.1. Staircase Algorithm
Given $A \in \mathbb{R}^{n,n}, B \in \mathbb{R}^{n,m}$ the method computes $P, Q \in \mathcal{U}_n(\mathbb{R})$ such that PAP^T, PBQ are in the form (3.7).
Step 0: Perform an SVD of B.

$$B = U_B \begin{bmatrix} \Sigma_B & 0 \\ 0 & 0 \end{bmatrix} V_B^T$$

with Σ_B an $n_1 \times n_1$ invertible matrix. Set $P := U_B^T$, $Q := V_B$ and

$$A := U_B^T A U_B = \begin{bmatrix} A_{1,1} & A_{1,2} \\ A_{2,1} & A_{2,2} \end{bmatrix}, \ B := U_B^T B V_B = \begin{bmatrix} \Sigma_B & 0 \\ 0 & 0 \end{bmatrix}$$

with $A_{1,1}$ of size $n_1 \times n_1$.
Step 1: Perform an SVD of $A_{2,1}$:

$$A_{2,1} = U_{2,1} \begin{bmatrix} \Sigma_{2,1} & 0 \\ 0 & 0 \end{bmatrix} V_{2,1}^T \text{ with } \Sigma_{2,1} \text{ an } n_2 \times n_2 \text{ nonsingular diagonal matrix.}$$

Set

$$P_2 := \begin{bmatrix} V_{2,1}^T & 0 \\ 0 & U_{2,1}^T \end{bmatrix}, \ P := P_2 P$$

and

$$A := P_2 A P_2^T =: \begin{bmatrix} A_{1,1} & A_{1,2} & A_{1,3} \\ A_{2,1} & A_{2,2} & A_{2,3} \\ 0 & A_{3,2} & A_{3,3} \end{bmatrix}, \ B := P_2 B =: \begin{bmatrix} B_1 & 0 \\ 0 & 0 \\ 0 & 0 \end{bmatrix},$$

where $A_{2,1} = [\Sigma_{2,1} \quad 0]$ with $\Sigma_{2,1}$ a nonsingular $n_2 \times n_2$ diagonal matrix and $B_1 := V_{2,1}^T \Sigma_B$ nonsingular.

Step 2:
$i = 3$
DO WHILE $(n_{i-1} > 0 \quad$ OR $\quad A_{i,i-1} \neq 0)$.
    Perform an SVD of $A_{i,i-1}$:

$$A_{i,i-1} = U_{i,i-1} \begin{bmatrix} \Sigma_{i,i-1} & 0 \\ 0 & 0 \end{bmatrix} V_{i,i-1}^T$$

with $\Sigma_{i,i-1}$ $n_i \times n_i$ a nonsingular diagonal matrix.
Set

$$P_i := \begin{bmatrix} I_{n_1} & & & & \\ & \ddots & & & \\ & & I_{n_{i-2}} & & \\ & & & V_{i,i-1}^T & \\ & & & & U_{i,i-1}^T \end{bmatrix}, \quad P := P_i P,$$

and

$$A := P_i A P_i^T =: \begin{bmatrix} A_{1,1} & \cdots & & & A_{1,i+1} \\ A_{2,1} & \ddots & & & A_{2,i+1} \\ & \ddots & \ddots & & \vdots \\ & & A_{i,i-1} & A_{i,i} & \vdots \\ 0 & & & A_{i+1,i} & A_{i+1,i+1} \end{bmatrix},$$

    where $A_{i,i-1} = [\Sigma_{i,i-1} \quad 0]$.
    $i := i + 1$
END
$s := i$.

It is clear that this algorithm terminates with $n_{i-1} = 0$ or $A_{i,i-1} = 0$ after at most $n - 1$ steps. The major difficulty in the method is to decide whether singular values are 0 or not, i.e. several rank decisions have to be made. This is a difficult decision, since in the presence of roundoff errors even 0 singular values are perturbed to small nonzero values. One of the more reliable ways to decide the rank of a matrix $M \in \mathbb{R}^{m,n}$, $m \geq n$, is as follows. Use a reliable numerical procedure to calculate the singular values $\sigma_1 \geq \sigma_2 \geq \ldots \geq \sigma_n \geq 0$, and consider a singular value σ_j to be zero if $\sigma_j \leq \mu\sigma_1$ where μ bounds the relative error in M. (If errors come only from rounding errors, then μ may be taken to be a modest multiple of the unit round-off.) The number of remaining nonzero singular values is then taken to be the *numerical rank* of the matrix.

As a direct consequence of Lemma 3.2 we have the following theorem.

Theorem 3.7.

(i) *System (3.1) is (completely) controllable if and only if in the staircase form (3.7) of (A, B) we have $n_s = 0$.*

(ii) *System (3.1) is stabilizable if and only if in the staircase form (3.7) of (A, B) all eigenvalues of $A_{s,s}$ have negative real-part.*

Proof.   (i) We may assume w.l.o.g. that (A, B) is in staircase form

$$
\begin{bmatrix}
\lambda I - A_{1,1} & -A_{1,2} & \cdots & & \cdots & -A_{1,s} & B_1 & 0 \\
-A_{2,1} & \lambda I - A_{2,2} & -A_{2,3} & & & \vdots & 0 & 0 \\
& \ddots & \ddots & \ddots & & -A_{s-2,s} & 0 & 0 \\
& & -A_{s-1,s-2} & \lambda I - A_{s-1,s-1} & -A_{s-1,s} & \vdots & \vdots \\
& & & 0 & \lambda I - A_{s,s} & 0 & 0
\end{bmatrix}
\begin{matrix}
n_1 \\ n_2 \\ \vdots \\ \vdots \\ n_s
\end{matrix}
$$

Since all matrices $B_1, A_{2,1}, \cdots, A_{s-1,s-2}$ have full rank, we have that rank $([\lambda I - A, B]) = n$ for all $\lambda \in \mathbb{C}$ if and only if rank $(\lambda I - A_{s,s}) = n_s$ for all $\lambda \in \mathbb{C}$, which is the case if and only if $n_s = 0$.

(ii) We have rank $([\lambda I - A, B]) = n$ for all $\lambda \in \mathbb{C}$, $\Re(\lambda) \geq 0$ if and only if rank $(\lambda I - A_{s,s}) = n$ for all $\lambda \in \mathbb{C}$, $\Re(\lambda) \geq 0$ which is equivalent to all eigenvalues of $A_{s,s}$ having negative real part. $\square$

This method allows us to check the system properties of stabilizability and controllability. But, due to the difficult rank decisions, it is preferable to compute either the distance to uncontrollability or the distance to unstabilizability, i.e. the smallest perturbations that make the system non-controllable or non-stabilizable [21,24].

3.3   Optimal Control

Typically there are many different ways to achieve a certain control goal, such as stabilizing a system or moving it into a desired state. The freedom in the choice of control can then be used to optimize some performance measure. This could be robustness of the resulting system under perturbations or just control costs. Consider again system (3.1) and try to minimize some cost functional. An example of such a functional is

$$
\begin{aligned}
&\mathcal{S}\left(x(t), u(t)\right) \\
&= \tfrac{1}{2}\Big\{ \left(x(t_f)^T M x(t_f)\right) \\
&\quad + \int_{t_0}^{t_f} \left\{ x(t)^T Q x(t) + u(t)^T R u(t) + x(t)^T S u(t) + u(t)^T S^T x(t) \right\} \, dt \Big\} \\
&\qquad t_0 < t_f \leq \infty, \hspace{4cm} (3.8)
\end{aligned}
$$

where $M = M^T, Q = Q^T \in \mathbb{R}^{n,n}$, M, Q positive semidefinite, $S \in \mathbb{R}^{n,m}$, $R = R^T \in \mathbb{R}^{m,m}$ positive definite. This functional represents for example energy costs in applying the controller. The solution to the optimal control problem to minimize (3.8) subject to (3.1) is given by the following theorem.

Theorem 3.8. *Consider the optimal control problem to minimize (3.8) sub-ject to (3.1). Let $u_* \in U_m := \{u(t) \in \mathbb{R}^m, u(t)$ piecewise continuous on $[t_0, t_f]\}$ be the optimal control and let $x_*(t) \in \mathbb{R}^n$ be the closed loop solution, i.e.*

$$\dot{x}_*(t) = Ax_*(t) + Bu_*(t), \quad x_*(t_0) = x^0. \tag{3.9}$$

Then there exists a costate function (Lagrange multiplier) $\mu(t) \in \mathbb{R}^n$ so that $x_*(t), \mu(t), u_*(t)$ solves the boundary value problem

$$\begin{bmatrix} A & 0 & B \\ Q & A^T & S \\ S^T & B^T & R \end{bmatrix} \begin{bmatrix} x_*(t) \\ \mu(t) \\ u_*(t) \end{bmatrix} = \begin{bmatrix} I_n & 0 & 0 \\ 0 & -I_n & 0 \\ 0 & 0 & 0 \end{bmatrix} \begin{bmatrix} \dot{x}_*(t) \\ \dot{\mu}(t) \\ \dot{u}_*(t) \end{bmatrix}, \tag{3.10}$$

$$x_*(t_0) = x^0, \quad \mu(t_f) = Mx_*(t_f). \tag{3.11}$$

Proof. This is a simplified version of the proof of the Pontryagin maximum principle [47]. Let u_* be the optimal control and consider a perturbation

$$u(t) = u_*(t) + \varepsilon v(t)$$

with $u(t) \in U_m$. Then from (3.9) we have

$$\dot{x}(t) = Ax(t) + Bu_*(t) + \varepsilon Bv(t).$$

By Theorem 3.1 this has the solution

$$x(t) = e^{A(t-t_0)}x^0 + \int_{t_0}^t e^{A(t-s)} B\left(u_*(s) + \varepsilon v(s)\right) ds$$

$$= x_*(t) + \varepsilon \int_{t_0}^t e^{A(t-s)} Bv(s) \, ds = x_*(t) + \varepsilon\varphi(t),$$

where $\varphi(t)$ solves the differential equation

$$\dot{\varphi}(t) = A\varphi(t) + Bv(t), \quad \varphi(t_0) = 0.$$

Introducing $\mu(t) \in \mathbb{R}^n$ and the Hamiltonian function

$$H(x,\mu,u) = x^T(t)Qx(t) + x(t)^T Su(t) + u(t)^T S^T x(t)$$
$$+ u(t)^T Ru(t) + \mu(t)^T (Ax(t) + Bu(t)) + (Ax(t) + Bu(t))^T \mu(t),$$

we can write $\mathcal{S}(x,u)$ as

$$\mathcal{S}(x,u) = \tfrac{1}{2}\left\{ x^T(t_f)Mx(t_f) + \int_{t_0}^{t_f} (H(x,\mu,u) - \mu^T\dot{x} - \dot{x}^T\mu) \, dt \right\}$$

and analogous for u_*, x_*

$$\mathcal{S}(x_*,u_*) = \tfrac{1}{2}\left\{ x_*^T(t_f)Mx_*(t_f) + \int_{t_0}^{t_f} (H(x_*,\mu,u_*) - \mu^T\dot{x}_* - \dot{x}_*^T\mu) \, dt \right\}.$$

Then we get

$$S(x,u) - S(x_*,u_*) = \tfrac{1}{2}\Big\{ \big(x(t)^T M x(t) - x_*(t)^T M x_*(t)\big)\,|_{t=t_f}$$

$$+ \int_{t_0}^{t_f} \big(H(x,\mu,u) - H(x_*,\mu,u_*)\big)\, dt$$

$$+ \int_{t_0}^{t_f} \big(\mu^T \underbrace{(\dot{x}_* - \dot{x})}_{-\varepsilon\dot\varphi} + \underbrace{(\dot{x}_* - \dot{x})^T}_{-\varepsilon\dot\varphi^T}\mu\big)\, dt\Big\}$$

$$\underbrace{\qquad\qquad\qquad\qquad\qquad}_{=-2\varepsilon\mu^T\dot\varphi}$$

and

$$\tfrac{1}{2}\big(H(x,\mu,u) - H(x_*,\mu,u_*)\big)$$

$$= \tfrac{1}{2}\Big\{ x^T Q x + x^T S u + u^T S^T x + u^T R u + \mu^T(Ax+Bu) + (Ax+Bu)^T\mu$$

$$-x_*^T Q x_* - x_*^T S u_* - u_*^T S^T x_* - u_*^T R u_* - \mu^T(Ax_*+Bu_*)$$

$$-(Ax_*+Bu_*)^T\mu\Big\}.$$

Since $u = u_* + \varepsilon v$ and $x = x_* + \varepsilon\varphi$, we get

$$\tfrac{1}{2}\big(H(x,\mu,u) - H(x_*,\mu,u_*)\big)$$

$$= \tfrac{1}{2}\Big\{ x_*^T Q x_* + 2\varepsilon x_*^T Q\varphi + \varepsilon^2\varphi^T Q\varphi - x_*^T Q x_*$$

$$+\varepsilon^2\varphi^T S v + x_*^T S u_* + \varepsilon(x_*^T S v + \varphi^T S u_*) - x_*^T S u_*$$

$$+\varepsilon^2 v^T S^T\varphi + u_*^T S^T x_* + \varepsilon(u_*^T S^T\varphi + v^T S^T x_*) - u_*^T S^T x_*$$

$$+\varepsilon^2 v^T R v + u_*^T R u_* + 2\varepsilon u_*^T R v - u_*^T R u_*$$

$$+\varepsilon\mu^T A\varphi + \varepsilon\mu^T B v + \varepsilon\varphi^T A^T\mu + \varepsilon v^T B^T\mu\Big\}$$

$$= \varepsilon(x_*^T Q\varphi + x_*^T S v + u_*^T S^T\varphi + u_*^T R v + \mu^T A\varphi + \mu^T B v) + \mathcal{O}(\varepsilon^2)$$

$$= \varepsilon\left\{ [x_*^T Q + u_*^T S^T + \mu^T A]\varphi + [x_*^T S + u_*^T R + \mu^T B]v\right\} + \mathcal{O}(\varepsilon^2).$$

Using partial integration we obtain

$$-\int_{t_0}^{t_f} \varepsilon\mu^T\dot\varphi\, dt = -\varepsilon\mu^T\varphi\big|_{t_0}^{t_f} + \varepsilon\int_{t_0}^{t_f} \dot\mu^T\varphi\, dt$$

$$= -\varepsilon\mu^T(t_f)\varphi(t_f) + \varepsilon\int_{t_0}^{t_f} \dot\mu^T\varphi\, dt$$

and

$$x^T M x - x_*^T M x_* = 2\varepsilon x_*^T M\varphi + \mathcal{O}(\varepsilon^2).$$

This then gives

$$S(x,u) - S(x_*,u_*)$$

$$= \varepsilon\left\{ \int_{t_0}^{t_f} \big([x_*^T Q + u_*^T S^T + \mu^T A]\varphi + \dot\mu^T\varphi + [x_*^T S + u_*^T R + \mu^T B]v\big)\, dt\right.$$

$$\left. -\mu^T(t_f)\varphi(t_f) + x_*^T(t_f)M\varphi(t_f)\right\} + \mathcal{O}(\varepsilon^2).$$

Since $S(x,u) - S(x_*, u_*) \geq 0$ for all sufficiently small (positive or negative) ε it follows that the factor of ε has to vanish for all v and associated φ. We choose μ as solution of

$$-\dot{\mu}(t) = A^T \mu(t) + Q x_* + S u_*$$

with end condition

$$\mu(t_f) = M x_*(t_f).$$

Then only the middle term remains, i.e.

$$\int_{t_0}^{t_f} [x_*^T S + u_*^T R + \mu^T B] v \, dt = 0 \quad \text{for all } v \in U_m$$

which implies that

$$x_*^T S + u_*^T R + \mu^T B \equiv 0 \quad \text{for all } t \in [t_0, t_f].$$

We thus obtain the boundary value problem (3.10), (3.11). □

The boundary value problem also leads to a sufficient condition.

Theorem 3.9. *Let x_*, μ, u_* be such that $\begin{bmatrix} x_* \\ \mu \\ u_* \end{bmatrix}$ solves (3.10), (3.11). If $\mathcal{R} = \begin{bmatrix} Q & S \\ S^T & R \end{bmatrix}$ and M is positive semidefinite, then*

$$S(x,u) \geq S(x_*, u_*)$$

for all x, u that satisfy (3.1).

Proof. Define

$$\Phi(s) = S\left(s x_*(t) + (1-s)x(t), s u_*(t) + (1-s)u(t)\right).$$

Then the assertion is equivalent to the statement that $\Phi(s)$ has a minimum at $s = 1$ for all $x(t), u(t)$ that satisfy (3.1). Since $\Phi(s)$ is quadratic in s, the minimum happens at $s = 1$ if and only if

$$\left.\frac{d\Phi}{ds}\right|_{s=1} = 0, \quad \left.\frac{d^2\Phi}{ds^2}\right|_{s=1} \geq 0.$$

(Normally we would need $\left.\frac{d^2\Phi}{ds^2}\right|_{s=1} > 0$, but since the functional is quadratic it suffices to have ≥ 0.) Now we have

$$\left.\frac{d\Phi}{ds}\right|_{s=1}$$
$$= (x_* - x)^T M x_*\big|_{t=t_f}$$
$$+ \frac{1}{2}\int_{t_0}^{t_f}\{(x_* - x)^T Q x_* + x_*^T Q(x_* - x) + u_*^T S^T(x_* - x) + (u_* - u)^T S^T x_*$$
$$+ (u_* - u)^T R u_* + u_*^T R(u_* - u) + (x_* - x)^T S u_* + x_*^T S(u_* - u)\} \, dt.$$

Multiplying the second equation of (3.10) from the left by x_*^T and inserting the other equations we obtain

$$
\begin{aligned}
x_*^T Q x_* &= -x_*^T A^T \mu - x_*^T S u_* - x_*^T \dot\mu \\
(\text{ 1. eq. of (3.10) }) \quad &= u_*^T B^T \mu - \dot x_*^T \mu - x_*^T S u_* - x_*^T \dot\mu \\
(\text{ 3. eq. of (3.10) }) \quad &= -u_*^T S^T x_* - u_*^T R u_* - \dot x_*^T \mu - x_*^T S u_* - x_*^T \dot\mu.
\end{aligned}
$$

Likewise after multiplication with x^T

$$
\begin{aligned}
x^T Q x_* &= -x^T A^T \mu - x^T S u_* - x^T \dot\mu \\
(\text{ 1. eq. } x) \quad &= u^T B^T \mu - \dot x^T \mu - x^T S u_* - x^T \dot\mu \\
(\text{ 2. eq. for } B^T \mu) \quad &= -u^T S^T x_* - u^T R u_* - \dot x^T \mu - x^T S u_* - x^T \dot\mu.
\end{aligned}
$$

Substituting yields

$$
\begin{aligned}
\left. \frac{d\Phi}{ds} \right|_{s=1} &= (x_* - x)^T M x_* \big|_{t=t_f} + \frac{1}{2} \int_{t_0}^{t_f} (x^T \dot\mu + \dot x^T \mu - x_*^T \dot\mu - \dot x_*^T \mu)\, dt \\
&= (x_* - x)^T M x_* \big|_{t=t_f} + x^T \mu \big|_{t=t_0}^{t=t_f} - x_*^T \mu \big|_{t=t_0}^{t=t_f}.
\end{aligned}
$$

For $t = t_0$ we have $x(t_0) = x_*(t_0)$ and for $t = t_f$ we have $\mu(t_f) = M x_*(t_f)$ and hence $\left. \frac{d\Phi}{ds} \right|_{s=1} = 0$. Also we have

$$
\begin{aligned}
\left. \frac{d^2\Phi}{ds^2} \right|_{s=1} &= (x_* - x)^T M (x_* - x) \big|_{t=t_f} \\
&+ \int_{t_0}^{t_f} \left[(x_* - x)^T, (u_* - u)^T \right] \begin{bmatrix} Q & S \\ S^T & R \end{bmatrix} \begin{bmatrix} x_* - x \\ u_* - u \end{bmatrix} dt \geq 0,
\end{aligned}
$$

since M and $\mathcal{R} = \begin{bmatrix} Q & S \\ S^T & R \end{bmatrix}$ are positive semidefinite. $\square$

We have thus reduced the solution of the optimal control problem to the solution of a boundary value problem. Returning to our discussion at the beginning of this article, we can solve this boundary value problem using the transformation to generalized Schur form via the QZ-algorithm, see [57]. Note that in this case the matrix E is definitely singular, but since R is positive definite, the following simplification is often used. The third equation of (3.10) yields that

$$
u(t) = -R^{-1}\left(S^T x(t) + B^T \mu(t) \right).
$$

Inserting this into the other equations gives the reduced boundary value problem that now has $E = I$,

$$
\begin{bmatrix} A - B R^{-1} S^T & -B R^{-1} B^T \\ -(Q - S R^{-1} S^T) & -A^T + S R^{-1} B^T \end{bmatrix} \begin{bmatrix} x \\ \mu \end{bmatrix} = \begin{bmatrix} \dot x \\ \dot\mu \end{bmatrix} \qquad (3.12)
$$

with boundary conditions

$$x(t_0) = x^0 \text{ and } \mu(t_f) = Mx(t_f). \tag{3.13}$$

To solve the eigenvalue problem we could apply the QR-algorithm. This is the standard procedure that is used in many software packages [4]. However, in general it is better to solve the generalized eigenvalue problem directly, since the inversion of R and the formation of the blocks of (3.12) may lead to large numerical errors from which the method may not recover. So instead of using the QZ-algorithm it is even better to use an algorithm that preserves the structure of the problem. We will discuss such algorithms below. Before we do this we make a small detour and analyze the structure of the eigenvalue problems associated with (3.12) and (3.10) in more detail.

4   Hamiltonian Matrices and Riccati Equations

In the last section we have seen that optimal control problems lead to linear boundary value problems (3.10) and (3.12) with a special symmetry structure. It would be ideal if we could make use of this symmetry structure when designing numerical methods for the solution of the eigenvalue problem.

4.1   The Hamiltonian Schur Form

Let us have a closer look at this structure. We first study (3.12). Since $\mathcal{R} = \begin{bmatrix} Q & S \\ S^T & R \end{bmatrix}$ is symmetric positive semidefinite with R symmetric positive definite, it follows that $Q - SR^{-1}S^T$ is symmetric positive semidefinite and $BR^{-1}B^T$ is also symmetric, positive semidefinite of rank m. Hence the boundary value problem (3.12) has the form

$$\begin{bmatrix} \dot{x} \\ \dot{\mu} \end{bmatrix} = \begin{bmatrix} F & G \\ H & -F^T \end{bmatrix} \begin{bmatrix} x \\ \mu \end{bmatrix} = \mathcal{H} \begin{bmatrix} x \\ \mu \end{bmatrix}$$

with G and H symmetric negative semidefinite. A matrix of the form $\mathcal{H} = \begin{bmatrix} F & G \\ H & -F^T \end{bmatrix}$ with G and H symmetric is called a *Hamiltonian matrix*. For the *indefinite J-inner product*, defined via

$$\langle x, y \rangle_J = x^T J y \text{ with } J = \begin{bmatrix} 0 & I_n \\ -I_n & 0 \end{bmatrix}$$

we have

$$\langle x, \mathcal{H}y \rangle_J = -\langle x\mathcal{H}^T, y \rangle_J.$$

The set $\mathcal{H}_{2n}(\mathbb{R})$ of Hamiltonian matrices in $\mathbb{R}^{2n,2n}$ is a *Lie–algebra* with matrix addition and *Lie–multiplication* $[\mathcal{H}_1, \mathcal{H}_2] = \mathcal{H}_1 \mathcal{H}_2 - \mathcal{H}_2 \mathcal{H}_1$, since

$$\begin{bmatrix} F_1 & G_1 \\ H_1 & -F_1^T \end{bmatrix} \begin{bmatrix} F_2 & G_2 \\ H_2 & -F_2^T \end{bmatrix} - \begin{bmatrix} F_2 & G_2 \\ H_2 & -F_2^T \end{bmatrix} \begin{bmatrix} F_1 & G_1 \\ H_1 & -F_1^T \end{bmatrix}$$

$$= \begin{bmatrix} F_1 F_2 + G_1 H_2 & F_1 G_2 - G_1 F_2^T \\ H_1 F_2 - F_1^T H_2 & H_1 G_2 + F_1^T F_2^T \end{bmatrix} - \begin{bmatrix} F_2 F_1 + G_2 H_1 & F_2 G_1 - G_2 F_1^T \\ H_2 F_1 - F_2^T H_1 & H_2 G_1 + F_2^T F_1^T \end{bmatrix}$$

$$= \begin{bmatrix} F_1 F_2 - F_2 F_1 + G_1 H_2 - G_2 H_1 & F_1 G_2 - G_1 F_2^T + G_2 F_1^T - F_2 G_1 \\ H_1 F_2 - F_1^T H_2 - H_2 F_1 + F_2^T H_1 & H_1 G_2 + F_1^T F_2^T - H_2 G_1 - F_2^T F_1^T \end{bmatrix} \in \mathcal{H}_{2n}.$$

The *Lie-group* (with the usual matrix multiplication) associated with this Lie-algebra are the *symplectic matrices* $S_{2n}(\mathbb{R}) := \{ S \in \mathbb{R}^{2n,2n} \mid S^T J S = J \}$, which are *orthogonal with respect to the J-inner product*, i.e.

$$\langle x, Sy \rangle_J = x^T J S y = x^T S^{-T} J y = \langle S^{-1} x, y \rangle_J$$

and we have that all $S \in S_{2n}(\mathbb{R})$ satisfy $|\det S| = 1$. But it should be noted that symplectic matrices are not norm bounded. Consider the matrix $\begin{bmatrix} 1 & v \\ 0 & 1 \end{bmatrix} \in S_2(\mathbb{R})$ with $v \in \mathbb{R}$, which can have arbitrarily large norm even though the determinant is 1. Hamiltonian matrices have the following symmetry in eigenvalues and eigenvectors.

Theorem 4.1. *Let* $z^{(0)}, z^{(1)}, \ldots, z^{(k)} \in \mathbb{C}^{2n}$ *and* $\lambda \in \mathbb{C}$ *so that for* $\mathcal{H} \in \mathcal{H}_{2n}(\mathbb{R})$ *(or* $\mathcal{H} \in \mathcal{H}_{2n}(\mathbb{C})$*)*

  (i) $(\mathcal{H} - \lambda I) z^{(0)} = 0$,
  (ii) $(\mathcal{H} - \lambda I) z^{(j)} = z^{(j-1)}$, $j = 1, \cdots, k$.

Then for $w^{(j)} = J z^{(j)}$, $j = 0, \cdots, k$ *we have*

  (iii) $w^{(0)H}(\mathcal{H} + \bar{\lambda} I) = 0$,
  (iv) $w^{(j)H}(\mathcal{H} + \bar{\lambda} I) = -w^{(j-1)H}$, $j = 1, \cdots, k$.

Proof.

$$(\mathcal{H} - \lambda I) z^{(0)} = 0 \iff (J\mathcal{H} - \lambda J) z^{(0)} = 0 \iff (-\mathcal{H}^H J - \lambda J) z^{(0)} = 0$$

$$\iff (\mathcal{H}^H + \bar{\lambda} I) w^{(0)} = 0 \iff w^{(0)H}(\mathcal{H} + \bar{\lambda} I) = 0$$

and analogously

$$(\mathcal{H} - \lambda I) z^{(j)} = z^{(j-1)} \iff (J\mathcal{H} - \lambda J) z^{(j)} = J z^{(j-1)}$$

$$\iff (-\mathcal{H}^H J - \lambda J) z^{(j)} = w^{(j-1)}$$

$$\iff w^{(j)H}(\mathcal{H} + \bar{\lambda} I) = -w^{(j-1)H}. \quad \square$$

This means that every eigenvalue (except for those eigenvalues that are on the imaginary axis) λ has a partner $-\bar{\lambda}$ and in the real case one even has quadruples $\lambda, \bar{\lambda}, -\lambda, -\bar{\lambda}$ with related left and right eigenvectors. But even more can be said about the Hamiltonian matrices $\mathcal{H}$ arising in our boundary value problem (3.12).

Lemma 4.1. *Let* $\mathcal{H} = \begin{bmatrix} F & G \\ H & -F^T \end{bmatrix} \in \mathcal{H}_{2n}(\mathbb{R})$, $\lambda \in \sigma(\mathcal{H})$, $z = \begin{bmatrix} z_1 \\ z_2 \end{bmatrix}$ *so that* $\mathcal{H}z = \lambda z$. *Then*

$$z_2^H G z_2 + z_1^H H z_1 = (\lambda + \bar{\lambda}) z_1^H z_2 = 2\Re(\lambda) z_1^H z_2. \qquad (4.1)$$

Proof. From $\mathcal{H}z = \lambda z$ it follows that

$$F z_1 + G z_2 = \lambda z_1,$$
$$H z_1 - F^T z_2 = \lambda z_2$$

and hence

$$z_2^H F z_1 + z_2^H G z_2 = \lambda z_2^H z_1,$$
$$z_1^H H z_1 - z_1^H F^T z_2 = \lambda z_1^H z_2.$$

Conjugation of the second equation yields

$$z_1^H H z_1 - z_2^H F z_1 = \bar{\lambda} z_2^H z_1$$

and addition of the equations gives

$$z_2^H G z_2 + z_1^H H z_1 = (\lambda + \bar{\lambda}) z_1^H z_2. \qquad \square$$

Using this Lemma we have the following result.

Theorem 4.2. *Let* $\mathcal{H} = \begin{bmatrix} F & G \\ H & -F^T \end{bmatrix} \in \mathcal{H}_{2n}(\mathbb{R})$ *and* $\mathcal{H} \begin{bmatrix} z_1 \\ z_2 \end{bmatrix} = \lambda \begin{bmatrix} z_1 \\ z_2 \end{bmatrix}$.

(i) If $\Re(\lambda) = 0$, *then*
$$z_2^H G z_2 + z_1^H H z_1 = 0.$$

(ii) If G, H *are negative semidefinite and* $\Re(\lambda) = 0$, *then*
$$G z_2 = 0 \text{ and } H z_1 = 0.$$

(iii) Let $\mathcal{H}$ *be as in (3.12),* $\mathcal{R} = \begin{bmatrix} Q & S \\ S^T & R \end{bmatrix}$ *be positive definite and* (A, B) *stabilizable. Then* $\mathcal{H}$ *has no eigenvalue with real part 0.*

Proof. (i) is clear from Lemma 4.1 and (ii) follows from (i). For (iii) suppose that there exists λ with $\Re(\lambda) = 0$ and $\mathcal{H} \begin{bmatrix} z_1 \\ z_2 \end{bmatrix} = \lambda \begin{bmatrix} z_1 \\ z_2 \end{bmatrix}$. By (i) we have $z_2^H G z_2 + z_1^H H z_1 = 0$. Since $G = -BR^{-1}B^T$ and $H = -(Q - SR^{-1}S^T)$ are negative semidefinite it follows by (ii) that $G z_2 = 0, H z_1 = 0$, i.e. $(Q - SR^{-1}S^T)z_1 = 0, BR^{-1}B^T z_2 = 0$ and since $Q - SR^{-1}S^T, R$ are positive definite, we have $z_1 = 0$ and $z_2^H B = 0$. Thus we have

$$z_2^H(A - BR^{-1}S^T) = z_2^H A = -\bar{\lambda} z_2^H.$$

But this is a contradiction to (A, B) being stabilizable. $\square$

The assumptions in this result can be relaxed, see [40] and some of the following results can also be proved for differential-algebraic systems and if eigenvalues are on the imaginary axis. In the following we discuss only the most simple case as it arises in our optimal control problem, that $\mathcal{H}$ has no purely imaginary eigenvalues. For the general case see [25,36].

Theorem 4.3. *Suppose that $\mathcal{H} \in \mathcal{H}_{2n}(\mathbb{R})$ and that $\mathcal{H}$ has no purely imaginary eigenvalues. Then there exists $S \in \mathcal{S}_{2n}(\mathbb{R})$, such that*

$$S^{-1}\mathcal{H}S = \begin{bmatrix} J_1 & 0 \\ 0 & -J_1^H \end{bmatrix}$$

where J_1 is in real Jordan form and has only eigenvalues with negative real part.

Proof. Theorem 4.1 implies that for all generalized eigenvectors we have that if z is a right generalized eigenvector to λ, then Jz is a left generalized eigenvector to $-\bar{\lambda}$. Therefore there exists a matrix V so that

$$V^{-1}\mathcal{H}V = \begin{bmatrix} J_1 & 0 \\ 0 & J_2 \end{bmatrix}$$

is in Jordan form, where J_1 has only eigenvalues in the left half plane and J_2 has only eigenvalues in the right half plane and we may assume w.l.o.g that $J_2 = -J_1^H$. We still need to show that we can achieve this *symplectic Jordan form* with a symplectic matrix S. Let

$$V = \begin{bmatrix} V_1 & V_2 \end{bmatrix}.$$

Then we set

$$W = \begin{bmatrix} -(JV_2)^H \\ (JV_1)^H \end{bmatrix} = J^H V^H J$$

and by Theorem 4.1

$$W\mathcal{H} = \begin{bmatrix} J_1 & 0 \\ 0 & -J_1^H \end{bmatrix} W.$$

Since the decomposition into eigen- and principle vectors is unique up to block permutations and since J_1 and J_1^H have no common eigenvalues, there exists a nonsingular block diagonal matrix

$$D = \begin{bmatrix} D_1 & 0 \\ 0 & D_2 \end{bmatrix}$$

so that

$$W = J^H V^H J = DV^{-1} \iff V^H JV = JD = \begin{bmatrix} 0 & D_2 \\ -D_1 & 0 \end{bmatrix}.$$

Since $V^H J V$ is skew Hermitian, we have $D_2 = D_1^H$. Then

$$S := V \begin{bmatrix} D_1^{-1} & 0 \\ 0 & I \end{bmatrix}$$

is symplectic. □

This structured Jordan form is unfortunately not computable in a numerically stable way, since arbitrarily small perturbations may change the Jordan structure and since symplectic matrices may have arbitrary large norm. To get numerical backward stability we need to use transformations from the *compact Lie group* of orthogonal (unitary) symplectic matrices given by $US_{2n}(\mathbb{C}) = \{Q \in S_{2n}(\mathbb{C}) \,|\, Q^H Q = I\}$ or $US_{2n}(\mathbb{R}) = \{Q \in S_{2n}(\mathbb{R}) \,|\, Q^T Q = I\}$ respectively. We have the following lemma.

Lemma 4.2. Let $Q = \begin{bmatrix} Q_{1,1} & Q_{1,2} \\ Q_{2,1} & Q_{2,2} \end{bmatrix} \in US_{2n}(\mathbb{R})$ with $Q_{ij} \in \mathbb{R}^{n,n}$. Then $Q_{1,2} = -Q_{2,1}, Q_{2,2} = Q_{1,1}$.

Proof. We have $Q \in S_{2n}(\mathbb{R})$ if and only if $Q^T J Q = J$ and $Q^T Q = I$. Then $JQ = Q^{-T} J = QJ$ implies

$$\begin{bmatrix} 0 & I \\ -I & 0 \end{bmatrix} \begin{bmatrix} Q_{1,1} & Q_{1,2} \\ Q_{2,1} & Q_{2,2} \end{bmatrix}$$
$$= \begin{bmatrix} Q_{2,1} & Q_{2,2} \\ -Q_{1,1} & -Q_{1,2} \end{bmatrix} = \begin{bmatrix} Q_{1,1} & Q_{1,2} \\ Q_{2,1} & Q_{2,2} \end{bmatrix} \begin{bmatrix} 0 & I \\ -I & 0 \end{bmatrix} = \begin{bmatrix} -Q_{1,2} & Q_{1,1} \\ -Q_{2,2} & Q_{2,1} \end{bmatrix}. \qquad \square$$

What can we achieve with orthogonal symplectic transformations?

Theorem 4.4.

(i) *Suppose that $\mathcal{H} \in \mathcal{H}_{2n}(\mathbb{R})$ ($\mathcal{H}_{2n}(\mathbb{C})$) has no eigenvalues with real part 0. Then there exists $Q \in US_{2n}(\mathbb{C})$, so that*

$$Q^H \mathcal{H} Q = \begin{bmatrix} T & N \\ 0 & -T^H \end{bmatrix}, \quad T, N \in \mathbb{C}^{n,n}$$

with T upper triangular. Here $N = N^H$ and T can be chosen so that all eigenvalues of T have negative real part. This form is called the Hamiltonian Schur form.

(ii) *Suppose that $\mathcal{H} \in \mathcal{H}_{2n}(\mathbb{R})$ has no eigenvalues with real part 0. Then there exists $Q \in US_{2n}(\mathbb{R})$, so that*

$$Q^T \mathcal{H} Q = \begin{bmatrix} T & N \\ 0 & -T^T \end{bmatrix} \quad T, N \in \mathbb{R}^{n,n},$$

with T quasi-upper triangular. Here $N = N^T$ and T can be chosen so that all eigenvalues of T have negative real part. This form is called the real Hamiltonian Schur form.

Proof. See the Laub trick in (5.1) below. $\quad\square$

This theorem immediately suggests that we may construct a numerical algorithm like the QR-algorithm, that exploits the structure and computes the Hamiltonian Schur form, see Section 5.

4.2 Solution of the Optimal Control Problem via Riccati Equations

We will now come back to the solution of the boundary value problems (3.12), (3.13). We make the ansatz: $\mu(t) = X(t)x(t)$. Then (3.13) implies that $\mu(t_f) = X(t_f)x(t_f) = Mx(t_f)$, that is $X(t_f) = M$. It follows that

$$\begin{bmatrix} \dot{x} \\ \dot{\mu} \end{bmatrix} = \begin{bmatrix} F & G \\ H & -F^T \end{bmatrix} \begin{bmatrix} x \\ \mu \end{bmatrix}$$

and

$$\begin{bmatrix} \dot{x} \\ X(t)\dot{x} + \dot{X}(t)x \end{bmatrix} = \begin{bmatrix} F & G \\ H & -F^T \end{bmatrix} \begin{bmatrix} x \\ X(t)x \end{bmatrix}.$$

Then

$$\dot{x} = (F + GX(t))x$$

and

$$X(t)\dot{x} + \dot{X}(t)x = Hx - F^T X(t)x$$

give

$$\dot{X}(t)x - Hx + F^T X(t)x + X(t)Fx + X(t)GX(t)x = 0.$$

Therefore, if $X(t)$ satisfies the initial value problem for the ordinary *matrix Riccati differential equation*

$$\dot{X}(t) = H - F^T X(t) - X(t)F - X(t)GX(t), \qquad (4.2)$$
$$X(t_f) = M,$$

then x solves the closed loop system

$$\dot{x} = (F + GX(t))\,x, \quad \mu = X(t)x. \qquad (4.3)$$

For the case $t_f = \infty$, $M = 0$ we obtain that $\mu(t) = Xx(t)$ where X is constant and satisfies the *algebraic Riccati equation*

$$H - F^T X - XF - XGX = 0. \qquad (4.4)$$

In both cases we obtain the optimal control u as a feedback

$$u(t) = -R^{-1}\left(S^T x(t) + B^T \mu(t)\right)$$
$$= -R^{-1}\left(S^T + B^T X\right)x(t),$$

where X solves either the algebraic Riccati equation (4.4) or the Riccati–differential equation (4.2).

The algebraic Riccati equation is not uniquely solvable, but since we want $x(t)$ to be asymptotically stable, we need that the closed loop solution

$$\dot{x} = Ax + Bu = \left[A - BR^{-1}(S^T + B^T X)\right] x$$

is asymptotically stable and hence $(A - BR^{-1}S^T) - BR^{-1}B^T X$ should have eigenvalues with negative real part. But since

$$\begin{bmatrix} F & G \\ H & -F^T \end{bmatrix} \begin{bmatrix} I \\ X \end{bmatrix} = \begin{bmatrix} I \\ X \end{bmatrix}(F + GX)$$

$$= \begin{bmatrix} I \\ X \end{bmatrix}(A - BR^{-1}S^T - BR^{-1}B^T X),$$

the columns of $\begin{bmatrix} I \\ X \end{bmatrix}$ must span the invariant subspace associated with the eigenvalues in the left half plane. This subspace is unique and we may get it from the Hamiltonian Jordan or Schur form. If

$$S = \begin{bmatrix} S_{1,1} & S_{1,2} \\ S_{2,1} & S_{2,2} \end{bmatrix} \in S_{2n}(\mathbb{R}), \quad S^{-1}\mathcal{H}S = \begin{bmatrix} T_{1,1} & T_{1,2} \\ 0 & -T_{1,1}^T \end{bmatrix}$$

with $T_{1,1}$ having all eigenvalues in the left half plane, then the columns of $\begin{bmatrix} S_{1,1} \\ S_{2,1} \end{bmatrix}$ span this subspace. We will show that then $S_{1,1}$ is invertible and $X = S_{2,1}S_{1,1}^{-1}$ is the symmetric, positive semidefinite solution of (4.3).

Lemma 4.3.

(i) Let X be a symmetric solution of the algebraic Riccati equation

$$0 = H - F^T X - XF - XGX. \tag{4.5}$$

Then the columns of $\begin{bmatrix} I \\ X \end{bmatrix}$ span an n–dimensional invariant subspace of

$$\mathcal{H} = \begin{bmatrix} F & G \\ H & -F^T \end{bmatrix}. \tag{4.6}$$

(ii) If X is symmetric, so that the columns of $\begin{bmatrix} I \\ X \end{bmatrix}$ span an invariant subspace of $\mathcal{H}$ as in (4.6), then X solves (4.5).

Proof. We have

$$\begin{bmatrix} F & G \\ H & -F^T \end{bmatrix} \begin{bmatrix} I \\ X \end{bmatrix} = \begin{bmatrix} F + GX \\ H - F^T X \end{bmatrix} = \begin{bmatrix} I \\ X \end{bmatrix} Z.$$

Thus $H - F^T X = XZ = X(F + GX)$ and X satisfies (4.5). $\square$

It remains to show that there exists an n-dimensional invariant subspace associated with the eigenvalues in the left half plane of the form $\begin{bmatrix} I \\ X \end{bmatrix}$ with X symmetric.

Theorem 4.5. *Let* $\mathcal{H} = \begin{bmatrix} F & G \\ H & -F^T \end{bmatrix} \in \mathcal{H}_{2n}(\mathbb{R})$ *be as in (3.12) and assume that* (A, B) *is stabilizable and* $\mathcal{R} = \begin{bmatrix} Q & S \\ S^T & R \end{bmatrix}$ *is positive definite. Let*

$$S \in \mathcal{S}_{2n}(\mathbb{R}), \ S^{-1}\mathcal{H}S = \begin{bmatrix} T_{1,1} & T_{1,2} \\ 0 & -T_{1,1}^T \end{bmatrix},$$

so that $T_{1,1}$ *has all its eigenvalues in the open left half plane. Then* $S_{1,1}$ *is invertible,* $S_{2,1}S_{1,1}^{-1}$ *is symmetric and* $X = S_{2,1}S_{1,1}^{-1}$ *is the unique positive semidefinite solution of (4.5).*

Proof. We have

$$\begin{bmatrix} F & G \\ H & -F^T \end{bmatrix} \begin{bmatrix} S_{1,1} & S_{1,2} \\ S_{2,1} & S_{2,2} \end{bmatrix} = \begin{bmatrix} S_{1,1} & S_{1,2} \\ S_{2,1} & S_{2,2} \end{bmatrix} \begin{bmatrix} T_{1,1} & T_{1,2} \\ 0 & -T_{1,1}^T \end{bmatrix}.$$

Hence

$$FS_{1,1} + GS_{2,1} = S_{1,1}T_{1,1},$$
$$HS_{1,1} - F^T S_{2,1} = S_{2,1}T_{1,1},$$

and thus $S_{1,1}^T H - S_{2,1}^T F = T_{1,1}^T S_{2,1}^T$.
This implies that

$$S_{2,1}^T F S_{1,1} + S_{2,1}^T G S_{2,1} = S_{2,1}^T S_{1,1} T_{1,1}$$
$$S_{1,1}^T H S_{1,1} - S_{2,1}^T F S_{1,1} = T_{1,1}^T S_{2,1}^T S_{1,1}$$

and thus

$$S_{2,1}^T G S_{2,1} + S_{1,1}^T H S_{1,1} = S_{2,1}^T S_{1,1} T_{1,1} + T_{1,1}^T S_{2,1}^T S_{1,1},$$

which is a Lyapunov equation. Since $S \in \mathcal{S}_{2n}(\mathbb{R})$ implies that

$$[I_n, \ 0] S^T J S \begin{bmatrix} I_n \\ 0 \end{bmatrix} = 0$$

we have that $S_{2,1}^T S_{1,1} = S_{1,1}^T S_{2,1}$. But since $T_{1,1}$ has only eigenvalues with negative real part and $S_{2,1}^T G S_{2,1} + S_{1,1}^T H S_{1,1}$ is negative semidefinite, it follows by the Lyapunov theorem [27] that $S_{2,1}^T S_{1,1}$ is unique and positive semidefinite. We still have to show that $S_{1,1}$ is invertible. Suppose that $w \neq 0$ so that $S_{1,1}w = 0$. Then

$$w^T S_{2,1}^T G S_{2,1} w + w^T S_{1,1}^T H S_{1,1} w = 0$$

implies that $GS_{2,1}w = 0$ and hence $B^T S_{2,1}w = 0$, since R is positive definite. But since $FS_{1,1} + GS_{2,1} = S_{1,1}T_{1,1}$ then $S_{1,1}(T_{1,1}w) = 0$ and thus $T_{1,1}w \in \text{Ker}(S_{1,1})$, i.e. $\text{Ker}(S_{1,1})$ is an invariant subspace of $T_{1,1}$. Since in every invariant subspace of a matrix there is at least one eigenvector z, there exists $z \in \text{Ker}(S_{1,1})$, $z \neq 0$ so that $T_{1,1}z = \lambda z$ for some eigenvalue λ of $T_{1,1}$. Hence

$$HS_{1,1}z - F^T S_{2,1}z = S_{2,1}T_{1,1}z,$$
$$-F^T S_{2,1}z = \lambda S_{2,1}z$$

and thus $(F^T + \lambda I)S_{2,1}z = 0$. Then $S_{2,1}z \neq 0$, since $\begin{bmatrix} S_{1,1} \\ S_{2,1} \end{bmatrix}$ has full rank. Therefore $S_{2,1}z$ is an eigenvector of F^T to $-\lambda, \Re(-\lambda) > 0$. But since $z \in \text{Ker}(S_{1,1})$, it follows that $GS_{2,1}z = 0$ and hence $B^T S_{2,1}z = 0$. Thus we have

$$\text{rank} \begin{bmatrix} F^T - (-\lambda)I \\ B^T \end{bmatrix} < n \quad \text{for some } \lambda \text{ with } \Re(-\lambda) > 0$$

and therefore, $(F, B) = (A - BR^{-1}S^T, B)$ and also (A, B) is not stabilizable, which is a contradiction. $\square$

In summary we have the following theorem which characterizes the solution of the linear quadratic optimal control problem.

Theorem 4.6. *Consider the optimal control problem (3.1), (3.8).*

(i) *If $t_f < \infty$, then the optimal control exists and is given by the linear feedback*

$$u(t) = -R^{-1}\left(S^T + B^T X(t)\right)x(t) \qquad (4.7)$$

where $X(t)$ is the unique solution of the Riccati differential equation

$$\dot{X}(t) = -(Q - SR^{-1}S^T) - (A - BR^{-1}S^T)^T X(t)$$
$$-X(t)(A - BR^{-1}S^T) + X(t)BR^{-1}B^T X(t) \qquad (4.8)$$
$$X(t_f) = M.$$

(ii) *If (A, B) is stabilizable, $t_f = \infty$, $M = 0$ and $\mathcal{R}$ positive definite, then the unique solution of of the optimal control problem given by (3.1), (3.8) is the linear feedback*

$$u(t) = -R^{-1}(S^T + B^T X)x(t) \qquad (4.9)$$

where X is the unique positive semidefinite solution of the algebraic Riccati equation

$$0 = (Q - SR^{-1}S^T) + (A - BR^{-1}S^T)^T X + X(A - BR^{-1}S^T) - XBR^{-1}B^T X. \qquad (4.10)$$

In this case the closed loop system satisfies

$$\lim_{t \to \infty} x(t) = 0. \qquad (4.11)$$

Proof. Most of the proof we have given already.

(i) From the theory of differential equations it follows that (4.8) has a unique solution. Thus we have a solution of the form (4.7) and Theorem 3.9 implies that we have a minimal solution. It remains to discuss the uniqueness of (3.12), (3.13). The solution of the boundary value problem is

$$\begin{bmatrix} x(t) \\ \mu(t) \end{bmatrix} = e^{\mathcal{H}(t-t_0)} \begin{bmatrix} q_0 \\ v_0 \end{bmatrix}$$

Since $Z(t) := e^{\mathcal{H}(t-t_0)}$ is nonsingular for all t it follows that in the partition

$$Z(t) = [Z_1(t),\ Z_2(t)] \text{ with } Z_i(t) \in \mathbb{C}^{2n,n} \quad i = 1, 2$$

$Z_i(t)$ has full rank. Hence

$$x(t_0) = Z_1(t_0)q_0 = x^0$$

and

$$\mu(t_f) = Mx(t_f) = Z_2(t_f)v_0 = MZ_1(t_f)q_0.$$

This proves (i).

(ii) From the stabilizability of (A, B) and the positive definiteness of $\mathcal{R}$ it follows by Theorem 4.2 that

$$\mathcal{H} = \begin{bmatrix} A - BR^{-1}S^T & -BR^{-1}B^T \\ -(Q - SR^{-1}S^T) & -(A - BR^{-1}S^T)^T \end{bmatrix}$$

has no purely imaginary eigenvalues. We have already shown that the feedback (4.9) with X solution of (4.10) leads to a solution of (3.12), (3.13) for $t_f = \infty$, $M = 0$ if $\lim_{t\to\infty} \mu(t) = 0$. The uniqueness follows from the uniqueness of the invariant subspace. □

This theorem implies that it suffices to solve a differential Riccati equation if $t_f < \infty$ or to compute the semidefinite solution of an algebraic Riccati equation to stabilize the system if $t_f = \infty$. In both cases the optimal control is given by

$$u(t) = -R^{-1}(S^T + B^T X)x(t). \tag{4.12}$$

But it should be noted that from a numerical analysis point of view computing the solution via the algebraic Riccati equation may be not advisable. First of all, the problem of solving the algebraic Riccati equation may have a much larger condition number than the boundary value problem itself. This situation becomes even more critical when one solves other optimal control problems such as the optimal H_∞ control problem, which is also commonly solved via the solution of Riccati equations. This problem becomes very ill-conditioned in the neighborhood of the optimal solution, see

[60]. The Riccati approach also fails when one studies the optimal control of differential-algebraic equations [40]. Thus from the point of view of accuracy and numerical robustness it is much better to solve the boundary value problem via the solution of the structured eigenvalue problem directly. This topic is discussed in the next section.

5 Numerical Solution of Hamiltonian Eigenvalue Problems

We have seen that to solve the boundary value problem, it suffices to compute the invariant subspace associated with the stable eigenvalues of a Hamiltonian matrix via the Hamiltonian Schur form. In this section we discuss numerical methods to compute the Hamiltonian Schur form of Hamiltonian matrices

$$\mathcal{H} = \begin{bmatrix} F & G \\ H & -F^T \end{bmatrix}$$

that have no purely imaginary eigenvalues.

One of the first and most simple ideas to do this is due to Laub [33]. This idea is often called the 'Laub-trick'. If we use the classical QR-algorithm to compute the real Schur form

$$\begin{bmatrix} Q_{1,1} & Q_{1,2} \\ Q_{2,1} & Q_{2,2} \end{bmatrix}^T \begin{bmatrix} F & G \\ H & -F^T \end{bmatrix} \begin{bmatrix} Q_{1,1} & Q_{1,2} \\ Q_{2,1} & Q_{2,2} \end{bmatrix} = \begin{bmatrix} T_{1,1} & T_{1,2} \\ 0 & T_{2,2} \end{bmatrix},$$

where all eigenvalues of $T_{1,1}$ have negative real part, then the analysis of the last section can be used to show that the matrix

$$\begin{bmatrix} Q_{1,1} & -Q_{2,1} \\ Q_{2,1} & Q_{1,1} \end{bmatrix}$$

is orthogonal and symplectic and it gives us the real Hamiltonian Schur form

$$\begin{bmatrix} Q_{1,1} & -Q_{2,1} \\ Q_{2,1} & Q_{1,1} \end{bmatrix}^T \begin{bmatrix} F & G \\ H & -F^T \end{bmatrix} \begin{bmatrix} Q_{1,1} & -Q_{2,1} \\ Q_{2,1} & Q_{1,1} \end{bmatrix} = \begin{bmatrix} T_{1,1} & \tilde{T}_{1,2} \\ 0 & -T_{1,1}^T \end{bmatrix}. \tag{5.1}$$

Furthermore, we may use this approach as well to compute

$$X = Q_{2,1} Q_{1,1}^{-1}$$

the desired solution of the algebraic Riccati equation and

$$\dot{x} = (Q_{1,1} T_{1,1} Q_{1,1}^{-1}) x = (F + GX) x$$

the closed loop system.

Hence, we get the solution of the Riccati equation as a by-product of the solution of the eigenvalue problem. Actually this is essentially the method that is currently implemented in MATLAB toolboxes [37,38,26] or the sub-routine library for control SLICOT [12] to solve algebraic Riccati equations.

This method is numerically backwards stable and therefore computes the invariant subspace of a matrix $\mathcal{H} + \Delta$, where Δ is a small perturbation. Unfortunately, since the QR-algorithm is not preserving the structure, this perturbation Δ is not Hamiltonian, so that in general the method computes the eigenvalues of a non-Hamiltonian matrix. This is not so much a problem if the subspace is computed with high relative accuracy, which is the case as long as the eigenvalues $\mathcal{H}$ are sufficiently far away from the imaginary axis. But if they come close to or are on the imaginary axis then this approach fails because small perturbations change the number of eigenvalues in the left (or right) half plane and it is then very difficult to decide what the correct invariant subspace is, for examples see [11]. It can be shown [25,48] that if there are eigenvalues on the imaginary axis, then *only* a structure preserving method for which the backward error Δ is Hamiltonian will allow us to compute the invariant subspace associated with eigenvalues in the closed left half plane accurately. Note that the structured perturbation analysis shows that if $\mathcal{H}$ has a pair of eigenvalues that are on the imaginary axis, and they form a 2×2 Jordan block, then the invariant subspace associated with the eigenvalues in the closed left half plane is unique. For more details, see [25]. But this also explains why a non-structure preserving method will have difficulties computing the subspace accurately. The classical perturbation theory for multiple eigenvalues [53] shows that if eigenvalues belong to a 2×2 Jordan block, then a perturbation of order $\sqrt{eps}$ has to be expected for the eigenvalues and the computation of the eigenvectors is an ill-conditioned problem. However, if the backward error is Hamiltonian then under some extra conditions the computation of the invariant subspace is still well-conditioned, see [25,48].

Due to its importance in control theory the solution of Hamiltonian eigenvalue problems has received a lot of attention in the last 20 years, starting with the pioneering work of Laub [33] and Paige/Van Loan [44] and Van Dooren [57]. Unfortunately, the construction of a completely satisfactory method is still an open problem. Such a method should be numerically backward stable, have a complexity of $O(n^3)$ or less and at the same time preserve the Hamiltonian structure. Many attempts have been made to tackle this problem, see [18,34,40] and the references therein, but it has been shown in [2] that a modification of a standard QR-like methods to solve this problem is in general hopeless, due to the missing reduction to a Hessenberg–like form. As we have seen, such a Hessenberg reduction is essential for the reduction of complexity from $\mathcal{O}(n^4)$ to $\mathcal{O}(n^3)$ and also for the deflation procedure, which is responsible for fast convergence.

For this reason other methods like the Orthogonal Symplectic Multi-shift method for the solution of Algebraic Riccati Equations, OSMARE of [1] were developed that do not follow the direct line of a standard QR-like method.

This method is backward stable and structure preserving. First it computes all the eigenvalues and then uses the computed eigenvalues as 'exact' shifts thus following the idea of Lemma 2.1. But it may suffer from a loss of convergence, in particular for large problems, due to the fact that it does not have a good deflation procedure incorporated. For small problems, say $n < 100$, it is still the best available method so far, see the comparison below.

A method to compute the eigenvalues of real Hamiltonian matrices was suggested in [58], which computes actually the eigenvalues of $\mathcal{H}^2$ and then takes square roots. If the matrix $\mathcal{H}$ is real then the matrix $\mathcal{H}^2$ is *skew Hamiltonian*, i.e. $J\mathcal{H}^2 = -(J\mathcal{H}^2)^T$. For real skew Hamiltonian matrices a nice Hessenberg-like reduction exist. It was shown in [44] that every real skew Hamiltonian matrix

$$\mathcal{W} = \begin{bmatrix} F & G \\ H & F^T \end{bmatrix}, \ G = -G^T, \ H = -H^T$$

can be transformed in a finite number of similarity transformations to the form

$$U^T \mathcal{W} U = \begin{bmatrix} \tilde{F} & \tilde{G} \\ \tilde{H} & \tilde{F}^T \end{bmatrix}$$

with $U \in \mathcal{U}_{2n}(\mathbb{R})$, $\tilde{F}$ upper Hessenberg and $\tilde{H} = \tilde{H}^T$ skew symmetric and diagonal. Hence $\tilde{H} = 0$ and the problem automatically deflates and reduces to an eigenvalue problem for $\tilde{F}$. This allows us to compute the *skew Hamiltonian Schur form* in a more efficient way than using the QR-algorithm on $\mathcal{H}$. This method is called the square reduced method $SQRED$. On the other hand, this method has two drawbacks. It does not compute the invariant subspace because the eigenvalues λ and $-\lambda$ are mapped both to λ^2 and hence it is difficult to separate the associated eigenvectors. Furthermore, the method may suffer from a loss of half of the possible accuracy as was shown in [58]. However, this loss of accuracy can be avoided by an idea of [14] which forms the basis for the following algorithms. This idea is based on the following two lemmas.

Lemma 5.1. *If $\mathcal{H} \in \mathcal{H}_{2n}(\mathbb{R})$, then there exist $U_1, U_2 \in \mathcal{US}_{2n}(\mathbb{R})$, which are computable in a finite number of steps, such that*

$$\mathcal{H} = U_2 \begin{bmatrix} H_{1,1} & H_{1,2} \\ 0 & -H_{2,2}^T \end{bmatrix} U_1^T,$$

where $H_{1,1}, H_{1,2}, H_{2,2} \in \mathbb{R}^{n,n}$, $H_{1,1}$ is upper triangular and $H_{2,2}$ upper Hessenberg.

Proof. The proof of this lemma is very technical but it provides a constructive procedure to compute this reduction, see [14]. □

The second lemma provides a method to compute the eigenvalues of a Hamiltonian matrix.

Lemma 5.2. *(Symplectic URV-Decomposition). If $\mathcal{H} \in \mathcal{H}_{2n}(\mathbb{R})$, then there exist $U_1, U_2 \in \mathcal{US}_{2n}(\mathbb{R})$ such that*

$$\mathcal{H} = U_2 \begin{bmatrix} H_t & H_r \\ 0 & -H_b^T \end{bmatrix} U_1^T,$$

where H_t is upper triangular and H_b is quasi upper triangular .
  Moreover,

$$\mathcal{H} = J\mathcal{H}^T J = U_1 \begin{bmatrix} H_b & H_r^T \\ 0 & -H_t^T \end{bmatrix} U_2^T.$$

The positive and negative square roots of the eigenvalues of $H_t H_b$ are the eigenvalues of $\mathcal{H}$.

Proof. The proof is immediate from

$$U_2^T \mathcal{H} U_1 U_1^T \mathcal{H} U_2 = U_2^T \mathcal{H} U_1 U_1^T J \mathcal{H}^T J U_2 = U_2^T \mathcal{H} U_1 J U_1^T \mathcal{H}^T U_2 J$$
$$= (U_2 \mathcal{H} U_1) J (U_2^T \mathcal{H} U_1)^T J. \qquad \square$$

This result would not help us a lot if we first had to compute the product $H_t H_b$ to compute its eigenvalues. But fortunately it is possible to compute the eigenvalues of a product of matrices without forming the product, by using the periodic Schur decomposition [15,29]. In our case this decomposition yields real orthogonal transformation matrices U, V such that in product $H_{1,1} H_{2,2}$

$$\hat{H} = U^T H_{1,1} V V^T H_{2,2}^T U, \qquad H_b^T = (U^T H_{2,2} V)^T$$

are quasi-upper triangular, while

$$\hat{H}_t := U^T H_{1,1} V$$

is upper triangular. The numerical method that is based on these observations was suggested in [14] and is called the *URVPSD*-algorithm
  Let us demonstrate the properties of this method by some numerical tests from [14] which compare the *URVPSD*-algorithm with *SQRED*, the square reduced method from [58] as implemented in [7] and the nonsymmetric eigensolver *DGEEVX* from LAPACK [3]. The test is performed for example 2 from [58].
  Table 5.1 shows the absolute errors in the eigenvalue approximations computed by the three methods.
  In Table 5.1 the loss of accuracy of the *SQRED*-method is obvious. Using the periodic Schur decomposition yields the exact eigenvalues with respect to machine precision as does the *QR*-algorithm implemented in LAPACK. However, there is a price to pay. Since compared with the method of [58] the complexity has grown and the implementation of the method needs more complex data structures, it turns out that the method is not as fast as the flop count suggests but it is still substantially faster than the *QR*-algorithm.

5.1   Subspace Computation

The symplectic URV-decomposition allows us to compute the eigenvalues of Hamiltonian matrices very accurately and it makes use of the Hamiltonian structure. These eigenvalues can then be used in the multishift method. But the method can also be employed to directly compute the invariant subspaces that are needed in the solution of the optimal control problem. This approach was suggested in [13] and it uses an embedding procedure that we first describe for general matrices $A \in \mathbb{R}^{n,n}$.

Consider the block matrix

$$B = \begin{bmatrix} 0 & A \\ A & 0 \end{bmatrix}. \tag{5.2}$$

With $\mathcal{X} = \frac{\sqrt{2}}{2} \begin{bmatrix} I_n & -I_n \\ I_n & I_n \end{bmatrix} \in \mathcal{U}_{2n}(\mathbb{R})$ we have $\mathcal{X}^T B \mathcal{X} = \begin{bmatrix} A & 0 \\ 0 & -A \end{bmatrix}$. We have the following relationship between the spectra of A and B.

$$\begin{aligned}
\sigma(B) &= \sigma(A) \cup \sigma(-A), \\
\sigma_0(B) &= \sigma_0(A) \cup \sigma_0(A), \\
\sigma_+(B) &= \sigma_+(A) \cup \sigma_+(-A) = \sigma_+(A) \cup (-\sigma_-(A)), \\
\sigma_-(B) &= \sigma_-(A) \cup \sigma_-(-A) = (-\sigma_+(A)) \cup \sigma_-(A) = -\sigma_+(B).
\end{aligned}$$

(Note that in the spectra we count eigenvalues with their algebraic multiplicities.) We obtain the following relations for the invariant subspaces of A and B.

Theorem 5.1. *[13] Let $A \in \mathbb{R}^{n,n}$ and $B \in \mathbb{R}^{2n,2n}$ be related as in (5.2) and let $\begin{bmatrix} Q_1 \\ Q_2 \end{bmatrix} \in \mathbb{R}^{2n,n}$ with $Q_1, Q_2 \in \mathbb{R}^{n,n}$, have orthonormal columns such that*

$$B \begin{bmatrix} Q_1 \\ Q_2 \end{bmatrix} = \begin{bmatrix} Q_1 \\ Q_2 \end{bmatrix} R,$$

where

$$\sigma_+(B) \subseteq \sigma(R) \subseteq \sigma_+(B) \cup \sigma_0(B).$$

λ	URVPSD	SQRED	LAPACK
1	0	0	7.8×10^{-16}
10^{-2}	5.5×10^{-16}	5.5×10^{-16}	5.0×10^{-17}
10^{-4}	1.6×10^{-18}	1.6×10^{-14}	2.6×10^{-18}
10^{-6}	1.0×10^{-18}	1.5×10^{-11}	8.4×10^{-18}
10^{-8}	3.1×10^{-17}	2.2×10^{-9}	4.7×10^{-17}

Table 5.1. Absolute errors $|\lambda - \tilde{\lambda}|$.

Then

$$\text{range}\{Q_1 + Q_2\} = \text{inv}_+(A) + \mathcal{N}_1, \quad where \quad \mathcal{N}_1 \subseteq \text{inv}_0(A),$$

and

$$\text{range}\{Q_1 - Q_2\} = \text{inv}_-(A) + \mathcal{N}_2, \quad where \quad \mathcal{N}_2 \subseteq \text{inv}_0(A).$$

Moreover, if we partition R as

$$R = \begin{bmatrix} R_{1,1} & R_{1,2} \\ 0 & R_{2,2} \end{bmatrix}, \quad where \quad \sigma(R_{1,1}) = \sigma_+(B),$$

and, accordingly, $Q_1 = \begin{bmatrix} Q_{1,1} & Q_{1,2} \end{bmatrix}$, $Q_2 = \begin{bmatrix} Q_{2,1} & Q_{2,2} \end{bmatrix}$, then

$$B \begin{bmatrix} Q_{1,1} \\ Q_{2,1} \end{bmatrix} = \begin{bmatrix} Q_{1,1} \\ Q_{2,1} \end{bmatrix} R_{1,1},$$

and there exists an orthogonal matrix Z such that

$$\frac{\sqrt{2}}{2}(Q_{1,1} + Q_{2,1}) = \begin{bmatrix} 0 & P_+ \end{bmatrix} Z, \qquad \frac{\sqrt{2}}{2}(Q_{1,1} - Q_{2,1}) = \begin{bmatrix} P_- & 0 \end{bmatrix} Z,$$

where P_+, P_- are orthogonal bases of $\text{inv}_+(A)$, $\text{inv}_-(A)$, respectively.

This Theorem shows how to compute the invariant subspaces $\text{inv}_+(A)$ and $\text{inv}_-(A)$ via the Schur form of B. For general matrices, this is not a suitable method because we can easily compute invariant subspaces directly from a reordering of the diagonal elements of the Schur form of A. However, for real Hamiltonian matrices, where we cannot compute the Hamiltonian Schur form easily, the situation is different.

Here the block matrix

$$B = \begin{bmatrix} 0 & \mathcal{H} \\ \mathcal{H} & 0 \end{bmatrix} \tag{5.3}$$

is just a permutation of a Hamiltonian matrix, since with $\mathcal{P} = \begin{bmatrix} I_n & 0 & 0 & 0 \\ 0 & 0 & I_n & 0 \\ 0 & I_n & 0 & 0 \\ 0 & 0 & 0 & I_n \end{bmatrix}$,

we have that

$$\tilde{B} := \mathcal{P}^T B \mathcal{P} = \begin{bmatrix} 0 & F & 0 & G \\ F & 0 & G & 0 \\ 0 & H & 0 & -F^T \\ H & 0 & -F^T & 0 \end{bmatrix}$$

is Hamiltonian. This double sized Hamiltonian matrix is very special because it always has a Hamiltonian Schur form even if eigenvalues on the imaginary axis occur, regardless of the properties of $\mathcal{H}$.

Theorem 5.2. *[13] Let $\mathcal{H}$ be Hamiltonian and $\mathcal{B}$ be related as in (5.3). Then there exists $\mathcal{U} \in \mathcal{US}_{2n}(\mathbb{R})$ such that*

$$\mathcal{U}^T \mathcal{B} \mathcal{U} = \hat{R} = \begin{bmatrix} R & D \\ 0 & -R^T \end{bmatrix}$$

is in real Hamiltonian Schur form.

Furthermore, if $\mathcal{H}$ has no purely imaginary eigenvalues, then no eigenvalues of R are in the left half plane. Moreover, $\mathcal{U} = \mathcal{PW}$ with $\mathcal{W} \in \mathcal{US}_{4n}(\mathbb{R})$ and

$$\hat{R} = \mathcal{W}^T \tilde{\mathcal{B}} \mathcal{W},$$

i.e. $\hat{R}$ *is the real Hamiltonian Schur form of the Hamiltonian matrix* $\tilde{\mathcal{B}}$.

Proof. We will make use of the symplectic URV-decomposition of $\mathcal{H}$. By Lemma 5.2 there exist $U_1, U_2 \in \mathcal{US}_{2n}(\mathbb{R})$, such that

$$\mathcal{H} = U_2 \begin{bmatrix} H_t & H_r \\ 0 & -H_b^T \end{bmatrix} U_1^T \text{ and } \mathcal{H} = U_1 \begin{bmatrix} H_b & H_r^T \\ 0 & -H_t^T \end{bmatrix} U_2^T,$$

where H_t is upper triangular and H_b is quasi-upper triangular. Taking $\hat{\mathcal{U}} := \text{diag}(U_1, U_2)$, we have

$$\mathcal{B}_1 := \hat{\mathcal{U}}^T \mathcal{B} \hat{\mathcal{U}} = \left[\begin{array}{cc|cc} 0 & 0 & H_b & H_r^T \\ 0 & 0 & 0 & -H_t^T \\ \hline H_t & H_r & 0 & 0 \\ 0 & -H_b^T & 0 & 0 \end{array} \right].$$

Using the block form of $\mathcal{P}$,

$$\mathcal{B}_2 := \mathcal{P}^T \mathcal{B}_1 \mathcal{P} = \left[\begin{array}{cc|cc} 0 & H_b & 0 & H_r^T \\ H_t & 0 & H_r & 0 \\ \hline 0 & 0 & 0 & -H_t^T \\ 0 & 0 & -H_b^T & 0 \end{array} \right]$$

is Hamiltonian and block upper triangular. Let $U_3 = \begin{bmatrix} U_{1,1} & U_{1,2} \\ U_{2,1} & U_{2,2} \end{bmatrix}$ be such that

$$U_3^T \begin{bmatrix} 0 & H_b \\ H_t & 0 \end{bmatrix} U_3 =: \begin{bmatrix} \Sigma & \Gamma \\ 0 & -\Delta \end{bmatrix},$$

is in real Schur form with $\Sigma, \Delta \in \mathbb{R}^{n,n}$ quasi-upper triangular and

$$\sigma(\Sigma) = \sigma(\Delta), \qquad \sigma_-(\Sigma) = \emptyset.$$

Then

$$\mathcal{B}_3 := \begin{bmatrix} U_3 & 0 \\ 0 & U_3 \end{bmatrix}^T \mathcal{B}_2 \begin{bmatrix} U_3 & 0 \\ 0 & U_3 \end{bmatrix} = \left[\begin{array}{cc|cc} \Sigma & \Gamma & \Pi_1 & \Pi_2 \\ 0 & -\Delta & \Pi_2^T & \Pi_3 \\ \hline 0 & 0 & -\Sigma^T & 0 \\ 0 & 0 & -\Gamma^T & \Delta^T \end{array} \right].$$

Note that $\mathcal{B}_3$ is already in real Hamiltonian Schur form. However, the order of the eigenvalues on the block diagonal may not be as we require. But using the reordering procedure of Byers [19,20], there exists $\mathcal{V} :=$
$$
\begin{bmatrix}
I_n & 0 & 0 & 0 \\
0 & V_1 & 0 & V_2 \\
\hline
0 & 0 & I_n & 0 \\
0 & -V_2 & 0 & V_1
\end{bmatrix} \in
$$
$\mathcal{US}(\mathbb{R})$ such that

$$
\left[
\begin{array}{cc|cc}
\Sigma & \tilde{\Gamma} & \tilde{\Pi}_1 & \tilde{\Pi}_2 \\
0 & \tilde{\Delta} & \tilde{\Pi}_2^T & \tilde{\Pi}_3 \\
\hline
0 & 0 & -\Sigma^T & 0 \\
0 & 0 & -\tilde{\Gamma}^T & -\tilde{\Delta}^T
\end{array}
\right]
$$

is in real Hamiltonian Schur form with the required eigenvalue reordering and $\hat{\mathcal{U}} := \mathrm{diag}(U_3, U_3)\mathcal{V}$. The remaining assertions follow, since $\mathcal{W} = \mathcal{P}^T \mathcal{U} = \mathcal{P}^T \hat{\mathcal{U}} \mathcal{P} \tilde{\mathcal{U}}$. □

If we partition $\mathcal{U} := \begin{bmatrix} \mathcal{U}_{1,1} & \mathcal{U}_{1,2} \\ \mathcal{U}_{2,1} & \mathcal{U}_{2,2} \end{bmatrix}$, $\mathcal{U}_{ij} \in \mathbb{R}^{2n,2n}$, then using the structures of the matrices $\hat{\mathcal{U}}$, $\mathcal{P}$, U_3 and $\mathcal{V}$ we obtain

$$
\mathcal{U}_{1,1} = U_2 \begin{bmatrix} U_{1,1} & U_{1,2}V_1 \\ 0 & -U_{1,2}V_2 \end{bmatrix} \quad \text{and} \quad \mathcal{U}_{2,1} = U_1 \begin{bmatrix} U_{2,1} & U_{2,2}V_1 \\ 0 & -U_{2,2}V_2 \end{bmatrix}.
$$

By Theorem 5.1 we have

$$
\mathrm{range}\{\mathcal{U}_{1,1} - \mathcal{U}_{2,1}\} = \mathrm{inv}_-(\mathcal{H}) + \mathcal{N}_1 \quad \text{and} \quad \mathrm{range}\{\mathcal{U}_{1,1} + \mathcal{U}_{2,1}\} = \mathrm{inv}_+(\mathcal{H}) + \mathcal{N}_2,
$$

where $\mathcal{N}_1$, $\mathcal{N}_2 \subset \mathrm{inv}_0(\mathcal{H})$. The construction in the proof of Theorem 5.2 can be directly translated into a numerical algorithm for computing the invariant subspace of a Hamiltonian matrix $\mathcal{H}$ associated with the eigenvalues in the left half plane, see [13]. Let us denote this method as the *Lagrange subspace algorithm LSA*.

The computational cost for this algorithm is in the range of the costs for the general QR-algorithm. For a comparison of the accuracy we present some partial results from a Table in [13], that gives a comparison for some MATLAB implementations of solution methods for algebraic Riccati equations. Table 5.2 shows a comparison of relative errors for the subspace Algorithm LSA, the Riccati solver ARE which is an implementation of the Laub trick in the MATLAB Control Toolbox [37], $CARE$ which is the Riccati solver contained in the MATLAB LMI Toolbox [26], $ARESOLV$, the implementation of the Laub trick in the MATLAB Robust Control Toolbox [38] and $OSMARE$, the multishift method from [1]. As test cases the Riccati benchmark collection [11] was used. In general, the LSA-algorithm produces errors of the same order as the best of the other methods. For problems of larger dimension it produces the best results while the multishift method suffers from convergence problems. However, the biggest advantage

	LSA	ARE	CARE	ARESOLV	OSMARE
1	0	2.1×10^{-16}	2.4×10^{-15}	2.0×10^{-15}	7.4×10^{-17}
2	4.7×10^{-15}	1.4×10^{-15}	4.5×10^{-15}	5.5×10^{-15}	1.3×10^{-15}
7	8.3×10^{-5}	2.2×10^{-5}	2.0×10^{-4}	8.9×10^{-5}	5.9×10^{-5}
9	4.1×10^{-14}	1.2×10^{-14}	2.5×10^{-11}	3.8×10^{-12}	1.6×10^{-16}
10	1.6×10^{-16}	7.5×10^{-16}	6.1×10^{-11}	5.2×10^{-16}	1.2×10^{-11}
11	2.1×10^{-8}	1.6×10^{-8}	2.7×10^{-8}	1.2×10^{-8}	6.3×10^{-16}
12	5.7×10^{-4}	7.0×10^{-4}	3.8×10^{-4}	1.9×10^{-3}	9.5×10^{-4}
17	8.3×10^{-7}	1.1×10^{-6}	1.1×10^{-6}	1.1×10^{-6}	6.6×10^{-9}

Table 5.2. Relative Errors for the compared MATLAB functions. For Example 17: $|x_{1,n} - 1|$.

of the LSA-algorithm is that it can also be applied to matrices with eigenvalues on the imaginary axis. Furthermore, due to a recent embedding trick suggested in [9] and the construction of a structured QZ-type method for skew-Hamiltonian/Hamiltonian pencils in [10], the method can also be used for the computation of deflating subspaces as they arise in the control of descriptor systems.

6   Large Scale Problems

Large scale Hamiltonian eigenvalue problems or generalized skew Hamiltonian/Hamiltonian eigenvalue problems have recently been studied in [30,41]. The motivation there was the computation of singularity exponents for elasticity problems, the analysis of gyroscopic systems and the optimal control of parabolic partial differential equations.

In order to solve these problems, in [41] a structure preserving version $SHIRA$ of the implicitly restarted Arnoldi method IRA was developed. In [42] this method was extended to general polynomial eigenvalue problems, where the coefficient matrices alternate between symmetric and skew symmetric matrices. Let us briefly describe the idea of the $SHIRA$ algorithm of [41]. This method computes a small number of eigenvalues of real large scale generalized eigenvalue problems for skew-Hamiltonian/Hamiltonian pencils of the form

$$\lambda \mathcal{N} - \mathcal{H} = \lambda \begin{bmatrix} F_1 & G_1 \\ H_1 & F_1^T \end{bmatrix} - \begin{bmatrix} F_2 & G_2 \\ H_2 & -F_2^T \end{bmatrix}, \qquad (6.1)$$

where $G_1 = -G_1^T$, $H_1 = -H_1^T$, $G_2 = G_2^T$ and $H_2 = H_2^T$.

If the skew-Hamiltonian matrix $\mathcal{N}$ in the pencil $\lambda \mathcal{N} - \mathcal{H}$ in (6.1) is invertible and given in the factored form $\mathcal{N} = \mathcal{Z}_1 \mathcal{Z}_2$ with $\mathcal{Z}_2^T J = \pm J \mathcal{Z}_1$, then the pencil is equivalent to standard eigenvalue problem for the Hamiltonian matrix

$$\mathcal{W} = \pm \mathcal{Z}_1^{-1} \mathcal{H} \mathcal{Z}_2^{-1}.$$

The factorization of $\mathcal{N}$ can be determined (if the storage capacity allows) via a Cholesky-like decomposition for skew-symmetric matrices developed in [16], see also [8].

Suppose that one wishes to compute the eigenvalues of $\mathcal{W}$ nearest to some focal point τ. Then typically one would use shift-and-invert of the form $(\mathcal{W} - \tau I)^{-1}$ to get the eigenvalues near τ via a Krylov subspace method. But this matrix is no longer Hamiltonian. In order to preserve the structure, [41] suggest a rational transformation with four shifts $(\tau, -\tau, \bar{\tau}, -\bar{\tau})$ given by

$$R_1(\tau, \mathcal{W}) = (\mathcal{W} - \tau I)^{-1}(\mathcal{W} + \tau I)^{-1}(\mathcal{W} - \bar{\tau} I)^{-1}(\mathcal{W} + \bar{\tau} I)^{-1}. \qquad (6.2)$$

If the focal point τ is either real or purely imaginary, one may use the simpler transformation

$$R_2(\tau, \mathcal{W}) = (\mathcal{W} - \tau I)^{-1}(\mathcal{W} + \tau I)^{-1}. \qquad (6.3)$$

If the matrix $\mathcal{W}$ is real and Hamiltonian, then both matrices R_1 and R_2 are real and skew-Hamiltonian and hence the eigenvalues have algebraic multiplicity of at least two, if they are not purely imaginary. For a real skew-Hamiltonian matrix $\mathcal{K}$ and a given vector q_1, a classical Arnoldi iteration applied to $\mathcal{K}$ would generate the Krylov space

$$\mathcal{V}(\mathcal{K}, q_1, k) = \text{span}\{[q_1, \mathcal{K}q_1, \ldots, \mathcal{K}^{k-1}q_1]\}.$$

Using an appropriate orthogonal basis of this space given by the columns of an orthogonal matrix Q_k, one produces a 'Ritz'-projection

$$K_k = Q_k^T \mathcal{K} Q_k.$$

To obtain a structure-preserving method, one needs a 'Ritz'-projection that is again skew-Hamiltonian. For this it is necessary to have an isotropic subspace $\mathcal{V}$, i.e. a subspace for which $x^T J y = 0$ for all $x, y \in \mathcal{V}$, see [41]. Let $V_k \in \mathbb{R}^{2n,k}$ and let

$$Q^T V_k = \begin{bmatrix} R_k \\ 0 \\ T_k \\ 0 \end{bmatrix}, \qquad (6.4)$$

be the symplectic QR-factorization [17] of V_k, where $Q \in \mathcal{US}_{2n}(\mathbb{R})$ and where R_k and $T_k \in \mathbb{R}^{k \times k}$ are upper and strictly upper triangular, respectively. If V_k is isotropic and of full column rank, then it is shown in [17] that R_k^{-1} exists and that $T_k = 0$. This means, in particular, that if V_n is of full column rank, then there exists $\mathcal{Q} = [Q_n \,|\, JQ_n] \in \mathcal{US}_{2n}(\mathbb{R})$, such that

$$\begin{bmatrix} Q_n^T \\ Q_n^T J^T \end{bmatrix} \mathcal{K} \, [Q_n \ JQ_n] = \begin{bmatrix} H_n & K_n \\ 0 & H_n^T \end{bmatrix}.$$

Here $H_n = Q_n^T \mathcal{K} Q_n$ is an upper-Hessenberg matrix and $K_n = Q_n^T \mathcal{K} J Q_n$ is skew-symmetric. This idea is the basis for the structure preserving Arnoldi algorithm SHIRA introduced in [41], which is given by the recursion

$$X S \mathcal{K} Q_\ell = Q_\ell H_\ell + q_{\ell+1} h_{\ell+1,\ell} e_\ell^T, \tag{6.5}$$

where H_ℓ is the leading $\ell \times \ell$ principal submatrix of H_n. Since Q_ℓ is orthogonal and $Q_\ell^T J Q_\ell = 0$ for $\ell < n$, it is easily seen that $[Q_\ell | J Q_\ell] \in \mathbb{R}^{2n,\ell}$ is orthogonal. This implies that

$$\begin{bmatrix} Q_\ell^T \\ Q_\ell^T J^T \end{bmatrix} \mathcal{K} [Q_\ell \mid J Q_\ell] = \begin{bmatrix} H_\ell & G_\ell \\ 0 & H_\ell^T \end{bmatrix} \in \mathbb{R}^{2\ell,2\ell}$$

is skew-Hamiltonian and $G_\ell = Q_\ell^T \mathcal{K} J Q_\ell$ is skew-symmetric.

The transformations $R_1(\tau, \mathcal{W})$ and $R_2(\tau, \mathcal{W})$ can be viewed as a structure preserving shift-and-invert strategy to accelerate the convergence of the desired eigenvalues. In exact arithmetic, the values $t_{i,\ell} = (J q_i)^T \mathcal{K} q_\ell$ in the ℓ-th step of the SHIRA algorithm are zero, but in practice roundoff errors cause them to be nonzero. It has been shown in [41] how to design an isotropic orthogonalization scheme to ensure that the spaces $\mathrm{span}\{q_1, \ldots, q_\ell\}$ are isotropic to working precision. Similar to the IRA-algorithm the $SHIRA$-algorithm also uses implicit restarts. Due to the rational transformations (6.2) or (6.3) again the eigenvectors associated with pairs of eigenvalues λ and $-\lambda$ in the real case or λ and $-\bar{\lambda}$ in the complex case are both mapped to the same invariant subspace associated with $|\lambda|^2$. In [30] a special subspace extraction procedure was designed to get the desired subspace associated with the stable eigenvalues.

7   Conclusion

In this paper we have given a brief overview of classical eigenvalue methods and also some of the basics of linear control theory. We have demonstrated that control theory provides a rich source for eigenvalue problems that also have special structure. As it should be the goal of a good numerical method to exploit and retain the possible structures as much as possible, since they typically reflect properties of the underlying physical problem, we have discussed special methods to solve Hamiltonian eigenvalue problems. We have demonstrated that structure preservation is important but also presents a challenge in the development of methods.

Acknowledgment

In Sections 4, 5 and 6 we have summarized research work that was obtained in the last 20 years together with many colleagues and friends such as G. Ammar, P. Benner, A. Bunse-Gerstner, R. Byers, W.-W. Lin, D. Watkins and H. Xu. I am greatly thankful for this wonderful research cooperation.

References

1. G.S. Ammar, P. Benner, and V. Mehrmann. A multishift algorithm for the numerical solution of algebraic Riccati equations. *Electr. Trans. Num. Anal.*, 1:33–48, 1993.
2. G.S. Ammar and V. Mehrmann. On Hamiltonian and symplectic Hessenberg forms. *Linear Algebra Appl.*, 149:55–72, 1991.
3. E. Anderson, Z. Bai, C. Bischof, J. Demmel, J. Dongarra, J. Du Croz, A. Greenbaum, S. Hammarling, A. McKenney, S. Ostrouchov, and D. Sorensen. *LAPACK Users' Guide*. SIAM, Philadelphia, PA, second edition, 1995.
4. W.F. Arnold, III and A.J. Laub. Generalized eigenproblem algorithms and software for algebraic Riccati equations. *Proc. IEEE*, 72:1746–1754, 1984.
5. M. Athans and P.L. Falb. *Optimal Control*. McGraw-Hill, New York, 1966.
6. Z. Bai, J. Demmel, J. Dongarra, A. Ruhe, and H. van der Vorst. *Templates for the solution of algebraic eigenvalue problems*. SIAM, Philadelphia, PA. USA, 2000.
7. P. Benner, R. Byers, and E. Barth. HAMEV and SQRED: Fortran 77 subroutines for computing the eigenvalues of Hamiltonian matrices using Van Loan's square reduced method. Technical Report SFB393/96-06, Fakultät für Mathematik, TU Chemnitz–Zwickau, 09107 Chemnitz, FRG, 1996.
8. P. Benner, R. Byers, H. Fassbender, V. Mehrmann, and D. Watkins. Cholesky-like factorizations of skew-symmetric matrices. *Electr. Trans. Num. Anal.*, 11:85–93, 2000.
9. P. Benner, R. Byers, V. Mehrmann, and H. Xu. Numerical methods for linear quadratic and H_∞ control problems. In G. Picci und D.S. Gillian, editor, *Dynamical Systems, Control, Coding, Computer Vision. Progress in Systems and Control Theory, Vol. 25*, pages 203–222, Basel, 1999. Birkhäuser Verlag.
10. P. Benner, R. Byers, V. Mehrmann, and H. Xu. Numerical computation of deflating subspaces of skew Hamiltonian/Hamiltonian pencils. *SIAM J. Matrix Anal. Appl.*, 24:165–190, 2002.
11. P. Benner, A.J. Laub, and V. Mehrmann. Benchmarks for the numerical solution of algebraic Riccati equations. *IEEE Control Systems Magazine*, 7(5):18–28, 1997.
12. P. Benner, V. Mehrmann, V. Sima, S. Van Huffel, and A. Varga. SLICOT - a subroutine library in systems and control theory. *Applied and Computational Control, Signals, and Circuits*, 1:499–532, 1999.
13. P. Benner, V. Mehrmann, and H. Xu. A new method for computing the stable invariant subspace of a real Hamiltonian matrix. *J. Comput. Appl. Math.*, 86:17–43, 1997.
14. P. Benner, V. Mehrmann, and H. Xu. A numerically stable, structure preserving method for computing the eigenvalues of real Hamiltonian or symplectic pencils. *Numer. Math.*, 78(3):329–358, 1998.
15. A. Bojanczyk, G.H. Golub, and P. Van Dooren. The periodic Schur decomposition; algorithms and applications. In *Proc. SPIE Conference, vol. 1770*, pages 31–42, 1992.
16. J. R. Bunch. A note on the stable decomposition of skew-symmetric matrices. *Math. Comp.*, 38:475–479, 1982.
17. A. Bunse-Gerstner. Matrix factorization for symplectic QR-like methods. *Linear Algebra Appl.*, 83:49–77, 1986.

18. A. Bunse-Gerstner, R. Byers, and V. Mehrmann. Numerical methods for algebraic Riccati equations. In S. Bittanti, editor, *Proc. Workshop on the Riccati Equation in Control, Systems, and Signals*, pages 107–116, Como, Italy, 1989.

19. R. Byers. *Hamiltonian and Symplectic Algorithms for the Algebraic Riccati Equation.* PhD thesis, Cornell University, Dept. Comp. Sci., Ithaca, NY, 1983.

20. R. Byers. A Hamiltonian QR-algorithm. *SIAM J. Sci. Statist. Comput.*, 7:212–229, 1986.

21. R. Byers. Numerical stability and instability in matrix sign function based algorithms. In C.I. Byrnes and A. Lindquist, editors, *Computational and Combinatorial Methods in Systems Theory*, pages 185–200. Elsevier (North-Holland), New York, 1986.

22. E.A. Coddington and R. Carlson. *Linear ordinary differential equations.* SIAM, Philadelphia, PA, 1997.

23. J.W. Demmel and B. Kågström. Computing stable eigendecompositions of matrix pencils. *Linear Algebra Appl.*, 88:139–186, 1987.

24. L. Elsner and C. He. An algorithm for computing the distance to uncontrollability. *Sys. Control Lett.*, 17:453–464, 1991.

25. G. Freiling, V. Mehrmann, and H. Xu. Existence, uniqueness and parametrization of lagrangian invariant subspaces. *SIAM J. Matrix Anal. Appl.*, 23:1045–1069, 2002.

26. P. Gahinet, A. Laub, and A. Nemirovski. The LMI Control Toolbox. The MathWorks, Inc., 24 Prime Park Way, Natick, MA 01760, 1995.

27. F.R. Gantmacher. *Theory of Matrices*, volume 1. Chelsea, New York, 1959.

28. G.H. Golub and C.F. Van Loan. *Matrix Computations.* Johns Hopkins University Press, Baltimore, third edition, 1996.

29. J.J. Hench and A.J. Laub. Numerical solution of the discrete-time periodic Riccati equation. *IEEE Trans. Automat. Control*, 39:1197–1210, 1994.

30. T.-M. Hwang, W.-W. Lin, and V. Mehrmann. Numerical solution of quadratic eigenvalue problems for damped gyroscopic systems. *SIAM J. Sci. Statist. Comput.*, 2003. To appear.

31. T. Kailath. *Systems Theory.* Prentice-Hall, Englewood Cliffs, NJ, 1980.

32. H.W. Knobloch and H. Kwakernaak. *Lineare Kontrolltheorie.* Springer-Verlag, Berlin, 1985.

33. A.J. Laub. A Schur method for solving algebraic Riccati equations. *IEEE Trans. Automat. Control*, AC-24:913–921, 1979.

34. A.J. Laub. Invariant subspace methods for the numerical solution of Riccati equations. In S. Bittanti, A.J. Laub, and J.C. Willems, editors, *The Riccati Equation*, pages 163–196. Springer-Verlag, Berlin, 1991.

35. R.B. Lehouq, D.C. Sorensen, and C. Yang. *ARPACK Users' Guide.* SIAM, Philadelphia, PA, USA, 1998.

36. W.-W. Lin, V. Mehrmann, and H. Xu. Canonical forms for Hamiltonian and symplectic matrices and pencils. *Linear Algebra Appl.*, 301-303:469–533, 1999.

37. The MathWorks, Inc., Cochituate Place, 24 Prime Park Way, Natick, Mass, 01760. *The MATLAB Control Toolbox, Version 3.0b*, 1993.

38. The MathWorks, Inc., Cochituate Place, 24 Prime Park Way, Natick, Mass, 01760. *The MATLAB Robust Control Toolbox, Version 2.0b*, 1994.

39. MATLAB, Version 5. The MathWorks, inc., 24 Prime Park Way, Natick, MA 01760-1500, USA, 1996.

40. V. Mehrmann. *The Autonomous Linear Quadratic Control Problem, Theory and Numerical Solution.* Number 163 in Lecture Notes in Control and Information Sciences. Springer-Verlag, Heidelberg, July 1991.

41. V. Mehrmann and D. Watkins. Structure-preserving methods for computing eigenpairs of large sparse skew-Hamiltoninan/Hamiltonian pencils. *SIAM J. Sci. Comput.*, 22:1905–1925, 2001.

42. V. Mehrmann and D. Watkins. Polynomial eigenvalue problems with Hamiltonian structure. *Electr. Trans. Num. Anal.*, 2002. To appear.

43. V. Mehrmann and H. Xu. Numerical methods in control. *J. Comput. Appl. Math.*, 123:371–394, 2000.

44. C.C. Paige and C.F. Van Loan. A Schur decomposition for Hamiltonian matrices. *Linear Algebra Appl.*, 14:11–32, 1981.

45. B.N. Parlett. *The Symmetric Eigenvalue Problem.* Prentice-Hall, Englewood Cliffs, NJ, 1980.

46. P.H. Petkov, N.D. Christov, and M.M. Konstantinov. *Computational Methods for Linear Control Systems.* Prentice-Hall, Hertfordshire, UK, 1991.

47. L.S. Pontryagin, V. Boltyanskii, R. Gamkrelidze, and E. Mishenko. *The Mathematical Theory of Optimal Processes.* Interscience, New York, 1962.

48. A.C.M. Ran and L. Rodman. Stability of invariant Lagrangian subspaces i. *Operator Theory: Advances and Applications (I. Gohberg ed.)*, 32:181–218, 1988.

49. Y. Saad. *Numerical methods for large eigenvalue problems.* Manchester University Press, Manchester, M13 9PL, UK, 1992.

50. W. Schiehlen. *Multibody Systems Handbook.* Springer-Verlag, 1990.

51. V. Sima. *Algorithms for Linear-Quadratic Optimization,* volume 200 of *Pure and Applied Mathematics.* Marcel Dekker, Inc., New York, NY, 1996.

52. D.C. Sorensen. Implicit application of polynomial filters in a k-step Arnoldi method. *SIAM J. Matrix Anal. Appl.*, 13:357–385, 1992.

53. G.W. Stewart and J.-G. Sun. *Matrix Perturbation Theory.* Academic Press, New York, 1990.

54. F. Tisseur. Backward error analysis of polynomial eigenvalue problems. *Linear Algebra Appl.*, 309:339–361, 2000.

55. F. Tisseur and K. Meerbergen. The quadratic eigenvalue problem. *SIAM Rev.*, 43:234–286, 2001.

56. P. Van Dooren. The computation of Kronecker's canonical form of a singular pencil. *Linear Algebra Appl.*, 27:103–121, 1979.

57. P. Van Dooren. A generalized eigenvalue approach for solving Riccati equations. *SIAM J. Sci. Statist. Comput.*, 2:121–135, 1981.

58. C.F. Van Loan. A symplectic method for approximating all the eigenvalues of a Hamiltonian matrix. *Linear Algebra Appl.*, 61:233–251, 1984.

59. J.H. Wilkinson. *The Algebraic Eigenvalue Problem.* Oxford University Press, Oxford, 1965.

60. K. Zhou, J.C. Doyle, and K. Glover. *Robust and Optimal Control.* Prentice-Hall, Upper Saddle River, NJ, 1996.

Universitext

Printing and Binding: Strauss GmbH, Mörlenbach